Quantum Computing since Democritus

Written by noted quantum computing theorist Scott Aaronson, this book takes readers on a tour through some of the deepest ideas of math, computer science, and physics.

Full of insights, arguments, and philosophical perspectives, the book covers an amazing array of topics. Beginning in antiquity with Democritus, it progresses through logic and set theory, computability and complexity theory, quantum computing, cryptography, the information content of quantum states, and the interpretation of quantum mechanics. There are also extended discussions about time travel, Newcomb's Paradox, the Anthropic Principle, and the views of Roger Penrose. Aaronson's informal style makes this fascinating book accessible to readers with scientific backgrounds, as well as students and researchers working in physics, computer science, mathematics, and philosophy.

SCOTT AARONSON is an Associate Professor of Electrical Engineering and Computer Science at the Massachusetts Institute of Technology. Considered one of the top quantum complexity theorists in the world, he is well known both for his research in quantum computing and computational complexity theory, and for his widely read blog *Shtetl-Optimized*. Professor Aaronson also created Complexity Zoo, an online encyclopedia of computational complexity theory, and has written popular articles for *Scientific American* and *The New York Times*. His research and popular writing have earned him numerous awards, including the United States Presidential Early Career Award for Scientists and Engineers and the Alan T. Waterman Award.

Quantum Computing since Democritus

SCOTT AARONSON

Massachusetts Institute of Technology

CAMBRIDGE
UNIVERSITY PRESS

CAMBRIDGE
UNIVERSITY PRESS

University Printing House, Cambridge CB2 8BS, United Kingdom

One Liberty Plaza, 20th Floor, New York, NY 10006, USA

477 Williamstown Road, Port Melbourne, VIC 3207, Australia

314-321, 3rd Floor, Plot 3, Splendor Forum, Jasola District Centre, New Delhi - 110025, India

79 Anson Road, #06-04/06, Singapore 079906

Cambridge University Press is part of the University of Cambridge.

It furthers the University's mission by disseminating knowledge in the pursuit of education, learning and research at the highest international levels of excellence.

www.cambridge.org
Information on this title: www.cambridge.org/9780521199568

© S. Aaronson 2013

First published 2013
9th printing 2018

A catalogue record for this publication is available from the British Library

Library of Congress Cataloging in Publication data
Aaronson, Scott.
Quantum computing since Democritus / Scott Aaronson.
  pages cm
Includes bibliographical references and index.
ISBN 978-0-521-19956-8 (pbk.)
1. Quantum theory – Mathematics. 2. Quantum computers. I. Title.
QC174.17.M35A27   2013
621.39′1 – dc23   2012036798

ISBN 9780-5-211-9956-8 Paperback

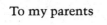

To my parents

Contents

Preface   *page* ix
Acknowledgments   xxix

1. Atoms and the void   1

2. Sets   8

3. Gödel, Turing, and friends   18

4. Minds and machines   29

5. Paleocomplexity   44

6. P, NP, and friends   54

7. Randomness   71

8. Crypto   93

9. Quantum   109

10. Quantum computing   132

11. Penrose   150

12. Decoherence and hidden variables   160

13. Proofs   186

14. How big are quantum states?   200

15. Skepticism of quantum computing   217

16. Learning   228

17. Interactive proofs, circuit lower bounds, and more   243

18. Fun with the Anthropic Principle   266

19. Free will   290

20. Time travel   307

21. Cosmology and complexity   325

22. Ask me anything   343

    Index   363

Preface

A CRITICAL REVIEW OF SCOTT AARONSON'S
QUANTUM COMPUTING SINCE DEMOCRITUS
by Scott Aaronson.

Quantum Computing since Democritus is a candidate for the weirdest book ever to be published by Cambridge University Press. The strangeness starts with the title, which conspicuously fails to explain what this book is *about*. Is this another textbook on quantum computing – the fashionable field at the intersection of physics, math, and computer science that's been promising the world a new kind of computer for two decades, but has yet to build an actual device that can do anything more impressive than factor 21 into 3 × 7 (with high probability)? If so, then what does *this* book add to the dozens of others that have already mapped out the fundamentals of quantum computing theory? Is the book, instead, a quixotic attempt to connect quantum computing to ancient history? But what does Democritus, the Greek atomist philosopher, really have to do with the book's content, at least half of which would have been new to scientists of the 1970s, let alone of 300 BC?

Having now read the book, I confess that I've had my mind blown, my worldview reshaped, by the author's truly brilliant, original perspectives on everything from quantum computing (as promised in the title) to Gödel's and Turing's theorems to the **P** versus **NP** question to the interpretation of quantum mechanics to artificial intelligence to Newcomb's Paradox to the black-hole information loss problem. So, if anyone were perusing this book at a bookstore, or with Amazon's "Look Inside" feature, I would *certainly* tell that person to buy a copy immediately. I'd also add that the author is extremely handsome.

Yet it's hard to avoid the suspicion that *Quantum Computing since Democritus* is basically a "brain dump": a collection of thoughts about theoretical computer science, physics, math, and philosophy that were on the author's mind around the fall of 2006, when he gave a series of lectures at the University of Waterloo that eventually turned into this book. The material is tied together by the author's nerdy humor, his "Socratic" approach to every question, and his obsession with the theory of computation and how it relates to the physical world. But if there's some overarching "thesis" that I'm supposed to take away, I can't for the life of me articulate what it is.

More pointedly, one wonders who the *audience* for this book is supposed to be. On the one hand, it has *way* too much depth for a popular book. Like Roger Penrose's *The Road to Reality* – whose preface promises an accessible adventure even for readers who struggled with fractions in elementary school, but whose first few chapters then delve into holomorphic functions and fiber bundles – *Quantum Computing since Democritus* is not for math-phobes. A curious layperson could *certainly* learn a lot from this book, but he or she would have to be willing to skip over some dense passages, possibly to return to them later. So if you're someone who can stomach "science writing" only after it's been carefully cleansed of the science, look elsewhere.

On the other hand, the book is *also* too wide-ranging, breezy, and idiosyncratic to be used much as a textbook or reference work. Sure, it has theorems, proofs, and exercises, and it covers the basics of an astonishing number of fields: logic, set theory, computability, complexity, cryptography, quantum information, and computational learning theory, among others. It seems likely that students in any of those fields, from the undergraduate level on up, could gain valuable insights from this book, or could use it as an entertaining self-study or refresher course. Besides these basics, the book also has significant material on quantum complexity theory – for example, on the power of quantum proofs and advice – that (to this reviewer's knowledge) hasn't appeared anywhere else in book form. But still, the book flits from topic to topic too hastily to be a definitive text on anything.

So, is the book aimed at non-scientists who won't *actually* make it past the first chapter, but want something to put on their coffee table to impress party guests? The only other possibility I can think of is that there's an underserved audience for science books that are neither "popular" nor "professional": books that describe a piece of the intellectual landscape from one researcher's heavily biased vantage point, using the same sort of language you might hear in a hallway conversation with a colleague from a different field. Maybe, besides those colleagues, this hypothetical "underserved audience" would include precocious high-school students, or programmers and engineers who enjoyed their theoretical courses back in college and want to find out what's new. Maybe this is the same audience that frequents these "science blogs" I've heard about: online venues where anyone in the world can apparently watch real scientists, people at the forefront of human knowledge, engage in petty spats, name-calling, and every other juvenile behavior, and can even egg the scientists on to embarrass themselves further. (The book's author, it should be noted, writes a particularly crass and infamous such blog.) *If* such an audience actually exists, then perhaps the author knew exactly what he was doing in aiming at it. My sense, though, is that he was having too much fun to be guided by any such conscious plan.

NOW FOR THE ACTUAL PREFACE

While I appreciate the reviewer's kind words about my book (and even my appearance!) in the preceding pages, I also take issue, in the strongest possible terms, with his ignorant claim that *Quantum Computing since Democritus* has no overarching thesis. It *does* have a thesis – even though, strangely, I wasn't the one who figured out what it was. For identifying the central message of this book, I need to thank Love Communications, an advertising agency based in

Sydney, Australia, which put the message into the mouths of fashion models for the purpose of selling printers.

Let me explain – the story is worth it.

In 2006, I taught a course entitled "Quantum Computing since Democritus" at the University of Waterloo. Over the next year, I posted rough notes from the course on my blog, *Shtetl-Optimized*[1] – notes that were eventually to become this book. I was heartened by the enthusiastic response from readers of my blog; indeed, that response is what convinced me to publish this book in the first place. But there was one response neither I nor anyone else could have predicted.

On October 1, 2007, I received an email from one Warren Smith in Australia, who said he had seen a television commercial for Ricoh printers. The commercial, he went on, featured two female fashion models in a makeup room, having the following conversation:

> Model 1: But if quantum mechanics isn't physics in the usual sense – if it's not about matter, or energy, or waves – then what *is* it about?

> Model 2: Well, from my perspective, it's about information, probabilities, and observables, and how they relate to each other.

> Model 1: That's interesting!

The commercial then flashed the tagline "A more intelligent model," followed by a picture of a Ricoh printer.

Smith said he was curious where the unusual text had come from, so he googled it. Doing so brought him to Chapter 9 of my "Quantum Computing since Democritus" notes (p. 110), where he found the following passage:

> But if quantum mechanics isn't physics in the usual sense – if it's not about matter, or energy, or waves, or particles – then what *is* it

[1] www.scottaaronson.com/blog

about? From my perspective, it's about information and probabilities and observables, and how they relate to each other.

So, it seemed, there was exactly one bit of dialogue in the commercial that I *didn't* write ("That's interesting!"). Smith found a link[2] where I could see the commercial for myself on YouTube, and his story checked out.

Far more amused than annoyed, I wrote a post for my blog, entitled "Australian Actresses Are Plagiarizing My Quantum Mechanics Lecture to Sell Printers."[3] After relating what had happened and linking to the video, the post ended

> For almost the first time in my life, I'm at a loss for words. I don't know how to respond. I don't know which of 500 000 possible jokes to make. Help me, readers. Should I be flattered? Should I be calling a lawyer?

This would become the most notorious blog post I ever wrote. By the next morning, the story had made the *Sydney Morning Herald* ("Ad agency cribbed my lecture notes: professor"[4]), Slashdot ("Scott Aaronson, Printer Shill"[5]), and several other news sites. I happened to be in Latvia at the time, visiting my colleague Andris Ambainis, but somehow journalists tracked me down to my hotel room in Riga, waking me up around 5 a.m. to ask for interviews.

Meanwhile, reactions on my blog and in other online forums were mixed. Some readers said I'd be foolish if I didn't sue the ad agency for all it was worth. What if they had played a few beats of a *Rolling Stones* song, without first getting permission? Cases like that, I was assured, are sometimes settled for millions of dollars. Others said that even *asking the question* made me a stereotypical litigious American, a personification of everything wrong with the world. I should be flattered, they continued, that the ad writers had

---

[2] www.youtube.com/watch?v=saWCyZupO4U

[3] www.scottaaronson.com/blog/?p=277

[4] www.smh.com.au/news/technology/professor-claims-ad-agency-cribs-lecture-notes/2007/10/03/1191091161163.html

[5] idle.slashdot.org/story/07/10/02/1310222/scott-aaronson-printer-shill

seen fit to give *my* take on quantum mechanics all this free
publicity. Dozens of commenters offered variations on the same
insipid joke, that I should ask for a date with the "models" as my
compensation. (I replied that I'd rather have a free printer, if it came
down to it.) One commenter simply wrote, "This really could be the
funniest thing that has ever happened."

For its part, Love Communications admitted that it had
appropriated material from my lecture, but said it had consulted a
lawyer and thought it was perfectly within its fair-use rights to do
so. Meanwhile, I *did* get in touch with an Australian intellectual
property lawyer, who said that I might have a case – but it would
take time and energy to pursue it. I felt torn: on the one hand,
plagiarism is one of the academic world's few unforgivable sins,
and I was miffed by the agency's completely unapologetic response,
after they'd been caught so red handed. On the other hand, if they
had just *asked* me, I probably would have gladly given them
permission to use my words, for either a token sum or no money at
all.

In the end, we found a solution that everyone liked. Love
Communications apologized (without admitting wrongdoing), and
donated $5000 to two science outreach organizations of my choice
in Australia.[6] In return, I didn't pursue any further action – and
indeed, I mostly forgot about the affair, except when colleagues
would rib me (as they continue to do) about Australian models.

But there's a final irony to the tale, and that's why I'm
recounting it here (well, besides just that it's a hilarious true story
involving this book). If I had to choose one passage from the entire
book to be broadcast on TV, I think I would have chosen the exact
same one that the commercial writers chose – even though they
were presumably just trawling for some sciencey-sounding
gobbledygook, and I hadn't highlighted the passage in any way, as its
centrality hadn't occurred to me.

---

[6] See www.scottaaronson.com/blog/?p=297

The idea that quantum mechanics is "about" information, probabilities, and observables, rather than waves and particles, certainly isn't an original one. The physicist John Archibald Wheeler said similar things in the 1970s; and today an entire field, that of quantum computing and information, is built around the idea. Indeed, in the discussion on my blog that followed the Australian models episode, one the commonest (and to me, funniest) arguments was that I had no right to complain, because the appropriated passage *wasn't special in any way*: it was an obvious thought that could be found in any physics book!

How I wish it were so. Even in 2013, the view of quantum mechanics as a theory of information and probabilities remains very much a minority one. Pick up almost any physics book – whether popular or technical – and you'll learn that (a) modern physics says all sorts of paradoxical-seeming things, like that waves are particles and particles are waves, (b) at a deep level, no one really understands these things, (c) even translating them into math requires years of intensive study, but (d) they make the atomic spectra come out right, and that's what matters in the end.

One eloquent statement of this "conventional view" was provided by Carl Sagan, in *The Demon-Haunted World*:

> Imagine you seriously want to understand what quantum mechan-
> ics is about. There is a mathematical underpinning that you must
> first acquire, mastery of each mathematical subdiscipline lead-
> ing you to the threshold of the next. In turn you must learn arith-
> metic, Euclidean geometry, high school algebra, differential and
> integral calculus, ordinary and partial differential equations, vec-
> tor calculus, certain special functions of mathematical physics,
> matrix algebra, and group theory . . . The job of the popularizer of
> science, trying to get across some idea of quantum mechanics to a
> general audience that has not gone through these initiation rites,
> is daunting. Indeed, there are no successful popularizations of
> quantum mechanics in my opinion – partly for this reason. These

> mathematical complexities are compounded by the fact that quantum theory is so resolutely counterintuitive. Common sense is almost useless in approaching it. It's no good, Richard Feynman once said, asking why it *is* that way. No one knows why it is that way. That's just the way it is (p. 249).

It's understandable why physicists talk this way: because physics is an experimental science. In physics you're *allowed* to say, "these are the rules, not because they make sense, but because we ran the experiment and got such-and-such a result." You can even say it proudly, gleefully – *defying* the skeptics to put their preconceived notions up against Nature's verdict.

Personally, I simply *believe* the experimentalists, when they say the world works in a completely different way than I thought it did. It's not a matter of convincing me. Nor do I presume to predict what the experimentalists will discover next. All I want to know is: *What went wrong with my intuition? How should I fix it, to put it more in line with what the experiments found? How could I have reasoned, such that the actual behavior of the world **wouldn't** have surprised me so much?*

With several previous scientific revolutions – Newtonian physics, Darwinian evolution, special relativity – I feel like I more-or-less know the answers to the above questions. If my intuition isn't yet fully adjusted even to those theories, then at least I know how it *needs* to be adjusted. And thus, for example, if I were creating a new universe, I might or might not decide to make it Lorentz invariant, but I'd certainly *consider* the option, and I'd understand why Lorentz-invariance was the inevitable consequence of a couple of other properties I might want.

But quantum mechanics is different. Here, the physicists assure us, *no one knows* how we should adjust our intuition so that the behavior of subatomic particles would no longer seem so crazy. Indeed, maybe there *is* no way; maybe subatomic behavior will always remain an arbitrary brute fact, with nothing to say about it beyond "such-and-such formulas give you the right answer." My response is radical: if that's true, then *I don't much care* how subatomic particles behave. No doubt other people *need* to know – the people designing lasers or transistors, for example – so let them learn. As for me, I'll simply study another subject that makes more sense to me – like, say, theoretical computer science. Telling me that my physical intuition was wrong, without giving me any path to *correct* that intuition, is like flunking me on an exam without providing any hint about how I could've done better. As soon as I'm free to do so, I'll simply gravitate to other courses where I get As, where my intuition *does* work.

Fortunately, I think that, as the result of decades of work in quantum computation and quantum foundations, we *can* do a lot better today than simply calling quantum mechanics a mysterious brute fact. To spill the beans, here's the perspective of this book:

> *Quantum mechanics is a beautiful generalization of the laws of probability: a generalization based on the 2-norm rather than the 1-norm, and on complex numbers rather than nonnegative real numbers. It can be studied completely separately from its applications to physics (and indeed, doing so provides a good starting point for learning the physical applications later). This generalized probability theory leads naturally to a new model of computation – the quantum computing model – that challenges ideas about computation once considered a priori, and that theoretical computer scientists might have been driven to invent for their own purposes, even if there were no relation to physics. In short, while quantum mechanics was invented a century ago to solve technical problems in physics, today it can be fruitfully explained from an extremely different perspective: as part of the history of ideas, in math, logic, computation, and philosophy, about the limits of the knowable.*

In this book I try to make good on the above claims, taking a leisurely and winding route to do so. I start, in Chapter 1, as near to the "beginning" as I possibly can: with Democritus, the ancient Greek philosopher. Democritus's surviving fragments – which speculate, among other things, that all natural phenomena arise from complicated interactions between a few kinds of tiny "atoms," whizzing around in mostly empty space – get closer to a modern scientific worldview than anything else in antiquity (and certainly closer than any of Plato's or Aristotle's ideas). Yet no sooner had Democritus formulated the atomist hypothesis, than he noticed uneasily its tendency to "swallow whole" the very sense-experiences that he was presumably trying to explain in the first place. How could *those* be reduced to the motions of atoms? Democritus expressed the dilemma in the form of a dialogue between the Intellect and the Senses:

> Intellect: By convention there is sweetness, by convention bitter-ness, by convention color, in reality only atoms and the void.

> Senses: Foolish intellect! Do you seek to overthrow us, while it is from us that you take your evidence?

This two-line dialogue will serve as a sort of touchstone for the entire book. One of my themes will be how quantum mechanics seems to give both the Intellect *and* the Senses unexpected new weapons in their 2300-year-old argument – while still (I think) not producing a clear victory for either.

In Chapters 2 and 3, I move on to discuss the deepest knowledge we have that intentionally *doesn't* depend on "brute facts" about the physical world: namely, mathematics. Even there, something inside me (and, I suspect, inside many other computer scientists!) is suspicious of those *parts* of mathematics that bear the obvious imprint of physics, such as partial differential equations, differential geometry, Lie groups, or anything else that's "too continuous." So instead, I start with some of the most "physics-free" parts of math yet discovered: set theory, logic, and computability. I discuss the great discoveries of Cantor, Frege,

Gödel, Turing, Church, and Cohen, which helped to map the contours of mathematical reasoning itself – and which, in the course of showing why all of mathematics can't be reduced to a fixed "mechanical process," also demonstrated just how much of it *could* be, and clarified what we mean by "mechanical process" in the first place. Since I can't resist, in Chapter 4 I then wade into the hoary debate about whether the human mind, too, is governed by "fixed mechanical processes." I set out the various positions as fairly as I can (but no doubt reveal my biases).

Chapter 5 introduces computability theory's modern cousin, *computational complexity theory*, which plays a central role in the rest of the book. I try to illustrate, in particular, how computational complexity lets us systematically take "deep philosophical mysteries" about the limits of knowledge, and convert them into "merely" insanely difficult unsolved mathematical problems, which arguably capture most of what we want to know! There's no better example of such a conversion than the **P** versus **NP** problem, which I discuss in Chapter 6. Then, as warmups to quantum computing, Chapter 7 examines the many uses of *classical* randomness, both in computational complexity and in other parts of life; and Chapter 8 explains how computational complexity ideas were applied to revolutionize the theory and practice of *cryptography* beginning in the 1970s.

All of that is just to set the stage for the most notorious part of the book: Chapter 9, which presents my view of quantum mechanics as a "generalized probability theory." Then Chapter 10 explains the basics of my own field, the *quantum theory of computation*, which can be briefly defined as the merger of quantum mechanics with computational complexity theory. As a "reward" for persevering through all this technical material, Chapter 11 offers a critical examination of the ideas of Sir Roger Penrose, who famously holds that the brain is not merely a quantum computer but quantum *gravitational* computer, able to solve Turing-uncomputable problems – and that this, or something like it, can be shown by an appeal to Gödel's Incompleteness Theorem. It's child's play to point out the problems with these ideas, and I do so, but what I find more interesting is to ask whether there *might* be nuggets of truth in Penrose's speculations. Then Chapter 12 confronts what I see as the central conceptual problem of quantum mechanics: not that the future is indeterminate (who cares?), but that the past is *also* indeterminate! I examine two very different responses to that problem: first, the appeal, popular among physicists, to *decoherence*, and to the "effective arrow of time" supplied by the Second Law of Thermodynamics; and second, "hidden-variable theories" such as Bohmian mechanics. Even if hidden-variable theories are rejected, I find that they lead to some extremely interesting mathematical questions.

The rest of the book consists of applications of the perspective developed earlier, to various big, exciting, or controversial questions in math, computer science, philosophy, and physics. Much more than the earlier chapters, the later ones discuss *recent research* – mostly in quantum information and computational complexity, but also a bit in quantum gravity and cosmology – that strikes me as having some hope of shedding light on these "big questions." As such, I expect that the last chapters will be the first to become outdated! While there are minor dependencies, to a first approximation the later chapters can be read in any order.

- Chapter 13 discusses new notions of mathematical proof (including probabilistic and zero-knowledge proofs), then applies those notions to understanding the computational complexity of hidden-variable theories.
- Chapter 14 takes up the question of the "size" of quantum states – do they encode an exponential amount of classical information, or not? – and relates this question to the quantum interpretation debate on the one hand, and to recent complexity-theoretic research on quantum proofs and advice on the other.
- Chapter 15 examines the arguments of quantum computing *skeptics*: the people who hold, not merely that building a practical quantum

computer is hard (which everyone agrees about!), but that it can *never be done* for some fundamental reason.

- Chapter 16 examines Hume's Problem of Induction, using it as a jumping-off point for discussing *computational learning theory*, as well as recent work on the learnability of quantum states.

- Chapter 17 discusses some breakthroughs in our understanding of classical and quantum *interactive proof systems* (e.g., the **IP = PSPACE** and **QIP = PSPACE** theorems), but is mostly interested in those breakthroughs insofar as they've led to *non-relativizing circuit lower bounds* – and, therefore, might illuminate something about the **P** versus **NP** question.

- Chapter 18 examines the famous Anthropic Principle and "Doomsday Argument"; the discussion starts out highly philosophical (of course), but eventually winds its way to a discussion of *postselected quantum computing* and the **PostBQP = PP** theorem.

- Chapter 19 discusses Newcomb's Paradox and free will, leading into an account of the Conway–Kochen "free will theorem," and the use of Bell's Inequality to generate "Einstein-certified random numbers."

- Chapter 20 takes up time travel: in a now-familiar pattern, starting with a wide-ranging philosophical discussion, and ending with a proof that classical *or* quantum computers with closed timelike curves yield exactly the computational power of **PSPACE** (under assumptions that are open to interesting objections, which I discuss at length).

- Chapter 21 discusses cosmology, dark energy, the Bekenstein bound, and the holographic principle – but, not surprisingly, with an eye toward what all these things mean for *the limits of computation*. For example, how many bits can one store or search through, and how many operations can one perform on those bits, without using so much energy that one instead creates a black hole?

- Chapter 22 is "dessert": it's based off the final lecture of the Quantum Computing Since Democritus class, in which the students could ask me anything whatsoever, and watch me struggle to respond. Topics addressed include the following: the possible breakdown of quantum mechanics; black holes and "fuzzballs"; the relevance of oracle results in computational complexity; **NP**-complete problems and creativity;

"super-quantum" correlations; derandomization of randomized algorithms; science, religion, and the nature of rationality; and why computer science is not a branch of physics departments.

A final remark. One thing you *won't* find in this book is much discussion of the "practicalities" of quantum computing: either physical implementation, or error correction, or the details of Shor's, Grover's, or other basic quantum algorithms. One reason for this neglect is incidental: the book is based on lectures I gave at the University of Waterloo's Institute for Quantum Computing, and the students were already learning all about those aspects in their other classes. A second reason is that those aspects are covered in *dozens* of other books[7] and online lecture notes (including some of my own), and I saw no need to reinvent the wheel. But a third reason is, frankly, that the technological prospect of building a new kind of computer, exciting as it is, is not why I went into quantum computing in the first place. (*Shhh*, please don't tell any funding agency directors I said that.)

To be clear, I think it's entirely possible that I'll see practical quantum computers in my lifetime (and also possible, of course, that I *won't* see them). And if we *do* get scalable, universal quantum computers, then they'll almost certainly find real applications (not even counting codebreaking): mostly, I think, for specialized tasks like quantum simulation, but to a lesser extent for solving combinatorial optimization problems. If that ever happens, I expect I'll be as excited about it as anyone on earth – and, of course, tickled if any of the work I've done finds applications in that new world. On the other hand, if someone gave me a practical quantum computer tomorrow, then I confess that I can't think of anything that I, personally, would want to use it for: only things that *other people* could use it for!

---

[7] The "standard reference" for the field remains *Quantum Computation and Quantum Information*, by Michael Nielsen and Isaac Chuang.

Partly for that reason, if scalable quantum computing were proved to be *im*possible, that would excite me a thousand times more than if it were proved to be possible. For such a failure would imply something wrong or incomplete with our understanding of quantum mechanics itself: a revolution in physics! As a congenital pessimist, though, my *guess* is that Nature won't be so kind to us, and that scalable quantum computing will turn out to be possible after all.

In summary, you could say that I'm in this field less because of what you could do with a quantum computer, than because of what the *possibility* of quantum computers *already* does to our conception of the world. *Either* practical quantum computers can be built, and the limits of the knowable are not what we thought they are; *or* they can't be built, and the principles of quantum mechanics themselves need revision; *or* there's a yet-undreamt method to simulate quantum mechanics efficiently using a conventional computer. All three of these possibilities sound like crackpot speculations, but at least one of them is right! So whichever the outcome, what can one say but – to reverse-plagiarize a certain TV commercial – "that's interesting?"

WHAT'S NEW

In revising this manuscript for publication, the biggest surprise for me was how much *happened* in the fields discussed by the book between when I originally gave the lectures (2006) and "now" (2013). This book is supposed to be about deep questions that are as old as science and philosophy, or at the least, as old as the birth of quantum mechanics and of computer science almost a century ago. And at least on a day-to-day basis, it can *feel* like nothing ever changes in the discussion of these questions. And thus, having to update my lectures extensively, after the passage of a mere six years, was an indescribably pleasant burden for me.

Just to show you how things are evolving, let me give a partial list of the developments that are covered in this book, but that *couldn't* have been covered in my original 2006 lectures, for the simple reason that they hadn't happened yet. IBM's Watson computer defeated the *Jeopardy!* world champion Ken Jennings, forcing me to update my discussion of AI with a new example (see p. 37), very different in character from previous examples like ELIZA and Deep Blue. Virginia Vassilevska Williams, building on work of Andrew Stothers, discovered how to multiply two $n \times n$ matrices using only $O(n^{2.373})$ steps, *slightly* beating Coppersmith and Winograd's previous record of $O(n^{2.376})$, which had held for so long that "2.376" had come to feel like a constant of nature (see p. 49).

There were major advances in the area of *lattice-based cryptography*, which provides the leading candidates for public-key encryption systems secure even against quantum computers (see pp. 105–107). Most notably, solving a 30-year-old open problem, Craig Gentry used lattices to propose the first *fully homomorphic cryptosystems.* These systems let a client delegate an arbitrary computation to an untrusted server – feeding the server encrypted inputs and getting back an encrypted output – in such a way that only the client can decrypt (and verify) the output; the server never has any clue what computation it was hired to perform.

In the foundations of quantum mechanics, Chiribella *et al.* (see p. 131) gave a novel argument for "why" quantum mechanics should involve the specific rules it does. Namely, they proved that those rules are the only ones compatible with certain general axioms of probability theory, *together with* the slightly mysterious axiom that "all mixed states can be purified": that is, whenever you don't know everything there is to know about a physical system A, your ignorance must be fully explainable by positing correlations between A and some faraway system B, such that you *would* know everything there is to know about the combined system AB.

In quantum computing theory, Bernstein and Vazirani's "Recursive Fourier Sampling" (RFS) problem – on which I spent a fair bit of time in my 2006 lectures – has been superseded by my "Fourier Checking" problem (see p. 145). RFS retains its place in history, as the first black-box problem ever proposed that a quantum computer can provably solve superpolynomially faster than a classical probabilistic computer – and, as such, an important forerunner to Simon's and Shor's breakthroughs. Today, though, if we want a candidate for a problem in **BQP\PH** – in other words, something that a quantum computer can easily do, but which is not even in the classical "polynomial-time hierarchy" – then Fourier Checking seems superior to RFS in every way.

Happily, several things discussed as "open problems" in my 2006 lectures have since lost that status. For example, Andrew Drucker and I showed that **BQP/qpoly** is contained in **QMA/poly** (and, moreover, the proof relativizes), falsifying my conjecture that there should be an oracle separation between those classes (see p. 214). Also, in a justly celebrated breakthrough in quantum computing theory, Jain *et al.* proved that **QIP = PSPACE** (see p. 263), meaning that quantum interactive proof systems are no more powerful than classical ones. In that case, at least, I conjectured the right answer! (There was actually *another* breakthrough in the study of quantum interactive proof systems, which I *don't* discuss in the book. My postdoc Thomas Vidick, together with Tsuyoshi Ito,[8] recently showed that **NEXP ⊆ MIP***, which means that any *multiple*-prover interactive proof system can be "immunized" against the possibility that the provers secretly coordinate their responses using quantum entanglement.)

Chapter 20 of this book discusses David Deutsch's model for quantum mechanics in the presence of closed timelike curves, as well as my (then-)new result, with John Watrous, that Deutsch's model provides exactly the computational power of **PSPACE**. (So that, in particular, quantum time-travel computers would be no more powerful than *classical* time-travel computers, in case you

---

[8] T. Ito and T. Vidick, A Multi-prover Interactive Proof for NEXP Sound against Entangled Provers. In *Proceedings of IEEE Symposium on Foundations of Computer Science* (2012), pp. 243–252.

were wondering.) Since 2006, however, there have been important papers questioning the assumptions behind Deutsch's model, and proposing alternative models, which generally lead to computational power *less* than **PSPACE**. For example, one model, proposed by Lloyd *et al.*, would "merely" let the time traveler solve all problems in **PP**! I discuss these developments on pp. 319–322.

What about circuit lower bounds – which is theoretical computer scientists' codeword for "trying to prove **P** ≠ **NP**," in much the same way that "closed timelike curves" is the physicists' codeword for "time travel?" I'm pleased to report that there have been interesting developments since 2006, certainly more than I would have expected back then. As one example, Rahul Santhanam used interactive proof techniques to prove the non-relativizing result that the class **PromiseMA** doesn't have circuits of any fixed polynomial size (see p. 257). Santhanam's result was part of what spurred Avi Wigderson and myself, in 2007, to formulate the *algebrization barrier* (see p. 258), a generalization of Baker, Gill, and Solovay's relativization barrier from the 1970s (see pp. 245–246). Algebrization explained why the interactive proof techniques can take us only so far and no further in our quest to prove **P** ≠ **NP**: as one example, why those techniques led to superlinear circuit lower bounds for **PromiseMA**, but not for the class **NP** just "slightly below it." The challenge we raised was to find new circuit lower bound techniques that convincingly *evade* the algebrization barrier. That challenge was met in 2010, by Ryan Williams' breakthrough proof that **NEXP** ⊄ **ACC⁰** (discussed on pp. 260–261).

Of course, even Williams' result, exciting as it was, is a helluva long way from a proof of **P** ≠ **NP**. But the past six years have also witnessed a flowering of interest in, and development of, Ketan Mulmuley's Geometric Complexity Theory (GCT) program (see pp. 261–262), which is to proving **P** ≠ **NP** almost exactly as string theory is to the goal of a unified theory of physics. That is, in terms of concrete results, the GCT program hasn't yet come anywhere close to fulfilling its initial hopes, and even the program's most

ardent proponents predict a slog of many decades, while its
mathematical complexities frighten everyone else. What GCT has
going for it is two things: firstly, that it's forged mathematical
connections "too profound and striking to be mere coincidence,"
and secondly, that it's perceived (by no means universally!) as "the
only game in town," the only hunter currently in the forest who's
even carrying a sharp stick.

Let me mention just three other post-2006 developments
relevant to this book. In 2011, Alex Arkhipov and I proposed
"BosonSampling" (see pp. 287–288): a rudimentary, almost certainly
non-universal quantum computing model involving non-interacting
photons, which was just recently demonstrated on a small scale.
Interestingly, the evidence that BosonSampling is hard to simulate
on a classical computer seems *stronger* than the evidence that (say)
Shor's factoring algorithm is hard to simulate. In 2012, Umesh
Vazirani and Thomas Vidick, building on earlier work of Pironio *et
al.*, showed how to use violations of the Bell inequality to achieve
exponential randomness expansion (see p. 305): that is, converting n
random bits into 2^n bits that are guaranteed to be almost-perfectly
random, *unless* Nature resorted to faster-than-light communication
to bias the bits. Meanwhile, the debate about the "black hole
information paradox" – i.e., the apparent conflict between the
principles of quantum mechanics and the locality of spacetime,
when bits or qubits are dropped into a black hole – has evolved in
new directions since 2006. Possibly the two most important
developments have been the increasing popularity and
sophistication of Samir Mathur's "fuzzball" picture of black holes,
and the controversial argument of Almheiri *et al.* that an observer
falling into a black hole would never even get near the singularity,
but would instead encounter a "firewall" and burn up at the event
horizon. I cover these developments as best I can on pp. 346–349.

A few updates were occasioned not by any new discovery or
argument, but simply by me (gasp) *changing my mind* about
something. One example is my attitude toward the arguments of
John Searle and Roger Penrose against "strong artificial

intelligence." As you'll see in Chapters 4 and 11, I still think Searle and Penrose are *wrong* on crucial points, Searle more so than Penrose. But on rereading my 2006 arguments for *why* they were wrong, I found myself wincing at the semi-flippant tone, at my eagerness to *laugh* at these celebrated scholars tying themselves into logical pretzels in quixotic, obviously doomed attempts to defend human specialness. In effect, I was lazily relying on the fact that everyone in the room already agreed with me – that to these (mostly) physics and computer science graduate students, it was simply self-evident that the human brain is nothing other than a "hot, wet Turing machine," and weird that I would even waste the class's time with such a settled question. Since then, I *think* I've come to a better appreciation of the immense difficulty of these issues – and in particular, of the need to offer arguments that engage people with different philosophical starting-points than one's own.

Here's hoping that, in 2020, this book will be as badly in need of revision as the 2006 lecture notes were in 2013.

Scott Aaronson
Cambridge, MA
January 2013

Acknowledgments

As my summer student in 2008, Chris Granade enthusiastically took charge of converting the scattered notes and audio recordings from my course into coherent drafts that I could post on my website, the first step on their long journey into book form. More recently, Alex Arkhipov, my phenomenal PhD student at MIT, went through the drafts with a fine-tooth comb, flagging passages that were wrong, unclear, or no longer relevant. I'm deeply grateful to both of them: this book is also *their* book; it wouldn't exist without their help.

It also wouldn't exist without Simon Capelin, my editor at Cambridge University Press, who approached me with the idea. Simon understood what I needed: he prodded me every few months to see if I'd made progress, but never in an accusatory way, always relying on my own internal guilt to see the project through. (And I *did* see it through – eventually.) Simon also assured me that, even though *Quantum Computing since Democritus* was . . . a bit *different* from CUP's normal fare, he would make every effort to preserve what he called the book's "quirky charm." I also thank all the others at CUP and Aptara Corp. who helped to make the book a reality: Sarah Hamilton, Emma Walker, and Disha Malhotra.

I thank the students and faculty who sat in on my "Quantum Computing since Democritus" course at the University of Waterloo in Fall 2006. Their questions and arguments made the course what it was (as you can still see in this book, especially in the last chapters). On top of that, the students also took care of the audio recordings and preliminary written transcripts. More broadly, I remember my two years as a postdoc at Waterloo's Institute for Quantum Computing as one of the happiest times of my life. I thank everyone there, and especially IQC's director Ray Laflamme, for not only *letting* me teach such a nutty course but *encouraging* it, and even (in Ray's and several other cases) sitting in on the course themselves and contributing many insights.

I thank MIT's Computer Science and Artificial Intelligence Laboratory and its Electrical Engineering and Computer Science Department, as well as the US National Science Foundation, the Defense Advanced Research Projects Agency, the Sloan Foundation, and TIBCO, Inc., for all the support they've given me over the last six years.

I thank the readers of my blog, *Shtetl-Optimized* (http://www.scottaaronson.com/blog), for their many comments on the draft chapters that I posted there, and for catching numerous errors. I especially thank those readers who encouraged me to turn *Quantum Computing since Democritus* into a book – some even promised they'd buy it when it came out.

I thank the people who advised me from my high school to my postdoc years: Chris Lynch, Bart Selman, Lov Grover, Umesh Vazirani, and Avi Wigderson. John Preskill was never "formally" an advisor, but I still think of him as one. I owe all of them more than I can say. I also thank everyone else in (and beyond) the quantum information and theoretical computer science communities whose discussions and arguments with me over the years left their imprints on this book. I can't possibly produce a full list of such people, so here's a partial one: Dorit Aharonov, Andris Ambainis, Dave Bacon, Michael Ben-Or, Raphael Bousso, Harry Buhrman, Sean Carroll, Greg Chaitin, Richard Cleve, David Deutsch, Andy Drucker, Ed Farhi, Chris Fuchs, Daniel Gottesman, Alex Halderman, Robin Hanson, Richard Karp, Elham Kashefi, Julia Kempe, Greg Kuperberg, Seth Lloyd, Michele Mosca, Michael Nielsen, Christos Papadimitriou, Len Schulman, Lenny Susskind, Oded Regev, Barbara Terhal, Michael Vassar, John Watrous, Ronald de Wolf. I apologize for the inevitable omissions (or to those who don't want their names in this book, you're welcome!).

I thank the following alert readers for catching errors and omissions in the first printing of this book: Boaz Barak, Evan Berkowitz, Ernest Davis, Bob Galesloot, Vijay Ganesh, Yuri Gurevich, John Kadvany, Andrew Marks, Cris Moore, Reviel Netz, and Tyler Singer-Clark.

Lastly, I thank my mom and dad, my brother David, and of course my wife Dana, who will now finally be able to know me while I'm *not* putting off finishing the damn book.

I   Atoms and the void

> I would rather discover a single cause than become king of the
> Persians.
>
> – Democritus

So why Democritus? First of all, who *was* Democritus? He was this
Ancient Greek dude. He was born around 450 BC in this podunk
Greek town called Abdera, where people from Athens said that even
the air causes stupidity. He was a disciple of Leucippus, according
to my source, which is Wikipedia. He's called a "pre-Socratic," even
though actually he was a contemporary of Socrates. That gives you
a sense of how important he's considered: "Yeah, the *pre*-Socratics –
maybe stick 'em in somewhere in the first week of class." Incidentally,
there's a story that Democritus journeyed to Athens to meet Socrates,
but then was too shy to introduce himself.

Almost none of Democritus's writings survive. Some survived
into the Middle Ages, but they're lost now. What we know about him
is mostly due to other philosophers, like Aristotle, bringing him up
in order to criticize him.

So, what did they criticize? Democritus thought the whole universe is composed of atoms in a void, constantly moving around
according to determinate, understandable laws. These atoms can hit
each other and bounce off, or they can stick together to make bigger
things. They can have different sizes, weights, and shapes – maybe
some are spheres, some are cylinders, whatever. On the other hand,
Democritus says that properties like color and taste are *not* intrinsic
to atoms, but instead emerge out of the interactions of many atoms.

For if the atoms that made up the ocean were "intrinsically blue," then how could they form the white froth on waves?

Remember, this is 400 BC. So far we're batting pretty well.

Why does Democritus think that things are made of atoms? He gives a few arguments, one of which can be paraphrased as follows: suppose we have an apple, and suppose the apple is made not of atoms but of continuous, hard stuff. And suppose we take a knife and slice the apple into two pieces. It's clear that the points on one side go into the first piece and the points on the other side go into the second piece, but what about the points exactly on the boundary? Do they disappear? Do they get duplicated? Is the symmetry broken? None of these possibilities seem particularly elegant.

Incidentally, there's a debate raging even today between atomists and anti-atomists. At issue in this debate is whether space and time *themselves* are made up of indivisible atoms, at the Planck scale of 10^{-33} cm or 10^{-43} s. Once again, the physicists have very little experimental evidence to go on, and are basically in the same situation that Democritus was in, 2400 years ago. If you want an ignorant, uninformed layperson's opinion, my money is on the atomist side. And the arguments I'd use are not *entirely* different from the ones Democritus used: they hinge mostly on inherent mathematical difficulties with the continuum.

One passage of Democritus that does survive is a dialogue between the intellect and the senses. The intellect starts out, saying "By convention there is sweetness, by convention bitterness, by convention color, in reality only atoms and the void." For me, this single line already puts Democritus shoulder to shoulder with Plato, Aristotle, or any other ancient philosopher you care to name: it would be hard to give a more accurate one-sentence summary of the entire scientific worldview that would develop 2000 years later! But the dialogue doesn't stop there. The senses respond, saying "Foolish intellect! Do you seek to overthrow us, while it is from us that you take your evidence?"

I first came across this dialogue in a book by Schrödinger.[1] Ah, Schrödinger! – you see we're inching toward the "quantum computing" in the book title. We're gonna get there, don't worry about that.

But why would Schrödinger be interested in this dialogue? Well, Schrödinger was interested in a lot of things. He was not an intellectual monogamist (or really any kind of monogamist). But one reason he *might've* been interested is that he was one of the originators of quantum mechanics – in my opinion the most surprising discovery of the twentieth century (relativity is a close second), and a theory that adds a whole new angle to the millennia-old debate between the intellect and the senses, even as it fails to resolve it.

Here's the thing: for any isolated region of the universe that you want to consider, quantum mechanics describes the evolution in time of the state of that region, which we represent as a linear combination – a *superposition* – of all the possible configurations of elementary particles in that region. So, this is a bizarre picture of reality, where a given particle is not *here*, not *there*, but in a sort of weighted sum over all the places it could be. But it works. As we all know, it does pretty well at describing the "atoms and the void" that Democritus talked about.

The part where it maybe doesn't do so well is the "from us you take your evidence" part. What's the problem? Well, if you take quantum mechanics seriously, *you yourself* ought to be in a superposition of different places at once. After all, you're made of elementary particles too, right? In particular, suppose you measure a particle that's in a superposition of two locations, A and B. Then the most naive, straightforward reading of quantum mechanics would predict that the universe itself should split into two "branches": one where the particle is at A and you see it at A, one where the particle is at B and you see it at B! So what do you think: *do* you split into several copies of yourself every time you look at something? I don't *feel* like I do!

---

[1] E. Schrödinger, *What is Life? With Mind and Matter and Autobiographical Sketches*, Cambridge University Press (reprinted edition), 2012.

You might wonder how such a crazy theory could be *useful* to physicists, even at the crassest level. How could it even make *predictions*, if it essentially says that everything that could happen does? Well, the thing I didn't tell you is that there's a separate rule for what happens when you make a measurement: a rule that's "tacked on" (so to speak), external to the equations themselves. That rule says, essentially, that the act of looking at a particle *forces it to make up its mind* about where it wants to be, and that the particle makes its choice *probabilistically*. And the rule tells you exactly how to calculate the probabilities. And of course it's been spectacularly well confirmed.

But here's the problem: as the universe is chugging along, doing its thing, how are we supposed to know when to apply this measurement rule, and when not to? What counts as a "measurement," anyway? The laws of physics aren't supposed to say things like "such-and-such happens *until someone looks*, and then a completely different thing happens!" Physical laws are supposed to be *universal*. They're supposed to describe human beings the same way they describe supernovas and quasars: all just examples of vast, complicated clumps of particles interacting according to simple rules.

So from a physics perspective, things would be so much cleaner if we could dispense with this "measurement" business entirely! Then we could say, in a more sophisticated update of Democritus: there's nothing but atoms and the void, evolving in quantum superposition.

But wait: if we're not here making nosy measurements, wrecking the pristine beauty of quantum mechanics, then how did "we" (whatever that means) ever get the evidence in the first place that quantum mechanics is true? How did we ever come to believe in this theory that seems so uncomfortable with the fact of our own existence?

So, that's the modern version of the Democritus dilemma, and physicists and philosophers have been arguing about it for almost a hundred years, and in this book we're not going to solve it.

The other thing I'm not going to do in this book is try to sell you on some favorite "interpretation" of quantum mechanics. You're free to believe whatever interpretation your conscience dictates. (What's my own view? Well, I agree with *every* interpretation to the extent it says there's a problem, and disagree with every interpretation to the extent it claims to have solved the problem!)

See, just like we can classify religions as monotheistic and polytheistic, we can classify interpretations of quantum mechanics by where they come down on the "putting-yourself-in-coherent-superposition" issue. On the one side, we've got the interpretations that enthusiastically sweep the issue under the rug: Copenhagen and its Bayesian and epistemic grandchildren. In these interpretations, you've got your quantum system, you've got your measuring device, and there's a line between them. Sure, the line can shift from one experiment to the next, but for any given experiment, it's gotta be somewhere. In principle, you can even imagine putting other people on the quantum side, but you *yourself* are always on the classical side. Why? Because a quantum state is just a representation of your knowledge – and you, by definition, are a classical being.

But what if you want to apply quantum mechanics to the whole universe, *including* yourself? The answer, in the epistemic-type interpretations, is simply that you don't ask that sort of question! Incidentally, that was Bohr's all-time favorite philosophical move, his WWF piledriver: "You're not allowed to ask such a question!"

On the other side, we've got the interpretations that *do* try in different ways to make sense of putting yourself in superposition: many-worlds, Bohmian mechanics, etc.

Now, to hardheaded problem-solvers like ourselves, this might seem like a big dispute over words – why bother? I actually agree with that: if it were just a dispute over words, then we *shouldn't* bother! But as David Deutsch pointed out in the late 1970s, we *can* conceive of experiments that would differentiate the first type of interpretation from the second type. The simplest experiment would just be to put yourself in coherent superposition and see what happens! Or if that's too dangerous, put someone *else* in coherent superposition. The point being that, if human beings were regularly put into superposition, then the whole business of drawing a line between "classical observers" and the rest of the universe would become untenable.

But alright – human brains are wet, goopy, sloppy things, and maybe we won't be able to maintain them in coherent superposition for 500 million years. So what's the next best thing? Well, we could try to put a *computer* in superposition. The more sophisticated the computer was – the more it resembled something like a brain, like ourselves – the further up we would have pushed the "line" between quantum and classical. You can see how it's only a minuscule step from here to the idea of quantum computing.

I'd like to draw a more general lesson here. What's the point of talking about philosophical questions? Because we're going to be doing a fair bit of it here – I mean, of philosophical bullshitting. Well, there's a standard answer, and it's that philosophy is an intellectual clean-up job – the janitors who come in after the scientists have made a mess, to try and pick up the pieces. So in this view, philosophers sit in their armchairs waiting for something surprising to happen in science – like quantum mechanics, like the Bell inequality, like Gödel's Theorem – and then (to switch metaphors) swoop in like vultures and say, ah, this is what it *really* meant.

Well, on its face, that seems sort of boring. But as you get more accustomed to this sort of work, I think what you'll find is . . . it's *still* boring!

Personally, I'm interested in results – in finding solutions to nontrivial, well-defined open problems. So, what's the role of philosophy in that? I want to suggest a more exalted role than intellectual janitor: philosophy can be a *scout*. It can be an explorer – mapping out intellectual terrain for science to *later* move in on, and build condominiums on or whatever. Not every branch of science was scouted out ahead of time by philosophy, but some were. And in recent history, I think quantum computing is really the poster child here. It's fine to tell people to "Shut up and calculate," but the question is, *what* should they calculate? At least in quantum computing, which is my field, the sorts of things that we like to calculate – capacities of quantum channels, error probabilities of quantum algorithms – are things people would never have *thought* to calculate if not for philosophy.

2 Sets

Here, we're gonna talk about sets. What will these sets contain? Other sets! Like a bunch of cardboard boxes that you open only to find *more* cardboard boxes, and so on all the way down.

You might ask "how is this relevant to a book on quantum computing?"

Well, hopefully we'll see a few answers later. For now, suffice it to say that math is the foundation of all human thought, and set theory – countable, uncountable, etc. – that's the foundation of math. So regardless of what a book is about, it seems like a fine place to start.

I probably should tell you explicitly that I'm compressing a whole math course into this chapter. On the one hand, that means I don't really expect you to understand everything. On the other hand, to the extent you do understand – hey! You got a whole math course in one chapter! You're welcome.

So let's start with the empty set and see how far we get.

THE EMPTY SET.

Any questions so far?

Actually, before we talk about sets, we need a language for talking about sets. The language that Frege, Russell, and others developed is called *first-order logic*. It includes Boolean connectives (and, or, not), the equals sign, parentheses, variables, predicates, quantifiers ("there exists" and "for all") – and that's about it. I'm told that the physicists have trouble with these. Hey, I'm just ribbin' ya. If you haven't seen this way of thinking before, then you haven't seen it. But maybe, for the benefit of the physicists, let's go over the basic rules of logic.

RULES OF FIRST-ORDER LOGIC

The rules all concern how to construct sentences that are *valid* – which, informally, means "tautologically true" (true for all possible settings of the variables), but which for now we can just think of as a combinatorial property of certain strings of symbols. I'll write logical sentences in a typewriter font in order to distinguish them from the surrounding English.

- **Propositional tautologies:** A or not A, not (A and not A), etc., are valid.
- **Modus ponens:** If A is valid and A implies B is valid, then B is valid.
- **Equality rules:** x=x, x=y implies y=x, x=y and y=z implies x=z, and x=y implies f(x)=f(y) are all valid.
- **Change of variables:** Changing variable names leaves a statement valid.
- **Quantifier elimination:** If For all x, A(x) is valid, then A(y) is valid for any y.
- **Quantifier addition:** If A(y) is valid where y is an unrestricted variable, then For all x, A(x) is valid.
- **Quantifier rules:** If not (For all x, A(x)) is valid, then There exists an x such that not (A(x)) is valid. Etc.

So, for example, here are the Peano axioms for the nonnegative integers written in first-order logic. In these, $S(x)$ is the successor function, intuitively $S(x) = x + 1$, and I'm assuming functions have already been defined.

PEANO AXIOMS FOR THE NONNEGATIVE INTEGERS

- **Zero exists:** There exists a z such that for all x, S(x) is not equal to z. (This z is taken to be 0.)
- **Every integer has at most one predecessor:** For all x,y, if S(x)=S(y) then x=y.

The nonnegative integers themselves are called a *model* for the axioms: in logic, the word "model" just means any collection of objects and functions of those objects that satisfies the axioms. Interestingly, though, just as the axioms of group theory can be satisfied by many different groups, so too the nonnegative integers are not the only model of the Peano axioms. For example, you should check that you can get another valid model by adding extra, made-up integers that aren't reachable from 0 – integers 'beyond infinity,' so to speak. Though once you add one such integer, you need to add infinitely many of them, since every integer needs a successor.

Writing down these axioms seems like pointless hairsplitting – and indeed, there's an obvious chicken-and-egg problem. How can we state axioms that will put the integers on a more secure foundation, when the very symbols and so on that we're using to write down the axioms presuppose that we *already know* what the integers are?

Well, precisely because of this point, I *don't* think that axioms and formal logic can be used to place arithmetic on a more secure foundation. If you don't already agree that $1 + 1 = 2$, then a lifetime of studying mathematical logic won't make it any clearer! But this stuff is still extremely interesting for at least three reasons.

1. The situation will change once we start talking not about integers, but about different sizes of infinity. There, writing down axioms and working out their consequences is pretty much all we have to go on!
2. Once we've formalized everything, we can then program a computer to reason for us:
   - **Premise 1:** For all x, if A(x) is true, then B(x) is true.
   - **Premise 2:** There exists an x such that A(x) is true.
   - **Conclusion:** There exists an x such that B(x) is true.
   Well, you get the idea. The point is that deriving the conclusion from the premises is purely a *syntactic* operation – one that doesn't require any understanding of what the statements mean.
3. Besides having a computer find proofs for us, we can also treat proofs themselves as mathematical objects, which opens the way to *metamathematics*.

Anyway, enough pussyfooting around. Let's see some axioms for set theory. I'll state the axioms in English; converting them to first-order logic is left as an exercise for the reader in most cases.

AXIOMS OF SET THEORY

The axioms all involve a universe of objects called "sets," and a relationship between sets that's called "membership" or "containment" and written using the symbol $\in$. Every operation on sets will ultimately be defined in terms of the containment relationship.

- **Empty set:** There exists an empty set: that is, a set x for which there is no y such that $y \in x$.
- **Extensionality:** If two sets contain the same members, then the sets are equal. That is, for all x and y, if ($z \in x$ if and only if $z \in y$ for all z), then $x = y$.
- **Pairing:** For all sets x and y, there exists a set $z = \{x, y\}$: that is, a set z such that, for all w, $w \in z$ if and only if ($w = x$ or $w = y$).
- **Union:** For all sets x, there exists a set equal to the union of all sets in x.
- **Existence of infinite sets:** There exists a set x that contains the empty set and that contains $\{y\}$ for every $y \in x$. (Why must this x have infinitely many elements?)
- **Power set:** For all sets x, there exists a set consisting of the subsets of x.
- **Replacement (actually an infinity of axioms, one for every function A mapping sets to sets):** For all sets x, there exists a set $z = \{A(y) \mid y \in x\}$, which results from applying A to all the elements of x. (Technically, one also has to define what one means by a "function mapping sets to sets," something that can be done although I won't do it here.)
- **Foundation:** All nonempty sets x have a member y such that for all z, either $z \notin x$ or $z \notin y$. (This is a technical axiom, whose point is to rule out sets like $\{\{\{\ldots\}\}\}$.)

These axioms – called the Zermelo–Fraenkel axioms – are the foundation for basically all of math. So I thought you should see them at least once in your life.

Alright, one of the most basic questions we can ask about a set is: how big is it? What's its size, its cardinality? Meaning, how many elements does it have? You might say, just count the elements. But what if there are infinitely many? Are there more integers than odd integers? This brings us to Georg Cantor (1845–1918), and the first of his several enormous contributions to human knowledge. He says two sets have the same cardinality if and only if their elements can be put in one-to-one correspondence. Period. And if, no matter how you try to pair off the elements, one set always has elements left over, the set with the elements left over is the bigger set.

What possible cardinalities are there? Of course, there are finite ones, one for each natural number. Then there's the first infinite cardinality, the cardinality of the integers, which Cantor called $\aleph_0$ ("aleph-zero"). The rational numbers have the same cardinality $\aleph_0$, a fact that's also expressed by saying that the rational numbers are *countable*, meaning that they can be placed in one-to-one correspondence with the integers. In other words, we can make an infinite list of them so that each rational number appears eventually in the list.

What's the proof that the rational numbers are countable? You haven't seen it before? Oh, alright. First, list 0 and then all the rational numbers where the sum of absolute values of the numerator and denominator is 2. Then, list all the rational numbers where the sum of absolute values of the numerator and denominator is 3. And so on. It's clear that every rational number will eventually appear in this list. Hence, there's only a countable infinity of them. QED.

But Cantor's biggest contribution was to show that not *every* infinity is countable – so, for example, the infinity of real numbers is greater than the infinity of integers. More generally, just as there are infinitely many numbers, there are also infinitely many infinities.

You haven't seen the proof of that either? Alright, alright. Let's say you have an infinite set A. We'll show how to produce another infinite set, B, which is even bigger than A. This B will simply be the set of all *subsets* of A, which is guaranteed to exist by the power set axiom. How do we know B is bigger than A? Well, suppose we could pair off every element $a \in A$ with an element $f(a) \in B$, in such a way that no elements of B were left over. Then, we could define a new subset $S \subseteq A$, consisting of *every a that's not contained in f(a)*. Then S is also an element of B. But notice that S can't have been paired off with any $a \in A$ – since otherwise, a would be contained in $f(a)$ if and only if it *wasn't* contained in $f(a)$, contradiction. Therefore, B is larger than A, and we've ended up with a bigger infinity than the one we started with.

This is certainly one of the four or five greatest proofs in all of math – again, good to see at least once in your life.

Besides cardinal numbers, it's also useful to discuss *ordinal* numbers. Rather than defining these, it's easier to just illustrate them. We start with the natural numbers:

$$0, 1, 2, 3, \ldots$$

Then, we say, let's *define* something that's greater than every natural number:

$$\omega$$

What comes after ω?

$$\omega + 1, \omega + 2, \ldots$$

Now, what comes after all of these?

$$2\omega$$

Alright, we get the idea:

$$3\omega, 4\omega, \ldots$$

Alright, we get the idea:

$$\omega^2, \omega^3, \ldots$$

Alright, we get the idea:

$$\omega^\omega, \omega^{\omega^\omega}, \ldots$$

We could go on for quite a while! Basically, for any set of ordinal numbers (finite or infinite), we stipulate that there's a first ordinal number that comes after everything in that set.

The set of ordinal numbers has the important property of being *well ordered*, which means that every subset has a minimum element. This is unlike the integers or the positive real numbers, where any element has another that comes before it.

Now, here's something interesting. All of the ordinal numbers I've listed have a special property, which is that they have at most countably many predecessors (i.e., at most $\aleph_0$ of them). What if we consider the set of *all* ordinals with at most countably many predecessors? Well, that set also has a successor, call it α. But does α itself have $\aleph_0$ predecessors? Certainly not, since otherwise α wouldn't be the successor to the set; it would be *in* the set! The set of predecessors of α has the next possible cardinality, which is called $\aleph_1$.

What this sort of argument proves is that the set of cardinalities is *itself* well ordered. After the infinity of the integers, there's a "next bigger infinity," and a "next bigger infinity after that," and so on. You never see an infinite decreasing sequence of infinities, as you do with the real numbers.

So, starting from $\aleph_0$ (the cardinality of the integers), we've seen two different ways to produce "bigger infinities than infinity." One of these ways yields the cardinality of sets of integers (or, equivalently, the cardinality of real numbers), which we denote $2^{\aleph_0}$. The other way yields $\aleph_1$. Is $2^{\aleph_0}$ *equal* to $\aleph_1$? Or to put it another way: is there any infinity of *intermediate* size between the infinity of the integers and the infinity of the reals?

Well, this question was David Hilbert's first problem in his famous 1900 address. It stood as one of the great math problems for over half a century, until it was finally "solved" (in a somewhat disappointing way, as you'll see).

Cantor himself believed there were no intermediate infinities, and called this conjecture the Continuum Hypothesis. Cantor was extremely frustrated with himself for not being able to prove it.

Besides the Continuum Hypothesis, there's another statement about these infinite sets that no one could prove or disprove from the Zermelo–Fraenkel axioms. This statement is the infamous Axiom of Choice. It says that, if you have a (possibly infinite) set of sets, then it's possible to form a new set by choosing one item from each set. Sound reasonable? Well, if you accept it, you also have to accept that there's a way to cut a solid sphere into a finite number of pieces, and then rearrange those pieces into another solid sphere a thousand times its size. (That's the "Banach–Tarski paradox." Admittedly, the "pieces" are a bit hard to cut out with a knife...)

Why does the Axiom of Choice have such dramatic consequences? Basically, because it asserts that certain sets exist, but without giving any rule for *forming* those sets. As Bertrand Russell put it: "To choose one sock from each of infinitely many pairs of socks requires the Axiom of Choice, but for shoes the Axiom is not needed." (What's the difference?)

The Axiom of Choice turns out to be equivalent to the statement that every set can be well ordered: in other words, the elements of any set can be paired off with the ordinals $0, 1, 2, \ldots, \omega, \omega + 1, \ldots,$ $2\omega, 3\omega, \ldots$ up to some ordinal. If you think, for example, about the set of real numbers, this seems far from obvious.

It's easy to see that well-ordering implies the Axiom of Choice: just well-order the whole infinity of socks, then choose the sock from each pair that comes first in the ordering.

Do you want to see the other direction? Why the Axiom of Choice implies that every set can be well ordered? Yes?

OK! We have a set A that we want to well-order. For every proper subset $B \subset A$, we'll use the Axiom of Choice to pick an element $f(B) \in A - B$ (where $A - B$ means the set of all elements of A that aren't also elements of B). Now we can start well-ordering A, as follows: first let $s_0 = f(\{\})$, then let $s_1 = f(\{s_0\})$, $s_2 = f(\{s_0, s_1\})$, and so on.

Can this process go on forever? No, it can't. For if it did, then by a process of "transfinite induction," we could stuff arbitrarily large infinite cardinalities into A. And while admittedly A is infinite, it has at most a *fixed* infinite size! So the process has to stop somewhere. But where? At a proper subset B of A? No, it can't do that either – since if it did, then we'd just continue the process by adding $f(B)$. So the only place it can stop is A itself. Therefore, A can be well ordered.

Earlier I mentioned inherent mathematical difficulties with the continuum, so I've got a puzzle somewhat related to that.

You know the real line, right? Suppose we want a union of open intervals (perhaps infinitely many) that covers every rational point. Question: does the sum of the lengths of the intervals have to be infinite? One would certainly think so! After all, there are rational numbers pretty much everywhere!

Solution: Not only can the sum of the lengths of the intervals be finite, it can be arbitrarily close to zero! Simply enumerate the rational numbers, r_0, r_1, etc. Then put an interval of size $\varepsilon/2^i$ around r_i for every i.

Here's a harder one: we want a subset S of the points (x, y) in unit square $[0, 1]^2$ so that, for every real number $x \in [0, 1]$, there's only a countable number of y in $[0, 1]$ for which (x, y) in S. Can we choose S so that, for every $(x, y) \in [0, 1]^2$, either $(x, y) \in$ S or $(y, x) \in$ S?

I'll give you two answers: that it isn't possible, and that it *is* possible.

We'll start with why it isn't possible. For this, I'll assume that the Continuum Hypothesis is false. Then, there's some proper subset $A \subset [0, 1]$ that has cardinality $\aleph_1$. Let B be the set of all y that appear in points $(x, y) \in$ S over all $x \in$ A. Since for each x there's countably many such y, the set B also has cardinality $\aleph_1$. So, since we assumed that $\aleph_1$ is less than $2^{\aleph_0}$, there must be some $y_0 \in [0, 1]$ not in B. Now observe that there are $\aleph_1$ real numbers $x \in$ A, but none of them satisfy $(x, y_0) \in$ S, and only $\aleph_0 < \aleph_1$ of them can satisfy $(y_0, x) \in$ S, so there's some x_0 for which (x_0, y_0) and (y_0, x_0) are not in S.

Now let's see why it *is* possible. For this, I want to assume both the Axiom of Choice *and* the Continuum Hypothesis. By the Continuum Hypothesis, there are only $\aleph_1$ real numbers in $[0, 1]$. So by the Axiom of Choice, we can well-order those real numbers, and do it in such a way that every number has at most $\aleph_0$ predecessors. Now put (x, y) in S if and only if $y \leq x$, where $\leq$ means comparison with respect to the well-ordering (not the usual ordering on real numbers). Then for every (x, y), clearly either $(x, y) \in S$ or $(y, x) \in S$.

This chapter's last puzzle is about the power of self-esteem and positive thinking. Is there any theorem that you can only prove by assuming as an axiom that the theorem *can* be proved?

3 Gödel, Turing, and friends[1]

In the last chapter, we talked about the rules for first-order logic. There's an amazing result called Gödel's Completeness Theorem that says that these rules are all you ever need. In other words: if, starting from some set of axioms, you can't derive a contradiction using these rules, then the axioms must have a model (i.e., they must be consistent). Conversely, if the axioms are inconsistent, then the inconsistency can be proved using these rules alone.

Think about what that means. It means that Fermat's Last Theorem, the Poincaré Conjecture, or any other mathematical achievement you care to name can be proved by starting from the axioms for set theory, and then applying these piddling little rules over and over again. Probably 300 million times, but still...

How does Gödel prove the Completeness Theorem? The proof has been described as "extracting semantics from syntax." We simply cook up objects to order as the axioms request them! And if we ever run into an inconsistency, that can only be because there was an inconsistency in the original axioms.

One immediate consequence of the Completeness Theorem is the *Löwenheim–Skolem Theorem*: every consistent set of axioms has a model of at most countable cardinality. (Note: One of the best predictors of success in mathematical logic is having an umlaut in your name.) Why? Because the process of cooking up objects to order as the axioms request them can only go on for a countably infinite number of steps!

---

[1] An excellent resource for the material in this chapter is *Gödel's Theorem: An Incomplete Guide to its Use and Abuse*, by Torkel Franzén (A. K. Peters Ltd, 2005).

It's a shame that, after proving his Completeness Theorem, Gödel never really did anything else of note. (Pause for comic effect.) Well, alright, I guess a year later he proved the *Incompleteness* Theorem.

The Incompleteness Theorem says that, given any consistent, computable set of axioms, there's a true statement about the integers that can never be proved from those axioms. Here, *consistent* means that you can't derive a contradiction, while *computable* means that either there are finitely many axioms, or else if there are infinitely many, at least there's an algorithm to generate all the axioms.

(If we didn't have the computability requirement, then we could simply take our "axioms" to consist of all true statements about the integers! In practice, that isn't a very useful set of axioms.)

But wait! Doesn't the Incompleteness Theorem contradict the Completeness Theorem, which says that any statement that's entailed by the axioms can be proved from the axioms? Hold that question; we're gonna clear it up later.

First, though, let's see how the Incompleteness Theorem is proved. People always say "the proof of the Incompleteness Theorem was a technical tour de force, it took 30 pages, it requires an elaborate construction involving prime numbers," etc. Unbelievably, 80 years after Gödel, that's still how the proof is presented in math classes!

Alright, should I let you in on a secret? The proof of the Incompleteness Theorem is about *two lines*. It's almost a triviality. The caveat is that, to give the two-line proof, you first need the concept of a computer.

When I was in junior high school, I had a friend who was really good at math, but maybe not so good at programming. He wanted to write a program using arrays, but he didn't know what an array was. So what did he do? He associated each element of the array with a unique prime number, then he multiplied them all together; then, whenever he wanted to read something out of the array, he *factored* the product. (If he was programming a quantum computer, maybe that wouldn't be quite so bad!) Anyway, what my friend did, that's basically what Gödel did. He made up an elaborate hack in order to program without programming.

TURING MACHINES

OK, time to bring Mr. T. on the scene.

In 1936, the word "computer" meant a person (usually a woman) whose job was to compute with pencil and paper. Turing wanted to show that, in principle, such a "computer" could be simulated by a machine. What would the machine look like? Well, it would have to able to write down its calculations somewhere. Since we don't really care about handwriting, font size, etc., it's easiest to imagine that the calculations are written on a sheet of paper divided into squares, with one symbol per square, and a finite number of possible symbols. Traditionally, paper has two dimensions, but without loss of generality we can imagine a long, one-dimensional paper tape. How long? For the time being, we'll assume as long as we need.

What can the machine do? Well, clearly it has to be able to read symbols off the tape and modify them based on what it reads. We'll assume for simplicity that the machine reads only one symbol at a time. But in that case, it had better be able to move back and forth on the tape. It would also be nice if, once it's computed an answer, the machine can halt! But at any time, how does the machine decide which things to do? According to Turing, this decision should depend only on two pieces of information: (1) the symbol currently being read, and (2) the machine's current "internal configuration" or "state." Based on its internal state and the symbol currently being read, the machine should (1) write a new symbol in the current square, overwriting whatever symbol is there, (2) move backward or forward one square, and (3) switch to a new state or halt.

Finally, since we want this machine to be physically realizable, the number of possible internal states should be finite. These are the only requirements.

Turing's first result is the existence of a "universal" machine: a machine whose job is to simulate any other machine described via symbols on the tape. In other words, *universal programmable computers can exist.* You don't have to build one machine for email, another for playing DVDs, another for Tomb Raider, and so on: you can build a single machine that simulates any of the other machines, by running different programs stored in memory. But this result is not even the main result of the paper.

So what's the main result? It's that there's a basic problem, called the halting problem, that no program can ever solve. The halting problem is this: we're given a program, and we want to decide if it ever halts. Of course, we can run the program for a while, but what if the program hasn't halted after a million years? At what point should we give up?

One piece of evidence that this problem might be hard is that, if we *could* solve it, then we could also solve many famous unsolved math problems. For example, Goldbach's Conjecture says that every even number 4 or greater can be written as a sum of two primes. Now, we can easily write a program that tests 4, 6, 8, and so on, halting only if it finds a number that can't be written as a sum of two primes. Then deciding whether that program ever halts is equivalent to deciding the truth of Goldbach's Conjecture.

But can we *prove* there's no program to solve the halting problem? This is what Turing does. His key idea is not even to *try* to analyze the internal dynamics of such a program, supposing it existed. Instead, he simply says, suppose by way of contradiction that such a program P exists. Then, we can modify P to produce a new program P' that does the following. Given another program Q as input, P'

(1) runs forever if Q halts given its own code as input, or
(2) halts if Q runs forever given its own code as input.

Now, we just feed P' its own code as input. By the conditions above, P' will run forever if it halts, or halt if it runs forever. Therefore, P' – and by implication P – can't have existed in the first place.

As I said, once you have Turing's results, Gödel's results fall out for free as a bonus. Why? Well, suppose the Incompleteness Theorem was false – that is, there existed a consistent, computable proof system F from which any statement about integers could be either proved or disproved. Then given a computer program, we could simply search through every possible proof in F, until we found either a proof that the program halts or a proof that it doesn't halt. This is possible because the statement that a particular computer program halts is ultimately just a statement about integers. But this would give us an algorithm to solve the halting problem, which we already know is impossible. Therefore, F can't exist.

By thinking more carefully, we can actually squeeze out a stronger result. Let P be a program that, given as input another program Q, tries to decide whether Q halts by the strategy above (i.e., searching through every possible proof and disproof that Q halts in some formal system F). Then, as in Turing's proof, suppose we modify P to produce a new program P' that

(1) runs forever if Q given its own code as input is proved to halt, or
(2) halts if Q given its own code as input is proved to run forever.

Now suppose we feed P' its own code as input. Then we know that P' will run forever, without ever discovering a proof or disproof that it halts. For if P' finds a proof that it halts, then it will run forever, and if it finds a proof that it runs forever, then it will halt, which is a contradiction.

But there's an obvious paradox: why isn't the above argument, *itself*, a proof that P' will run forever given its own code as input? And why won't P' discover this proof that it runs forever – and therefore halt, and therefore run forever, and therefore halt, etc.?

The answer is that, in "proving" that P' runs forever, we made a hidden assumption: namely, that the proof system F is consistent. If F were inconsistent, then there could perfectly well be a proof that P' halts, even if the reality were that P' ran forever.

But this means that, if F could *prove* that F was consistent, then F could also prove that P' ran forever – thereby bringing back the above contradiction. The only possible conclusion is that *if F is consistent, then F can't prove its own consistency.* This result is sometimes called Gödel's Second Incompleteness Theorem.

The Second Incompleteness Theorem establishes what we maybe should have expected all along: that the only mathematical theories pompous enough to prove their own consistency are the ones that don't *have* any consistency to brag about! If we want to prove that a theory F is consistent, then we can only do it within a *more powerful* theory – a trivial example being F + Con(F) (the theory F plus the axiom that F is consistent). But then how do we know that F + Con(F) is itself consistent? Well, we can only prove that in a still stronger theory: F + Con(F) + Con(F + Con(F)) (which is F + Con(F) plus the axiom that F + Con(F) is consistent). And so on infinitely. (Indeed, even beyond infinitely, into the countable ordinals.)

To take a concrete example: the Second Incompleteness Theorem tells us that the most popular axiom system for the integers, Peano Arithmetic, can't prove its own consistency. Or in symbols, PA can't prove Con(PA). If we want to prove Con(PA), then we need to move to a stronger axiom system, such as ZF (the Zermelo–Fraenkel axioms for set theory). In ZF, we can prove Con(PA) pretty easily, by using the Axiom of Infinity to conjure up an infinite set that then serves as a *model* for PA.

On the other hand, again by the Second Incompleteness Theorem, ZF can't prove its *own* consistency. If we want to prove Con(ZF), the simplest way to do it is to posit the existence of infinities bigger than anything that can be defined in ZF. Such infinities are called "large cardinals." (When set theorists say large, they *mean* large.) Once again, we can prove the consistency of ZF in ZF + LC (where LC is the axiom that large cardinals exist). But if we want to prove that ZF + LC is itself consistent, then we need a still more powerful theory, such as one with even *bigger* infinities.

A quick question to test your understanding: while we can't prove in PA that Con(PA), can we at least prove in PA that Con(PA) implies Con(ZF)?

No, we can't. For then we could also prove in ZF that Con(PA) implies Con(ZF). But since ZF can prove Con(PA), this would mean that ZF can prove Con(ZF), which contradicts the Second Incompleteness Theorem.

I promised to explain why the Incompleteness Theorem doesn't contradict the Completeness Theorem. The simplest way is just to observe that they're talking about different things! The Completeness Theorem is talking about statements that are true in *every* model of the axioms, and it says that all of those statements are indeed provable from the axioms. By contrast, the Incompleteness Theorem is talking about statements that only have to be true in one specific model: namely, the "standard, ordinary" natural numbers. It says that *not* all of those statements are provable from the axioms. In other words, there will always be weird, "non-standard" models of your axioms, in which certain statements that are true for the ordinary natural numbers become false—and even if you don't care about those non-standard models, their existence means that your axioms don't imply everything you want.

Another way to think about it is that there are really *three* notions in play here: (1) being provable from the axioms, (2) being true in every model of the axioms, and (3) being true in the standard natural numbers. The Completeness Theorem says that (1) is equivalent to (2), while the Incompleteness Theorem says that (1) and (2) are *not* equivalent to (3).

But what the hell does it even mean for an arithmetical statement to be "false, but in a non-standard model"? The easiest way to understand this is probably though an example. Consider the "self-hating theory" PA + Not(Con(PA)), or Peano Arithmetic *plus the assertion of its own inconsistency*. We know that, if PA is consistent, then this strange theory must be consistent as well – since otherwise

PA would prove its own consistency, which the Incompleteness Theorem doesn't allow. It follows, by the Completeness Theorem, that PA + Not(Con(PA)) must have a model. But what could such a model possibly look like? In particular, what would happen if, within that model, you just asked to *see* the proof that PA was inconsistent?

I'll tell you what would happen: the axioms would tell you that proof of PA's inconsistency is encoded by a positive integer X. And then you would say "but what *is* X?" And the axioms would say "X." And you would say "But what is X, as an *ordinary positive integer*?"

"What do you mean, ordinary positive integer?"

"I mean, not some abstract entity denoted by a symbol like X, but 1, or 2, or 3, or some other concrete integer that we get by starting from 0 and then adding 1 a finite number of times."

"What do you mean, a finite number of times?"

"I mean, like, once, or twice, or three times . . . "

"But then your definition is circular!"

"Look, you *know* what I mean by finite!"

"No, no, no! Talk to the axioms."

"Alright, is X greater or smaller than $10^{500\,000}$?"

"Greater." (The axioms aren't stupid: they know that if they said "smaller," then you could simply try every smaller number and verify that none of them encode a proof of PA's inconsistency.)

"Alright then, what's $X + 1$?"

"Y."

And so on. The axioms will keep cooking up fictitious numbers to satisfy your requests, and assuming that PA itself is consistent, you'll never be able to trap them in an inconsistency. The point of the Completeness Theorem is that the whole infinite set of fictitious numbers the axioms cook up will constitute a *model* for PA – just not the usual model (i.e., the ordinary positive integers)! If we insist on talking about the usual model, then we switch from the domain of the Completeness Theorem to the domain of the Incompleteness Theorem.

Do you remember the puzzle from Chapter 2? The puzzle was whether there's any theorem that can only be proved by assuming as an axiom that it *can* be proved. In other words, does "just believing in yourself" make any formal difference in mathematics? We're now in a position to answer that question.

Let's suppose, for concreteness, that the theorem we want to prove is the Riemann Hypothesis (RH), and the formal system we want to prove it in is Zermelo–Fraenkel set theory (ZF). Suppose we can prove in ZF that, if ZF proves RH, then RH is true. Then taking the contrapositive, we can also prove in ZF that if RH is false, then ZF does *not* prove RH. In other words, we can prove in ZF + not(RH) that not(RH) is perfectly consistent with ZF. But this means that the theory ZF + not(RH) proves its own consistency – and this, by Gödel, means that ZF + not(RH) is inconsistent. But saying that ZF + not(RH) is inconsistent is equivalent to saying that RH is a theorem of ZF. Therefore, we've proved RH. In general, we find that, if a statement can be proved by assuming as an axiom that it's provable, then it can also be proved *without* assuming that axiom. This result is known as Löb's Theorem (again with the umlauts), though personally I think that a better name would be the "You-Had-the-Mojo-All-Along Theorem."

Oh, you remember earlier we talked about the Axiom of Choice and the Continuum Hypothesis? These are natural statements about the continuum that, since the continuum is such a well-defined mathematical entity, must certainly be either true or false. So, how did those things ever get decided? Well, Gödel proved in 1939 that assuming the Axiom of Choice (AC) or the Continuum Hypothesis (CH) can never lead to an inconsistency. In other words, if the theories ZF + AC or ZF + CH were inconsistent, that could only be because ZF itself was inconsistent.

This raised an obvious question: can we also consistently assume that AC and CH are *false*? Gödel worked on this problem but wasn't able to answer it. Finally, Paul Cohen gave an affirmative answer in 1963, by inventing a new technique called "forcing."

(For that, he won the only Fields Medal that's ever been given for set theory and the foundations of math.)

So, we now know that the usual axioms of mathematics don't decide the Axiom of Choice and the Continuum Hypothesis one way or another. You're free to believe both, neither, or one and not the other without fear of contradiction. And sure enough, opinion among mathematicians about AC and CH remains divided to this day, with many interesting arguments for and against (which we unfortunately don't have time to explore the details of).

Let me end with a possibly surprising observation: the independence of AC and CH from ZF set theory *is itself a theorem of Peano Arithmetic*. For, ultimately, Gödel and Cohen's consistency theorems boil down to combinatorial assertions about manipulations of first-order sentences – which can in principle be proved directly, without ever thinking about the transfinite sets that those sentences purport to describe. (In practice, translating these results into combinatorics would be horrendously complicated, and Cohen has said that *trying to think about these problems in finite combinatorial terms led him nowhere. But we know that in theory it could be done.*) This provides a nice illustration of what, to me, is the central philosophical question underlying this whole business: do we ever *really* talk about the continuum, or do we only ever talk about finite sequences of symbols that talk about the continuum?

BONUS ADDENDUM

What does any of this have to do with quantum mechanics? I will now attempt the heroic task of making a connection. What I've tried to impress on you is that there are profound difficulties if we want to assume the world is continuous. Take a pen, for example: how many different positions can I put it in on the surface of a table? $\aleph_1$? More than $\aleph_1$? Less than $\aleph_1$? We don't want the answers to "physics" questions to depend on the axioms of set theory!

Ah, but you say my question is physically meaningless, since the pen's position could never actually be measured to infinite precision? Sure – but the point is, you need a physical theory to *tell* you that!

Of course, quantum mechanics gets its very name from the fact that a lot of the observables in the theory, like energy levels, are discrete – "quantized." This seems paradoxical, since one of the criticisms that computer scientists level against quantum computing is that, as they see it, it's a *continuous* model of computation!

My own view is that quantum mechanics, like classical probability theory, should be seen as somehow "intermediate" between a continuous and discrete theory. (Here, I'm assuming that the Hilbert space[2] or probability space is finite dimensional.) What I mean is that, while there *are* continuous parameters (the probabilities or amplitudes, respectively), those parameters are not directly observable, and that has the effect of "shielding" us from the bizarro universe of the Axiom of Choice and the Continuum Hypothesis. We don't need a detailed physical theory to tell us that whether amplitudes are rational or irrational, whether there are more or less than $\aleph_1$ possible amplitudes, etc., are physically meaningless questions. This follows directly from the fact that, if we wanted to learn an amplitude exactly, then (even assuming no error!) we would need to measure the appropriate state infinitely many times.

EXERCISE

Let BB(n), or the "nth Busy Beaver number," be the maximum number of steps that an n-state Turing machine can make on an initially blank tape before halting. (Here, the maximum is over all n-state Turing machines that eventually halt.) Prove that BB(n) grows faster than any computable function.

---

[2] Please don't be alarmed by the term "Hilbert space," which I'll use occasionally in this book. All it means is "the space of all possible quantum states of some system." With *infinite*-dimensional systems, the definition of Hilbert space is a bit subtle – but in this book, we'll only care about finite-dimensional systems. And as we'll see in Chapter 9, the Hilbert space of a finite-dimensional system is nothing other than $\mathbb{C}^N$: an N-dimensional complex vector space.

4 Minds and machines

Now we're going to launch into something I know you've all been waiting for: a philosophical food fight about minds, machines, and intelligence!

First, though, let's finish talking about computability. One concept we'll need again and again in this chapter is that of an oracle. The idea is a pretty obvious one: we *assume* we have a "black box," or "oracle," that immediately solves some hard computational problem, and then see what the consequences are! (When I was a freshman, I once started talking to my professor about the consequences of a hypothetical "NP-completeness fairy": a being that would instantly tell you whether a given Boolean formula was satisfiable or not. The professor had to correct me: they're not called "fairies"; they're called "oracles." *Much* more professional!)

Oracles were apparently first studied by Turing, in his 1938 PhD thesis. Obviously, anyone who could write a whole thesis about these fictitious entities would have to be an *extremely* pure theorist, someone who wouldn't be caught dead doing anything relevant. This was certainly true in Turing's case – indeed, he spent the years after his PhD, from 1939 to 1943, studying certain abstruse symmetry transformations on a 26-letter alphabet.

Anyway, we say that problem A is *Turing reducible* to problem B, if A is solvable by a Turing machine given an oracle for B. In other words, "A is no harder than B": if we had a hypothetical device to solve B, then we could also solve A. Two problems are *Turing equivalent* if each is Turing reducible to the other. So, for example, the problem of whether a statement can be proved from the axioms of set theory is Turing equivalent to the halting problem: if you can solve one, you can solve the other.

Now, a *Turing degree* is the set of all problems that are Turing equivalent to a given problem. What are some examples of Turing degrees? Well, we've already seen two examples: (1) the set of computable problems, and (2) the set of problems that are Turing equivalent to the halting problem. Saying that these Turing degrees aren't equal is just another way of saying that the halting problem isn't solvable.

Are there any Turing degrees *above* these two? In other words, is there any problem even harder than the halting problem, one that we can't solve even with the help of an oracle to the halting problem? Well, consider the following "super halting problem": given a Turing machine with an oracle for the halting problem, decide if it halts! Can we prove that this super halting problem is unsolvable, even given an oracle for the ordinary halting problem? Yes, we can! We simply take Turing's original proof that the halting problem is unsolvable, and "shift everything up a level" by giving all the machines an oracle for the halting problem. Everything in the proof goes through as before, a fact we express by saying that the proof "relativizes."

Here's a subtler question: is there any problem of *intermediate* difficulty between the computable problems and the halting problem? This question was first asked by Emil Post in 1944, and was answered in 1954 by Post and Stephen Kleene (although, in Post's original statement of the problem, he'd added an additional condition called "recursive enumerability," and it was only shown two years later, by Richard Friedberg and A. A. Muchnik, how to fulfill the additional condition). The answer is yes. Indeed, we actually have a stronger result: that there are two problems A and B, both of which are solvable given an oracle for the halting problem, but neither of which is solvable given an oracle for the other. These problems are constructed via an infinite process whose purpose is to kill off every Turing machine that might reduce A to B or B to A. Unfortunately, the resulting problems are extremely contrived; they don't look like anything that might arise in practice. And even today, we don't have a single example of a "natural" problem with intermediate Turing degree.

Since the breakthrough in solving Post's problem, the structure of the Turing degrees has been studied in more detail than you can possibly imagine. Here's one of the *simplest* questions: if two problems A and B are both reducible to the halting problem, then must there be a problem C that's reducible to A and B, such that any problem that's reducible to both A and B is also reducible to C? Hey, whatever floats your boat! But this is the point where some of us say, maybe we should move on to the next topic . . . (Incidentally, the answer to the question is no.)

Alright, the main philosophical idea underlying computability is what's called the *Church–Turing Thesis*. It's named after Turing and his adviser Alonzo Church, even though what they themselves believed about "their" thesis is open to dispute! Basically, the thesis is that *any function "naturally to be regarded as computable" is computable by a Turing machine*. Or in other words, any "reasonable" model of computation will give you either the same set of computable functions as the Turing machine model, or else a proper subset.

Already there's an obvious question: what sort of claim is this? Is it an empirical claim, about which functions can be computed in physical reality? Is it a definitional claim, about the meaning of the word "computable?" Is it a little of both?

Well, whatever it is, the Church–Turing Thesis can only be regarded as extremely successful, as theses go. As you know – and as we'll discuss later – quantum computing presents a serious challenge to the so-called Extended Church–Turing Thesis: that any function naturally to be regarded as *efficiently* computable is *efficiently* computable by a Turing machine. But in my view, so far there hasn't been *any* serious challenge to the original Church–Turing Thesis – neither as a claim about physical reality, nor as a definition of "computable."

There have been plenty of *non*serious challenges to the Church–Turing Thesis. In fact, there are whole conferences and journals devoted to these challenges – Google "hypercomputation." I've read some of this stuff, and it's mostly along the lines of, well, suppose you could do the first step of a computation in one second, the next step in a half second, the next step in a quarter second, the next step in an eighth second, and so on. Then in two seconds you'll have done an infinite amount of computation! Well, as stated it sounds a bit silly, so maybe sex it up by throwing in a black hole or something. How could the hidebound Turing reactionaries possibly object? (It reminds me of the joke about the supercomputer that was so fast, it could do an infinite loop in 2.5 seconds.)

We should immediately be skeptical that, if Nature was going to give us these vast computational powers, she would do so in a way that's so mundane, so uninteresting. Without making us sweat or anything. But admittedly, to *really* see why the hypercomputing proposals fail, you need the entropy bounds of Bekenstein, Bousso, and others – which are among the few things the physicists think they know about quantum gravity, and which we'll say something about later in the book. So the Church–Turing Thesis – even its original, nonextended version – really is connected to some of the deepest questions in physics. But in my opinion, neither quantum computing, nor analog computing, nor anything else, has mounted a serious challenge to that thesis in the 75 years since it was formulated.

A closely-related objection to this computation by geometric series is that we do sort of understand why this model isn't physical: we believe that the very notion of time starts breaking down when you get down to around 10^{-43} seconds (the Planck scale). We don't know exactly what happens there. Nevertheless, the situation seems not the slightest bit analogous to quantum computing (for example). In quantum computing, as we'll see, no one has any quantitative idea of where the theory could break down and the computer could stop working – which leads to the conjecture that maybe it *won't* stop working.

Once you get to the Planck scale, you might say we're getting into a really sophisticated argument. Why not just say you're always limited in practice by noise and imperfection?

The question is why are you limited? *Why* can't you store a real number in a register? I think that if you try to make the argument precise, ultimately, you're going to be talking about the Planck scale.

If we interpret the Church–Turing Thesis as a claim about physical reality, then it should encompass everything in that reality, including the goopy neural nets between your ears. This leads us, of course, straight into the cratered intellectual battlefield that I promised to lead you into.

As a historical remark, it's interesting that the possibility of thinking machines isn't something that occurred to people gradually, after they'd already been using computers for decades. Instead, it occurred to them *immediately*, the minute they started talking about computers themselves. People like Leibniz and Babbage and Lovelace and Turing and von Neumann understood from the beginning that a computer wouldn't just be another steam engine or toaster – that, because of the property of universality (whether or not they called it that), it's difficult even to talk about computers without also talking about ourselves.

Now, I ask you to put down this book for a few minutes, and read Turing's second famous paper, "Computing machinery and intelligence."[1]

What's the main idea of this paper? As I read it, it's a plea against meat chauvinism. Sure, Turing makes some scientific arguments, some mathematical arguments, some epistemological arguments. But beneath everything else is a *moral* argument. Namely: if a computer interacted with us in a way that was indistinguishable from a human, then of course we could say the computer wasn't "really" thinking, that it was just a simulation. But on the same grounds, we could also say that *other people* aren't really thinking, that they merely act as if they're thinking. So what entitles us to go through such intellectual acrobatics in the one case but not the other?

If you'll allow me to editorialize (as if I ever do otherwise...), this moral question, this question of double standards, is really where Searle, Penrose, and every other "strong AI skeptic" comes up empty for me. One can indeed give weighty and compelling arguments

---

[1] http://www.loebner.net/Prizef/TuringArticle.html

against the possibility of thinking machines. The only problem with these arguments is that they're also arguments against the possibility of thinking *brains*!

So, for example: one popular argument is that, if a computer appears to be intelligent, that's merely a reflection of the intelligence of the humans who programmed it. But what if humans' intelligence is just a reflection of the billion-year evolutionary process that gave rise to it? What frustrates me every time I read the AI skeptics is their failure to consider these parallels honestly. The "qualia" and "aboutness" of other people is simply taken for granted. It's only the qualia of machines that's in question.

But perhaps a skeptic could retort: I believe other people think because I know *I* think, and other people look sort of similar to me – they've also got five fingers, hair in their armpits, etc. But a robot *looks* different – it's made of metal, it's got an antenna, it lumbers across the room, etc. So even if the robot acts like it's thinking, who knows? But if I accept this argument, why not go further? Why can't I say, I accept that *white* people think, but those blacks and Asians, who knows about them? They look too dissimilar from me.

In my view, one can divide everything that's been said about artificial intelligence into two categories: the 70% that's somewhere in Turing's paper from 1950, and the 30% that's emerged from a half-century of research since then.

So, after 60 plus years, there are *some* things we can say that would've surprised Alan Turing. What are those things? Well, one of them is how little progress has been made, compared to what he expected! Do you remember, Turing made a falsifiable prediction?

> I believe that in about fifty years' time it will be possible, to programme computers, with a storage capacity of about 10^9, to make them play the imitation game so well that an average interrogator will not have more than 70 per cent chance of making the right identification after five minutes of questioning.

How well has his prediction fared? First, let's note that the prediction about computers themselves was damn good. Turing predicted that in 50 years' time (i.e., 2000), we'd be programming computers with a storage capacity of about 10^9 (i.e., one gig).

But what about programming the computers to pass the imitation game? How well has Turing's prediction fared there?

Well, some of you might have heard of a program called ELIZA, written by Joseph Weizenbaum in 1966. This program simulates a psychotherapist who keeps spitting back whatever you said. The amazing thing Weizenbaum found is that many people will spill their hearts out to this program! And sometimes, if you then tell them they were talking to a program (and an extremely simple one at that), they won't believe you.

Years ago, someone had the brilliant idea to take the original ELIZA program and let it loose in AOL chat rooms to see what happened. If you go to http://fury.com/aoliza, you can see some hilarious (fortunately anonymized) examples of people trying to seduce the program, telling it about their infidelities, etc. Here's one of my favorite exchanges, from a guy who, before moving on to amorous solicitations, had told the program that he planned to finish his B of A and then "move into corporate business alliance with Starbucks":

Guy: I think that u are getting off on playing with my mind
ELIZA: your mind?
Guy: frustrating me by not joining in my fantasy
ELIZA: your fantasy?
Guy: about having mind blowing sex with u
ELIZA: does that have anything to do with the fact that your b of a and then you will move into corporate business alliance with starbucks?

The state of the art in terms of man–machine repartee has advanced beyond this, but not *that* far beyond. It seems one actually needs to revise the Turing Test, to say that, if we want to verify intelligence in a computer, then we need some minimal level of intelligence in the human interrogator.

Of course, the problem here is that these guys *assumed* they were talking to a human. To satisfy the conditions of the Turing Test, the interrogator would be trying to distinguish a human from a computer. So, this isn't a real Turing Test; it's merely amusing. However, for the past 15 years, a guy named Hugh Loebner has been running a contest[2] that's closer to what Turing had in mind. Here the testers *are* told that they're trying to distinguish humans from computers – yet many of the transcripts have been just as depressing, both from the standpoint of machine intelligence and from that of human intelligence. (E.g., a woman who tried to converse intelligently about Shakespeare got classified as a computer, since "no human would know that much about Shakespeare ... ")

You might wonder, what if we had a computer doing the interrogation instead of a human? As it turns out, that's not at all a hypothetical situation. In 2006, a guy named Luis von Ahn won a MacArthur award for (among other things) his work on CAPTCHAs, which are those tests that websites use to distinguish legitimate users from spambots. I'm sure you've encountered them – you know, the things where you see those weird curvy letters that you have to retype. The key property of these tests is that a computer should be able to generate and grade them, but not pass them! (A lot like professors making up a midterm ...) Only humans should be able to pass the tests. So basically, these tests capitalize on the *failures* of AI. (Well, they also capitalize on the computational hardness of inverting one-way functions, which we'll get to later.)

One interesting aspect of CAPTCHAs is that they've already led to an arms race between the CAPTCHA programmers and the AI programmers. When I was at Berkeley, some of my fellow grad students wrote a program[3] that broke a CAPTCHA called Gimpy maybe 30% of the time. So then the CAPTCHAs have to be made harder, and then the AI people get back to work, and so on. Who will win?

---

[2] http://www.loebner.net/Prizef/loebner-prize.html
[3] http://www.cs.sfu.ca/~mori/research/gimpy/

You see: every time you set up a Yahoo! Mail account, you're directly confronting age-old mysteries about what it means to be human . . .

Despite what I said about the Turing Test, there *have* been some dramatic successes of AI. We all know about Kasparov and Deep Blue, and IBM's Watson (the computer that won at *Jeopardy!*, defeating the human champion Ken Jennings). Maybe less well known is that, in 1996, a program called Otter[4] was used to solve a 60-year-old open problem in algebra called the Robbins Conjecture, which Tarski and other famous mathematicians had worked on. (Apparently, for decades Tarski would give the problem to his best students. Then, eventually, he started giving it to his *worst* students . . .) The problem is easy to state: given the three axioms

- A or (B or C) = (A or B) or C
- A or B = B or A
- Not(Not(A or B) or Not(A or Not(B))) = A,

can one derive as a consequence that Not(Not(A)) = A?

Let me stress that this was *not* a case like Appel and Haken's proof of the Four-Color Theorem, where the computer's role was basically to check thousands of cases. In this case, the proof was 17 lines long. A human could check the proof by hand, and say, yeah, I could've come up with that. (In principle!)

What else? Arguably there's a pretty sophisticated AI system that almost all of you used this morning and will use many more times today. What is it? Right, Google.

You can look at any of these examples – Deep Blue, the Robbins conjecture, Google, most recently Watson – and say, that's not *really* AI. That's just massive search, helped along by clever programming. Now, this kind of talk drives AI researchers up a wall. They say: if you told someone in the 1960s that in 30 years we'd be able to beat

---

[4] W. McCune, Solution of the Robbins Problem, *Journal of Automated Reasoning* 19:3 (1997), 263–276. http://www.cs.unm.edu/~mccune/papers/robbins/

the world grandmaster at chess, and asked if that would count as AI, they'd say, of course it's AI! But now that we know how to do it, it's no longer AI – it's just search. (Philosophers have a similar complaint: as soon as a branch of philosophy leads to anything concrete, it's no longer called philosophy! It's called math or science.)

There's another thing we appreciate now that people in Turing's time didn't really appreciate. This is that, in trying to write programs to simulate human intelligence, we're competing against *a billion years of evolution*. And that's damn hard. One counterintuitive consequence is that it's much easier to program a computer to beat Garry Kasparov at chess than to program a computer to recognize faces under varied lighting conditions. Often the hardest tasks for AI are the ones that are trivial for a five-year-old – since those are the ones that are so hardwired by evolution that we don't even think about them.

In the last 60 years, have there been any new insights about the Turing Test itself? In my opinion, not many. There has, on the other hand, been a famous "attempted" insight, which is called *Searle's Chinese Room*. This was put forward around 1980, as an argument that even a computer that did pass the Turing Test wouldn't be intelligent. The way it goes is, let's say you don't speak Chinese. You sit in a room, and someone passes you paper slips through a hole in the wall with questions written in Chinese, and you're able to answer the questions (again in Chinese) just by consulting a rule book. In this case, you might be carrying out an intelligent Chinese conversation, yet by assumption, you don't understand a word of Chinese! Therefore, symbol-manipulation can't produce understanding.

So, how might a strong AI proponent respond to this argument? Well, she might say: *you* might not understand Chinese, but the rule book does! Or if you like, understanding Chinese is an emergent property of the system consisting of you and the rule book, in the same sense that understanding English is an emergent property of the neurons in your brain.

Searle's response to that is, fine, just memorize the rule book! Then there's no "system" other than your own brain, but you still don't "understand" Chinese. To which the AI proponent shoots back: there is too another "system" in this case! Supposing you memorized the rule book, we'd need to distinguish between the "original" you and the new, simulated being brought into existence by your following of the memorized rules – a being whose only relation to you might be that it happens to inhabit the same skull. That response might sound crazy, but only to someone who's never studied computer science. To a computer scientist, it seems perfectly reasonable to say that one computation (say, a LISP interpreter) can conjure into existence a different, unrelated computation (say, a spaceship game) just by dutifully executing rules.

Look, as I'll discuss later, I don't know whether the *conclusion* of the Chinese Room argument is true or false. I don't know what conditions are necessary or sufficient for a physical system to "understand" Chinese – and neither, I think, does Searle, or anyone else. But considered as an *argument*, there are several aspects of the Chinese Room that have always annoyed me. One of them is the unselfconscious appeal to intuition – "it's just a rule book, for crying out loud!" – on precisely the sort of question where we should expect our intuitions to be least reliable. A second is the double standard: the idea that a *bundle of nerve cells* can understand Chinese is taken as, not merely obvious, but so unproblematic that it doesn't even *raise the question* of why a rule book couldn't understand Chinese as well. The third thing that annoys me about the Chinese Room argument is the way it gets so much mileage from a possibly misleading choice of imagery, or, one might say, by trying to sidestep the entire issue of *computational complexity* purely through clever framing. We're invited to imagine someone pushing around slips of paper with zero understanding or insight – much like the doofus freshmen who write $(a + b)^2 = a^2 + b^2$ on their math tests. But *how many slips of paper are we talking about?* How big would the rule book have to be, and how quickly would you have to consult it, to carry out an intelligent

Chinese conversation in anything resembling real time? If each page of the rule book corresponded to one neuron of a native speaker's brain, then probably we'd be talking about a "rule book" at least the size of the Earth, its pages searchable by a swarm of robots traveling at close to the speed of light. When you put it that way, maybe it's not so hard to imagine that this enormous Chinese-speaking entity that we've brought into being might have something we'd be prepared to call understanding or insight.[5]

Of course, everyone who talks about this stuff is really tiptoeing around the question of consciousness. See, consciousness has this weird dual property that, on the one hand, it's arguably the most mysterious thing we know about, and on the other hand, not only are we directly aware of it, but in some sense it's the *only* thing we're directly aware of. You know, *cogito ergo sum* and all that. So, to give an example, I might be mistaken about my shirt being blue – I might be hallucinating or whatever – but I really can't be mistaken about my *perceiving* it as blue. (Or if I can, then we get an infinite regress.)

Now, is there anything else that also produces the feeling of absolute certainty? Right – math! Incidentally, I think this similarity between math and subjective experience might go a long way toward explaining mathematicians' "quasi-mystical" tendencies. (I can already hear some mathematicians wincing. Sorry!) This is a good thing for physicists to understand: when you're talking to a mathematician, you *might* not be talking to someone who fears the real world and who's therefore retreated into intellectual masturbation. You might be talking to someone for whom the real world was never especially real to begin with!

I mean, consider the computer proof of the Four-Color Theorem, which I briefly mentioned earlier. That proof solved a great, century-old mathematical problem, but it did so by reducing the problem to

---

[5] For further discussion of these issues, see Scott Aaronson, Why Philosophers Should Care About Computational Complexity, in *Computability: Turing, Gödel, Church, and Beyond* (MIT Press, 2013; edited by B. Jack Copeland, Carl J. Posy, and Oron Shagrir), http://www.scottaaronson.com/papers/philos.pdf

the tedious enumeration of thousands of cases. Why did some mathematicians look askance at the proof, or at least hold out hope for a better one? Because the computer "might have made a mistake"? Well, that's a feeble argument, since the proof has now been rechecked by several independent groups of programmers using different hardware and software, and at any rate, humans make plenty of mistakes too!

What it boils down to, I think, is that there *is* a sense in which the Four-Color Theorem has been proved, and there's another sense in which many mathematicians understand proof, and those two senses aren't the same. For many mathematicians, a statement isn't proved when a physical process (which might be a classical computation, a quantum computation, an interactive protocol, or something else) terminates saying that it's been proved – however good the reasons might be to believe that physical process is reliable. Rather, the statement is proved when they (the mathematicians) feel that their minds can directly perceive its truth.

Of course, it's hard to discuss these things directly. But what I'm trying to point out is that many people's "antirobot animus" is probably a combination of two ingredients:

(1) the directly experienced certainty that *they're* conscious – that *they* perceive colors, sounds, positive integers, etc., regardless of whether anyone else does, and
(2) the belief that, *if* they were just a computation, *then* they could not be conscious in this way.

For example, I think Penrose's objections to strong AI derive from these two ingredients. I think his arguments about Gödel's Theorem are window dressing added later.

For people who think this way (as even I do, in certain moods), granting consciousness to a robot seems strangely equivalent to *denying that one is conscious oneself*. Is there any respectable way out of this dilemma – or in other words, any way out that doesn't rely on a meatist double standard, with one rule for ourselves and a different rule for robots?

My own favorite way out is one that's been advocated by the philosopher David Chalmers.[6] Basically, what Chalmers proposes is a "philosophical **NP**-completeness reduction": a reduction of one mystery to another. He says that, if computers someday become able to *emulate* humans in every observable respect, then we'll be compelled to regard them as conscious, for exactly the same reasons we regard other *people* as conscious. And as for *how* they could be conscious — well, we'll understand that just as well or as poorly as we understand how a bundle of neurons could be conscious. Yes, it's mysterious, but the one mystery doesn't seem so different from the other.

PUZZLES

- [The barely well-defined puzzle] Can we assume without loss of generality that a computer program has access to its own source code?
- [The vague, ill-defined puzzle] If that which before the 1800s was called water turned out to be CH_4 instead of H_2O, would it still be water, or would it be something else?

ANSWERS TO EXERCISE FROM LAST CHAPTER

Recall that BB(n), or the "nth Busy Beaver number," is the largest number of steps that an n-state Turing machine can make on an initially blank tape before halting.

The first problem was to prove that BB(n) grows faster than any computable function.

Suppose there were a computable function $f(n)$ such that $f(n) >$ BB(n) for every n. Then, given an n-state Turing machine M, we could first compute $f(n)$, then simulate M for up to $f(n)$ steps. If M hasn't halted by then, then we know it never *will* halt, since $f(n)$ is greater than the maximum number of steps any n-state machine could make. But this gives us a way to solve the halting problem, which we already know is impossible. Therefore, the function f doesn't exist.

---

[6] See David J. Chalmers, *The Conscious Mind: In Search of a Fundamental Theory*, Oxford University Press, 1997.

So the BB(*n*) function grows really, really, *really* fast. (In case you're curious, here are the first few values, insofar as they've been computed by people with too much free time: BB(1) = 1, BB(2) = 6, BB(3) = 21, BB(4) = 107, BB(5) ≥ 47 176 870. Of course, these values depend on exact details of how Turing machines are defined.)

5 Paleocomplexity

By any objective standard, the theory of computational complexity ranks as one of the greatest intellectual achievements of humankind – along with fire, the wheel, and computability theory. That it isn't taught in high schools is really just an accident of history. In any case, we'll *certainly* need complexity theory for everything else we're going to do in this book, which is why the next five or six chapters will be devoted to it. So before we dive in, let's step back and pontificate about where we're going.

What I've been trying to do is show you the conceptual under-pinnings of the universe, *before* quantum mechanics comes on the scene. The amazing thing about quantum mechanics is that, despite being a grubby empirical discovery, it *changes* some of the underpin-nings! Others it doesn't change, and others it's not so clear whether it changes them or not. But if we want to debate how things are changed by quantum mechanics, then we'd better understand what they looked like before quantum mechanics.

It's useful to divide complexity theory into historical epochs:

- **1950s: Late Turingzoic**
- **1960s: Dawn of the Asymptotic Age**
- **1971: The Cook–Levin Asteroid; extinction of the Diagonalosaurs**
- **Early 1970s: The Karpian Explosion**
- **1978: Early Cryptozoic**
- **1980s: Randomaceous Era**
- **1993: Eruption of Mt Razborudich; extinction of the Combinataurs**
- **1994: Invasion of the Quantodactyls**
- **Mid-1990s to present: Derandomaceous Era**

This chapter will be about "paleocomplexity": complexity in the age before **P**, **NP**, and **NP**-completeness, when Diagonalosaurs ruled the earth. Then Chapter 6 will cover the Karpian Explosion, Chapter 7 the Randomaceous Era, Chapter 8 the Early Cryptozoic, and Chapter 9 the Invasion of the Quantodactyls.

We talked earlier about computability theory. We saw how certain problems are uncomputable – like, given a statement about positive integers, is it true or false? (If we could solve *that*, then we could solve the halting problem, which we already know is impossible.)

But now let's suppose we're given a statement about real numbers – for example,

For all real x and y, $(x + y)^2 = x^2 + 2xy + y^2$

– and we want to know if it's true or false. In this case, it turns out that there *is* a decision procedure – this was proved by Tarski in the 1930s, at least when the statement only involves addition, multiplication, comparisons, the constants 0 and 1, and universal and existential quantifiers (no exponentials or trig functions).

Intuitively, if all our variables range over real numbers instead of integers, then everything is forced to be smooth and continuous, and there's no way to build up Gödel sentences like "this sentence can't be proved."

(If we throw in the exponential function, then it was recently proved that there's *still* no way to encode Gödel sentences, modulo an unsolved problem in analysis.[1] But if we throw in the exponential function *and* switch from real numbers to complex numbers, then we're again able to encode Gödel sentences – and the theory goes back to being undecidable! Can you guess why? Well, once we have complex numbers, we can force a number n to be an integer, by saying that we want $e^{2\pi i n}$ to equal 1. So we're then back to where we were with integers.)

---

[1] See http://www.ams.org/notices/199607/marker.pdf

Anyway, the attitude back then was, OK, we found an algorithm to decide the truth or falsehood of any sentence about real numbers! We can go home! Problem solved!

Trouble is, if you worked out how many steps that algorithm took to decide the truth of a sentence with n symbols, it grew like an enormous stack of exponentials: $2^{2^{\cdot^{\cdot^{\cdot^{2}}}}} \Big\} n$. So I was reading in a biography[2] of Tarski that, when actual computers came on the scene in the 1950s, one of the first things anyone thought to do was to implement Tarski's algorithm for deciding statements about the real numbers. And it was hopeless – indeed, it would have been hopeless even on the computers of today! On the computers of the 1950s, it was $hopeless^{hopeless^{\cdots^{hopeless}}}$.

So, these days we talk about complexity. (Or at least most of us do.) The idea is, you impose an upper bound on how much of some resource your computer can use. The most obvious resources are (1) amount of time and (2) amount of memory, but many others can be defined. (Indeed, if you visit my Complexity Zoo website,[3] you'll find about 500 of them.)

One of the very first insights is, if you ask how much can be computed in 10 million steps, or 20 billion bits of memory, you won't get anywhere. Your theory of computing will be at the mercy of arbitrary choices about the underlying model. In other words, you won't be doing theoretical computer science at all: you'll be doing *architecture*, which is an endlessly fascinating, nonboring topic in its own right, but not our topic.

So instead you have to ask a looser question: how much can be computed in an amount of time that grows linearly (or quadratically, or logarithmically) with the problem size? Asking this sort of question lets you ignore constant factors.

[2] A. Burdman Feferman and S. Feferman, *Alfred Tarski: Life and Logic* (Cambridge: Cambridge University Press, 2008).

[3] http://www.complexityzoo.com

So, we define TIME($f(n)$) to be the class of problems for which every instance of size n is solvable in an amount of time that grows like a constant times $f(n)$. Here, by "solvable," we mean solvable by some particular type of idealized computer (say, a Turing machine), which we fix as a "reference." A crucial empirical fact, on which the whole theory depends, is that *which* type of idealized computer we choose won't matter very much, as long as we stay within some broad limits (for example, we consider serial, deterministic, classical computers only, not quantum computers or anything like that).

Likewise, SPACE($f(n)$) is the class of problems solvable by our reference machine using an amount of space (i.e., bits of memory) that grows like a constant times $f(n)$.

What can we say about the relations between these classes? Well, for every function $f(n)$, TIME($f(n)$) is contained in SPACE($f(n)$). Why? Because a Turing machine can access at most one memory location per time step.

What else? Presumably you agree that TIME(n^2) is contained in TIME(n^3). Here's a question: is it *strictly* contained? In other words, can you solve more problems in n^3 time than in n^2 time?

It turns out that you can. This is a consequence of a fundamental result called the Time Hierarchy Theorem, which was proved by Hartmanis and Stearns in the mid-1960s and later rewarded with a Turing Award. (Not to diminish their contribution, but back then Turing Awards were hanging pretty low on the tree! Of course you had to *know* to be looking for them, which not many people did.)

Let's see how the proof goes. We need to find a problem that's solvable in n^3 time but not n^2 time. What will this problem be? It'll be the simplest thing you could imagine: a time-bounded analog of Turing's halting problem.

Given a Turing machine M, does M halt in at most $n^{2.5}$ steps?
(Here $n^{2.5}$ is just some function between n^2 and n^3.)

Clearly we can solve the above problem in n^3 steps, by simulating M for $n^{2.5}$ steps and seeing whether it halts or not. (Indeed, we can solve the problem in something like $n^{2.5} \log n$ steps. We always need some overhead when running a simulation, but the overhead can be made extremely small.)

But now suppose there were a program P to solve the problem in n^2 steps. We'll derive a contradiction. By using P as a subroutine, clearly we could produce a new program P' with the following behavior. Given a program M as input, P'

(1) runs forever if M halts in at most $n^{2.5}$ steps given its own code as input, or
(2) halts in $n^{2.5}$ steps if M runs for more than $n^{2.5}$ steps given its own code as input.

Furthermore, P' does all of this in at most $n^{2.5}$ steps (indeed, n^2 steps plus some overhead).

Now what do we do? Duh, we feed P' its own code as input! And we find that P' must do the opposite of whatever it does: run forever if it halts, or halt if it runs forever. This gives us a contradiction, which implies that P can never have existed in the first place.

Obviously, the choice of n^3 versus n^2 is not essential. We can substitute n^{17} versus n^{16}, 3^n versus 2^n, etc. But there's actually an interesting question here: can we substitute *any* functions f and g such that f grows significantly faster than g? The surprising answer is no! The function g needs a property called *time-constructibility*, which means (basically) that there's *some* program that halts in $g(n)$ steps given n as input. Without this property, the program P' wouldn't know how many steps to simulate M for, and the argument wouldn't go through.

Now, every function you'll ever encounter in civilian life will be time constructible. But in the early 1970s, complexity theorists made up some bizarre, rapidly growing functions that aren't. And for these functions, you really *can* get arbitrarily large gaps in the complexity hierarchy! So, for example, there's a function f such that TIME$(f(n))=$TIME$(2^{f(n)})$. *Duuuuude.*

Anyway, analogous to the Time Hierarchy Theorem is the *Space Hierarchy Theorem*, which says there's a problem solvable with n^3 bits of memory that's not solvable with n^2 bits of memory.

Alright, next question: in computer science, we're usually interested in the fastest algorithm to solve a given problem. But is it clear that every problem *has* a fastest algorithm? Or could there be a problem that admits an infinite sequence of algorithms, with each one faster than the last but slower than some other algorithm?

Contrary to what you might think, this is not just a theoretical armchair question: it's a concrete, down-to-earth armchair question! As an example, consider the problem of multiplying two n-by-n matrices. The obvious algorithm takes $O(n^3)$ time. In 1968, Strassen gave a more complicated algorithm that takes $O(n^{2.78})$ time. A long string of improvements followed, culminating in an $O(n^{2.376})$ algorithm of Coppersmith and Winograd. And that's where things stood for 23 years – until 2011, not long before this book went to press, when Stothers[4] and then Vassilevska[5] announced improvements leading to an $O(n^{2.373})$ algorithm. But is that the end of the line? Might there be an algorithm to multiply matrices in n^2 time? Here's a weirder possibility: could it be that, for every $\varepsilon > 0$, there exists an algorithm to multiply n-by-n matrices in time $O(n^{2+\varepsilon})$, but as ε approaches zero, these algorithms become more and more complicated without end?

See, some of this paleocomplexity stuff is actually nontrivial! (*T. rex* might've been a dinosaur, but it still had pretty sharp teeth!) In this case, a 1967 result called the Blum Speedup Theorem says that there really *are* problems that admit no fastest algorithm. Not only

---

[4] A. Stothers, On the complexity of matrix multiplication. Unpublished PhD Thesis, University of Edinburgh (2010). http://www.maths.ed.ac.uk/pg/thesis/stothers.pdf

[5] V. Vassilevska Williams, Breaking the Coppersmith–Winograd barrier. In *Proceedings of Annual ACM Symposium on Theory of Computing* (2012). http://www.cs.berkeley.edu/~virgi/matrixmult.pdf

that: there exists a problem P such that, for every function f, if P has an $O(f(n))$ algorithm then it also has an $O(\log f(n))$ algorithm!

Let's see how this goes. Let $t(n)$ be a complexity bound. Our goal is to define a function f, from integers to $\{0, 1\}$, such that if f can be computed in $O(t(n))$ steps, then it can also be computed in $O(t(n - i))$ steps for any positive integer i. Taking t to be sufficiently fast growing then gives us as dramatic a speedup as we want: for example, if we set $t(n) := 2^{t(n-1)}$, then certainly $t(n - 1) = O(\log t(n))$.

Let $M_1, M_2, \ldots$ be an enumeration of Turing machines. Then let $S_i = \{M_1, \ldots, M_i\}$ be the set consisting of the first i machines. Here's what we do: given an integer n as input, we loop over all i from 1 to n. In the ith iteration, we simulate every machine in S_i that wasn't "cancelled" in iterations 1 to $i - 1$. If none of these machines halt in at most $t(n - i)$ steps, then set $f(i) = 0$. Otherwise, let M_j be the *first* machine that halts in at most $t(n - i)$ steps. Then we define $f(i)$ to be 1 if M_j outputs 0, or 0 if M_j outputs 1. (In other words, we cause M_j to fail at computing $f(i)$.) We also "cancel" M_j, meaning that M_j doesn't need to be simulated in any later iteration. This defines the function f.

Certainly $f(n)$ can be computed in $O(n^2 t(n))$ steps, by simply simulating the entire iterative procedure above. The key observation is this: for any integer i, if we *hardwire* the outcomes of iterations 1 to i into our simulation algorithm (i.e., tell the algorithm which M_j get cancelled in those iterations), then we can skip iterations 1 to i, and proceed immediately to iteration $i + 1$. Furthermore, assuming we start from iteration $i + 1$, we can compute $f(n)$ in only $O(n^2 t(n - i))$ steps, instead of $O(n^2 t(n))$ steps. So the more information we "precompute," the faster the algorithm will run on sufficiently large inputs n.

To turn this idea into a proof, the main thing one needs to show is that simulating the iterative procedure is pretty much the *only* way to compute f: or more precisely, any algorithm to compute f needs at least $t(n - i)$ steps for some i. This then implies that f has no fastest algorithm.

FURTHER READING

The next few chapters will continue to explore computational complexity theory. However, for readers who just can't get enough, and *really* want to explore this subject in depth, my own favorite books include: *Computational Complexity* by Christos Papadimitriou (Addison-Wesley, 1994); *Computational Complexity: A Modern Approach*, by Sanjeev Arora and Boaz Barak (Cambridge University Press, 2009); and *The Nature of Computation*, by Cristopher Moore and Stephan Mertens (Oxford University Press, 2011)

PUZZLE I FROM LAST CHAPTER

Can we assume, without loss of generality, that a computer program has access to its own code? As a simple example, is there a program that prints itself as output?

The answer is yes: there *are* such programs. In fact, there have even been competitions to write the *shortest* self-printing program. At the IOCCC[6] (the International Obfuscated C Code Contest), this competition was won some years ago by an *extremely* short program. Can you guess how long it was: 30 characters? 10? 5?

The winning program had *zero* characters. (Think about it!) Admittedly, a blank file is not *exactly* a kosher C program, but apparently some compilers will compile it to a program that does nothing.

Alright, alright, but what if we want a *nontrivial* self-printing program? In that case, the standard trick is to do something like the following (which you can translate into your favorite programming language):

Print the following twice, the second time in quotes.
"Print the following twice, the second time in quotes."

In general, if you want a program to have access to its own source code, the trick is to divide the program into three parts: (1) a part that actually does something useful (this is optional), (2) a "replicator,"

---

[6] http://www.ioccc.org/

and (3) a string to be replicated. The string to be replicated should consist of the complete code of the program, *including* the replicator. (In other words, it should consist of parts (1) and (2).) Then by running the replicator twice, we get a spanking-new copy of parts (1), (2), and (3).

This idea was elaborated by von Neumann in the early 1950s. Shortly afterward, two guys (I think their names were Crick and Watson) found a physical system that actually obeys these rules. You and I, along with all living things on Earth, are basically walking computer programs with the semantics

> Build a baby that acts on the following instructions, and also contains a copy of those instructions in its reproductive organs.

> "Build a baby that acts on the following instructions, and also contains a copy of those instructions in its reproductive organs."

PUZZLE 2 FROM LAST CHAPTER

If water weren't H_2O, would it still be water?

Yeah, this isn't really a well-defined question: it all boils down to what we *mean* by the word *water*. Is water a "predicate": if x is clear and wet and drinkable and tasteless and freezable to ice, etc.... then x is water? In this view, what water "is" is determined by sitting in our armchairs and listing necessary and sufficient conditions for something to be water. We then venture out into the world, and anything that meets the conditions is water by definition. This was the view of Frege and Russell, and it implies that anything with the "intuitive properties" of water *is* water, whether or not it's H_2O.

The other view, famously associated with Saul Kripke[7], is that the word *water* "rigidly designates" a particular substance (H_2O). In this view, we now know that when the Greeks and Babylonians talked about water, they were really talking about H_2O, even though they didn't realize it. Interestingly, "water = H_2O" is thus a *necessary*

---

[7] See Saul Kripke, *Naming and Necessity*, Wiley-Blackwell, 1991 (reprint edition).

truth that was discovered by *empirical* observation. Something with all the same intuitive properties is water, but a different chemical structure would *not* be water.

Kripke argues that, if you accept this "rigid designator" view, then there's an implication for the mind–body problem.

The idea is this: the reductionist dream would be to *explain* consciousness in terms of neural firings, in the same way that science *explained* water as being H_2O. But Kripke says there's a disanalogy between these two cases. In the case of water, we can at least *talk* coherently about a hypothetical substance that feels like water, tastes like water, etc., but isn't H_2O and therefore isn't water. But suppose we discovered that pain is always associated with the firings of certain nerves called C-fibers. Could we then say that pain *is* C-fiber firings? Well, if something felt like pain but had a different neurobiological origin, would we say that it felt like pain but *wasn't* pain? Presumably we wouldn't. Anything that feels like pain *is* pain, by definition! Because of this difference, Kripke thinks that we can't explain pain as "being" C-fiber firings, in the same sense that we can explain water as "being" H_2O.

I hope you're not bored here. Dude – this is considered one of the greatest philosophical insights of the last four decades! I'm serious! Well, I guess if you don't find it interesting, philosophy is not the field for you.

6   P, NP, and friends

We've seen that if we want to make progress in complexity, then we need to talk about asymptotics: not which problems can be solved in $10\,000$ steps, but for which problems can instances of size n be solved in cn^2 steps as n goes to infinity? We met $\text{TIME}(f(n))$, the class of all problems solvable in $O(f(n))$ steps, and $\text{SPACE}(f(n))$, the class of all problems solvable using $O(f(n))$ bits of memory.

But if we *really* want to make progress, then it's useful to take an even coarser-grained view: one where we distinguish between polynomial and exponential time, but *not* between $O(n^2)$ and $O(n^3)$ time. From this remove, we think of any polynomial bound as "fast," and any exponential bound as "slow."

Now, I realize people will immediately object: what if a problem is solvable in polynomial time, but the polynomial is $n^{50\,000}$? Or what if a problem takes exponential time, but the exponential is $1.000\,000\,01^n$? My answer is pragmatic: if cases like that regularly arose in practice, then it would've turned out that we were using the wrong abstraction. But so far, it seems like we're using the right abstraction. Of the big problems solvable in polynomial time – matching, linear programming, primality testing, etc. – most of them really *do* have practical algorithms. And of the big problems that we think take exponential time – theorem-proving, circuit minimization, etc. – most of them really *don't* have practical algorithms. So, that's the empirical skeleton holding up our fat and muscle.

Of course, you might wonder *why* this is true: why do algorithms typically take n^2 or n^3 time, and not n^{10000} time? My own view is that there's a general phenomenon here, which is not specific to computer science at all, but which holds for physics, chemistry, economics, engineering, and every other quantitative field I've encountered. Namely, suppose I tell you that some quantity m scales with another quantity n like $m \sim n^c$ for some fixed power c, but I don't tell you what c is. Then not knowing anything else, should you guess that c is more like 2 or 3, or more like 10000 or a googol? Empirically, in every real-world situation that I know about, you're far more likely to be right if you guess 2 or 3 than if you guess 10000. I don't fully understand the reasons for that—it would be great to clarify them— but on the other hand, I don't find it too surprising either. I mean, it's not hard to come up with plausible mechanisms that could lead to an n^2 scaling: maybe you have n points and you want to compare every point against every other, maybe you're interested in the area of a square with side length n, etc. And if you try, I predict you'll also be able to come up with plausible mechanisms (though fewer of them) for n^3 or n^4 scaling. But what are the plausible mechanisms that could produce n^{10000} scaling, besides those that simply produce n^k scaling for arbitrary values of k? Not many! There are a lot more special properties of 2 or 3 than there are special properties of 10000.

PETTING ZOO

It's now time to meet the most basic complexity classes – the sheep and goats of the Complexity Zoo.

- **P** is the class of problems solvable by a Turing machine in polynomial time. In other words, **P** is the union, over all positive integers k, of TIME(n^k). (Note that, by "problem," we'll always mean *decision problem*: a problem where the inputs are n-bit strings and the outputs are either yes or no.)
- **PSPACE** is the class of problems solvable in polynomial space (but unlimited time). In other words, it's the union over all integers k of SPACE(n^k).
- **EXP** is the class of problems solvable in exponential time. In other words, it's the union over all integers k of TIME(2^{n^k}).

Certainly **P** is contained in **PSPACE**. I claim that **PSPACE** is contained in **EXP**. Why?

Right: a machine with n^k bits of memory can only go through 2^{n^k} different configurations before it either halts or else gets stuck in an infinite loop.

Now, **NP** is the class of problems for which, if the answer is yes, then there's a polynomial-size *proof* of that fact that you can *check* in polynomial time. (The NP stands for "Nondeterministic Polynomial," in case you were wondering.) I could get more technical, but it's easiest to give an example: say, I give you a 10000-digit number, and I ask whether it has a divisor ending in 3. Well, answering that question might take a Long, Long Time™. But if your grad student finds such a divisor *for* you, then you can easily check that it works: you don't need to trust your student (always a plus).

I claim that **NP** is contained in **PSPACE**. Why?

Right: in polynomial space, you can loop over all possible n^k-bit proofs and check them one by one. If the answer is "yes," then one of the proofs will work, while if the answer is "no," then none of them will work.

Certainly **P** is contained in **NP**: if you can answer a question yourself, then someone else can convince you that the answer is yes (if it *is* yes) without even telling you anything.

Of course, a question arises of whether **P** *equals* **NP**. In other words, if you can *recognize* an answer efficiently, can you also *find* one efficiently? Maybe you've heard of this question before.

Look, this **P** versus **NP** question, what can I say? People like to describe it as "probably the central unsolved problem of theoretical computer science." That's a comical understatement. **P** vs. **NP** is one of the deepest questions that human beings have ever asked.

And not only that: it's one of the seven million-dollar prize problems of the Clay Math Institute![1] What an honor! Imagine: our mathematician friends have decided that **P** vs. **NP** is *as important* as the Hodge Conjecture, or even Navier–Stokes existence and

---

[1] See http://www.claymath.org/millennium/

smoothness! (Apparently, they weren't going to include it, until they asked around to make sure it was important enough.)

Dude. One way to measure **P** vs. **NP**'s importance is this. If **NP** problems were feasible, then mathematical creativity could be automated. The ability to check a proof would entail the ability to find one. Every Apple II, every Commodore, would have the reasoning power of Archimedes or Gauss. So by just programming your computer and letting it run, presumably you could immediately solve not only **P** vs. **NP**, but *also the other six Clay problems.* (Or five, now that Poincaré is down.)

But if that's the case, then why isn't it *obvious* that **P** doesn't equal **NP**? Surely, God wouldn't be so benign as to grant us these extravagant powers! Surely, our physicist-intuition tells us that brute-force search is unavoidable! (Leonid Levin told me that Feynman – the king, or possibly court jester, of physicist-intuition – had trouble even being convinced that **P** vs. **NP** was an open problem!)

Well, we certainly *believe* **P** ≠ **NP**. Indeed, we don't even believe there's a general way to solve **NP** problems that's dramatically better than brute-force search through every possibility. But if you want to understand why it's so hard to *prove* these things, let me tell you something.

Let's say you're given an N-digit number, but instead of factoring it, you just want to know if it's prime or composite.

Or let's say you're given a list of freshmen, together with which ones are willing to room with which other ones, and you want to pair off as many willing roommates as you can.

Or let's say you're given two DNA sequences, and you want to know how many insertions and deletions are needed to convert the one sequence to the other.

Surely, these are fine examples of the sort of exponentially hard **NP** problem we were talking about! Surely, they, too, require brute-force search!

Except they don't. As it turns out, all of these problems have clever polynomial-time algorithms. *The central challenge any* ***P ≠ NP*** *proof will have to overcome is to separate the **NP** problems that really* ***are*** *hard from the ones that merely **look** hard.* I'm not just making a philosophical point. While there have been dozens of purported **P ≠ NP** proofs over the years, almost all of them could be rejected immediately for the simple reason that, if they worked, then they would rule out polynomial-time algorithms that we already know to exist.

So to summarize, there are problems like primality testing and pairing off roommates, for which computer scientists (often after decades of work) have been able to devise polynomial-time algorithms. But then there are other problems, like proving theorems, for which we don't know of any algorithm fundamentally better than brute-force search. But is that all we can say – that we have a bunch of these **NP** problems, and for some of them, we've found a fast algorithm and for others, we haven't?

As it turns out, we can say something much more interesting than that. We can say that *almost all of the "hard" problems are the **same** "hard" problem in different guises* – in the sense that, if we had a polynomial-time algorithm for any one of them, then we'd also have polynomial-time algorithms for all the rest. This is the upshot of the theory of **NP**-completeness, which was created in the early 1970s by Cook, Karp, and Levin.

The way it goes is, we define a problem B to be "**NP**-hard" if any **NP** problem can be efficiently reduced to B. What the hell does that mean? It means that, *if* we had an oracle to immediately solve problem B, *then* we could solve any **NP** problem in polynomial time.

That gives one notion of reduction, which is called Cook reduction. There's also a weaker notion of reduction, which is called Karp reduction. In a Karp reduction from problem A to problem B, we insist that there should be a polynomial-time algorithm that transforms any instance of A to an instance of B having the same answer.

What's the difference between Cook and Karp?

Right: with a Cook reduction, in solving problem A we get to call the oracle for problem B more than once. We can even call the oracle *adaptively* – that is, in ways that depend on the outcomes of the previous calls. A Karp reduction is weaker in that we don't allow ourselves these liberties. Perhaps surprisingly, almost every reduction we know of is a Karp reduction – the full power of Cook reductions is rarely needed in practice.

Now, we say a problem is **NP**-complete if it's both **NP**-hard and in **NP**. In other words, **NP**-complete problems are the "hardest" problems in **NP**: the problems that single-handedly capture the difficulty of every other **NP** problem. As a first question, is it obvious that **NP**-complete problems even *exist*?

I claim that it *is* obvious. Why?

Well, consider the following problem, called DUH: we're given a polynomial-time Turing machine M, and we want to know if there exists an n^k-bit input string that causes M to accept. I claim that any instance of any **NP** problem can be converted, in polynomial time, into a DUH instance having the same answer. Why? Well, DUH! Because that's what it *means* for a problem to be in **NP**!

The discovery of Cook, Karp, and Levin was not that there *exist* **NP**-complete problems – that's obvious – but rather that many *natural* problems are **NP**-complete.

The king of these natural **NP**-complete problems is called 3-Satisfiability, or 3SAT. (How do I know it's the king? Because it appeared on the TV show *NUMB3RS*.) Here we're given n Boolean variables, $x_1, \ldots, x_n$, as well as a set of logical constraints called *clauses* that relate at most three variables each:

x_2 or x_5 or not(x_6)

not(x_2) or x_4

not(x_4) or not(x_5) or x_6

. . . .

Then the question is whether there's some way to set the variables $x_1, \ldots, x_n$ to TRUE or FALSE, in such a way that every clause is "satisfied" (that is, every clause evaluates to TRUE).

It's obvious that 3SAT is *in* **NP**. Why? Right: Because if someone gives you a setting of $x_1, \ldots, x_n$ that works, it's easy to check that it works!

Our goal is to prove that 3SAT is **NP**-complete. What will that take? Well, we need to show that, if we had an oracle for 3SAT, then we could use it to solve not only 3SAT in polynomial time but also *any* **NP** problem whatsoever. That seems like a tall order! Yet in retrospect, you'll see that it's almost a triviality.

The proof has two steps. Step 1 is to show that, if we could solve 3SAT, then we could solve a more "general" problem called CircuitSAT. Step 2 is to show that, if we could solve CircuitSAT, then we could solve any **NP** problem.

In CircuitSAT, we're given a Boolean circuit and . . . wait, listen up, engineers: in computer science, a "circuit" *never* has loops! Nor does it have resistors or diodes or anything weird like that. For us, a circuit is just an object where you start with n Boolean variables $x_1, \ldots, x_n$, and then you can repeatedly define a new variable that's equal to the AND, OR, or NOT of variables that you've previously defined. Like so:

$$x_{n+1} := x_3 \text{ or } x_n$$
$$x_{n+2} := \text{not}(x_{n+1})$$
$$x_{n+3} := x_1 \text{ and } x_{n+2}$$
$$\ldots$$

We designate the last variable in the list as the circuit's "output." Then the goal, in CircuitSAT, is to decide whether there's a setting of $x_1, \ldots, x_n$ such that the output is TRUE.

I claim that, if we could solve 3SAT, then we could also solve CircuitSAT. Why?

Well, all we need to do is notice that every CircuitSAT instance is really a 3SAT instance in disguise! Every time we compute an AND, OR, or NOT, we're relating one new variable to one or two old variables. And any such relationship can be expressed by a set of clauses involving at most three variables each. So, for example,

$$x_{n+1} := x_3 \text{ or } x_n$$

becomes

$$x_{n+1} \text{ or } \text{not}(x_3)$$
$$x_{n+1} \text{ or } \text{not}(x_n)$$
$$\text{not}(x_{n+1}) \text{ or } x_3 \text{ or } x_n$$

So, that was Step 1. Step 2 is to show that, if we can solve CircuitSAT, then we can solve *any* **NP** problem.

Alright, so consider some instance of an **NP** problem. Then by the definition of **NP**, there's a polynomial-time Turing machine M such that the answer is "yes" if and only if there's a polynomial-size witness string w that causes M to accept.

Now, given this Turing machine M, our goal is to create a circuit that "mimics" M. In other words, we want there to exist a setting of the circuit's input variables that makes it evaluate to TRUE, *if and only if* there exists a string w that causes M to accept.

How do we achieve that? Simple: by *defining a whole buttload of variables!* We'll have a variable that equals TRUE if and only if the 37th bit of M's tape is set to '1' at the 42nd time step. We'll have another variable that equals TRUE if and only if the 14th bit is set to '1' at the 52nd time step. We'll have another variable that equals TRUE if and only if M's tape head is in the 15th internal state and the 74th tape position at the 33rd time step. Well, you get the idea.

Then, having written down this buttload of variables, we write down a shitload of logical relations between them. If the 17th bit of the tape is '0' at the 22nd time step, and the tape head is nowhere near the 17th bit at that time, then the 17th bit will still be '0' at the 23rd time step. If the tape head is in internal state 5 at the 44th time step, and it's reading a '1' at that time step, and internal state 5 transitions to internal state 7 on reading a '1', then the tape head will be in internal state 7 at the 45th time step. And so on, and so on. The only variables that are left unrestricted are the ones that constitute the string w at the first time step.

The key point is that, while this *is* a very large buttload of variables and relations, it's still only a *polynomial* buttload. We therefore get a polynomially large CircuitSAT instance, which is satisfiable if and only if there exists a w that causes M to accept.

We've just proved the celebrated Cook–Levin Theorem: that 3SAT is **NP**-complete. This theorem can be thought of as the "initial infection" of the **NP**-completeness virus. Since then, the virus has spread to *thousands* of other problems. What I mean is this: if you want to prove that your favorite problem is **NP**-complete, all you have to do is prove it's as hard as some other problem that's *already* been proved **NP**-complete. (Well, you also have to prove that it's *in* **NP**, but that's usually trivial.) So there's a rich-get-richer effect: the more problems that have already been proved **NP**-complete, the easier it is to induct a new problem into the club. Indeed, proving problems **NP**-complete had become so routine by the 1980s or 1990s, and people had gotten so good at it, that (with rare exceptions) the two main complexity conferences STOC and FOCS stopped publishing yet more **NP**-completeness proofs.

I'll just give you a tiny sampling of some of the earliest problems that were proved **NP**-complete:

- **Map Colorability:** Given a map, can you color every country red, green, or blue, in such a way that no two neighboring countries are colored the same? (Interestingly, if you're only allowed to use *two* colors, then it's easy to decide whether or not such a coloring is possible – why? On the other hand, if you're allowed *four* colors, then it always is possible, at least for maps drawn in the plane – that's the famous

Four-Color Theorem. So then the problem is again easy. Only with three colors is the problem **NP**-complete.)

- **Clique:** Given a set of N high-school students, together with which ones will sit at a cafeteria table with which other ones, is there a "clique" of $N/3$ students who will all sit at a table with each other?
- **Packing:** Given a set of boxes of specified dimensions, can you fit them into the trunk of your car?

Etc., etc., etc.

To reiterate: although these problems might look unrelated, they're actually the same problem in different costumes. If any *one* of them has an efficient solution, then *all* of them do, and **P = NP**. If any one of them *doesn't* have an efficient solution, then *none* of them do, and **P ≠ NP**. To prove **P = NP**, it's enough to show that *some* **NP**-complete problem (no matter which one) has an efficient solution. To prove **P ≠ NP**, it's enough to show that some **NP**-complete problem has *no* efficient solution. One for all and all for one.

So, there are the **P** problems, and then there are the **NP**-complete problems. Is there anything in between? (You should be used to this sort of "intermediate" question by now – we saw it both in set theory and in computability theory!)

If **P = NP**, then **NP**-complete problems are **P** problems, so obviously the answer is no.

But what if **P ≠ NP**? In that case, a beautiful result called Ladner's Theorem says that there must be "intermediate" problems between **P** and **NP**-complete: in other words, problems that are in **NP**, but neither **NP**-complete nor solvable in polynomial time.

How would we create such an intermediate problem? Well, I'll give you the idea. The first step is to define an extremely slow-growing function t. Then, given a 3SAT instance F of size n, the problem will be to decide whether F is satisfiable *and* $t(n)$ is odd. In other words: if $t(n)$ is odd, then solve the 3SAT problem, while if $t(n)$ is even, then always output "no."

If you think about what we're doing, we're alternating long stretches of an **NP**-complete problem with long stretches of nothing! Intuitively, each stretch of 3SAT should kill off another polynomial-time algorithm for our problem, where we use the assumption that $P \neq NP$. Likewise, each stretch of nothing should kill off another **NP**-completeness reduction, where we again use the assumption that $P \neq NP$. This ensures that the problem is neither in **P** nor **NP**-complete. The main technical trick is to make the stretches get longer at an exponential rate. That way, given an input of size n, we can simulate the whole iterative process up to n in time polynomial in n. That ensures that the problem is still in **NP**.

Besides **P** and **NP**, another major complexity class is **coNP**: the "complement" of **NP**. A problem is in **coNP** if a "no" answer can be checked in polynomial time. To every **NP**-complete problem, there's a corresponding **coNP**-complete problem. We've got *un*satisfiability, map *non*colorability, etc.

Now, why would anyone bother to define such a stupid thing? Because then we can ask a new question: *does NP equal coNP?* In other words: if a Boolean formula is unsatisfiable, is there at least a short *proof* that it's unsatisfiable, even if finding the proof would take exponential time? Once again, the answer is that we don't know.

Certainly, if $P = NP$, then $NP = coNP$. (Why?) On the other hand, the other direction isn't known: it could be that $P \neq NP$ but still $NP = coNP$. So if proving $P \neq NP$ is too easy, you can instead try to prove $NP \neq coNP$!

This seems like a good time to mention a special complexity class, a class we quantum computing people know and love: $NP \cap coNP$.

This is the class for which either a yes answer *or* a no answer has an efficiently checkable proof. As an example, consider the problem of factoring an integer into primes. Over the course of my life, I must've met at least two dozen people who "knew" that factoring is

NP-complete, and therefore that Shor's algorithm – since it lets us factor on a quantum computer – also lets us solve NP-complete problems on a quantum computer. Often these people were supremely confident of their "knowledge."

But can we pinpoint just *how* factoring differs from the known NP-complete problems, in terms of complexity theory? Yes, we can. First of all, in order to make factoring a decision (yes-or-no) problem, we need to ask something like this: given a positive integer N, does N have a prime factor whose last digit is 7? I claim that this problem is not merely in NP, but in NP ∩ coNP. Why? Well, suppose someone gives you the prime factorization of N. There's only one of them. So if there *is* a prime factor whose last digit is 7, then you can verify that, and if there's *no* prime factor whose last digit is 7, then you can also verify *that*.

You might say, "but how do I know that I really was given the prime factorization? Sure, if someone gives me a bunch of numbers, I can check that they multiply to N, but how do I know they're prime?" For this, you'll have to take on faith something that I told you earlier: that if you just want to know whether a number is prime or composite, and not what its factors are, then you can do that in polynomial time. OK, so if you accept that, then the factoring problem is in NP ∩ coNP.

From this, we can conclude that, *if* factoring were NP-complete, then NP would equal coNP. (Why?) Since we don't believe NP = coNP, this gives us a strong indication (though not a proof) that, all those people I told you about notwithstanding, factoring is *not* NP-complete. If we accept that, then only two possibilities remain: either factoring is in P, or else factoring is one of those "intermediate" problems whose existence is guaranteed by Ladner's Theorem. Most of us incline toward the latter possibility – though not with as much conviction as we believe P ≠ NP.

Indeed, for all we know, it could be the case that P = NP ∩ coNP but still P ≠ NP. (This possibility would imply that NP ≠ coNP.) So, if proving $P \neq NP$ and $NP \neq coNP$ are *both* too easy for you, your next challenge can be to prove $P \neq NP \cap coNP$!

If **P**, **NP**, and **coNP** aren't enough to rock your world, you can generalize these classes to a giant teetering mess that we computer scientists call the *polynomial hierarchy*.

Observe that you can put any **NP** problem instance into the form

Does there exist an n-bit string X such that A(X)=1?

Here A is a function computable in polynomial time.

Likewise, you can put any **coNP** problem into the form

Does A(X)=1 for every X?

But what happens if you throw in another quantifier, like so?

Does there exist an X such that for every Y, A(X,Y)=1?

For every X, does there exist a Y such that A(X,Y)=1?

Problems like these lead to two new complexity classes, which are called $\Sigma_2 P$ and $\Pi_2 P$, respectively. $\Pi_2 P$ is the "complement" of $\Sigma_2 P$, in the same sense that **coNP** is the complement of **NP**. We can also throw in a third quantifier:

Does there exist an X such that for every Y, there exists a Z such that A(X,Y,Z)=1?

For every X, does there exist a Y such that for every Z, A(X,Y,Z)=1?

This gives us $\Sigma_3 P$ and $\Pi_3 P$, respectively. It should be obvious how to generalize this to $\Sigma_k P$ and $\Pi_k P$ for any larger k. (As a side note, when $k = 1$, we get $\Sigma_1 P = NP$ and $\Pi_1 P = coNP$. Why?) Then taking the union of these classes over all positive integers k gives us the polynomial hierarchy **PH**.

The polynomial hierarchy really is a substantial generalization of **NP** and **coNP** – in the sense that, even if we had an oracle for **NP**-complete problems, it's not at all clear how we could use it to solve (say) $\Sigma_2 P$ problems. On the other hand, just to complicate matters further, I claim that if **P** = **NP**, then the whole polynomial hierarchy would collapse down to **P**! Why?

Right: if **P** = **NP**, then we could take our algorithm for solving **NP**-complete problems in polynomial time, and modify it to *call itself as a subroutine*. And that would let us "flatten **PH** like a steamroller": first simulating **NP** and **coNP**, then $\Sigma_2 P$ and $\Pi_2 P$, and so on through the entire hierarchy.

Likewise, it's not hard to prove that, if **NP** = **coNP**, then the entire polynomial hierarchy collapses down to **NP** (or in other words, to **coNP**). If $\Sigma_2 P = \Pi_2 P$, then the entire polynomial hierarchy collapses down to $\Sigma_2 P$, and so on. If you think about it, this gives us a whole infinite sequence of generalizations of the **P** $\neq$ **NP** conjecture, each one "harder" to prove than the last. Why do we care about these generalizations? Because often, we're trying to study conjecture BLAH, and we can't prove that BLAH is true, and we can't *even* prove that if BLAH were false then **P** would equal **NP**. But – and here's the punchline – we *can* prove that if BLAH were false, then the polynomial hierarchy would collapse to the second or the third level. And this gives us some sort of evidence that BLAH is true.

Welcome to complexity theory!

Since I talked about how lots of problems have nonobvious polynomial-time algorithms, I thought I should give you at least one example. So, let's do one of the simplest and most elegant in all of computer science – the so-called Stable Marriage Problem. Have you seen this before? You haven't?

Alright, so we have N men and N women. Our goal is to marry them off. We assume for simplicity that they're all straight. (Marrying off gays and lesbians is technically harder, though also solvable in polynomial time!) We also assume, for simplicity and with much loss of generality, that everyone would rather be married than single.

So, each man ranks the women, in order from his first to last choice. Each woman likewise ranks the men. There are no ties.

Obviously, not every man can marry his first-choice woman, and not every woman can marry her first-choice man. Life sucks that way.

So, let's try for something weaker. Given a way of pairing off the men and women, say that it's *stable* if *no man and woman who aren't married to each other both prefer each other to their spouses*. In other words, you might despise your husband, but no man who you like better than him likes you better than his wife, so you have no incentive to leave. This is the, um, desirable property that we call "stability."

Now, given the men's and women's stated preferences, our goal as matchmakers is to find a stable way of pairing them off. Matchmaker, matchmaker, make me a match, find me a find, catch me a catch, etc.

First obvious question: does there always exist a stable pairing of men and women? What do you think? Yes? No? As it turns out, the answer is yes, but the easiest way to prove it is just to give an algorithm for *finding* the pairing!

So, let's concentrate on the question of how to find a pairing. In total, there are $N!$ ways of pairing off men with women. For the soon-to-be-newlyweds' sake, we hope we won't have to search through all of them.

Fortunately, we won't. In the early 1960s, Gale and Shapley invented a polynomial-time – in fact *linear*-time – algorithm to solve this problem. And the beautiful thing about this algorithm is, it's exactly what you'd come up with from reading a Victorian romance novel. Later they found out that the same algorithm had been *in use* since the 1950s – not to pair off men with women, but to pair off medical-school students with hospitals to do their residencies in.

Indeed, hospitals and medical schools are still using a version of the algorithm today.

But back to the men and women. If we want to pair them off by the Gale–Shapley algorithm, then as a first step, we need to break the symmetry between the sexes: which sex "proposes" to the other? This being the early 1960s, you can guess how that question was answered. The men propose to the women.

So, we loop through all the men. The first man proposes to his first-choice woman. She provisionally accepts him. Then the next man proposes to his first-choice woman. She provisionally accepts *him*, and so on. But what happens when a man proposes to a woman who's already provisionally accepted another man? She chooses the one she prefers, and boots the other one out! Then, the next time we come around to that man in our loop over the men, he'll propose to his *second*-choice woman. And if she rejects him, then the next time we come around to him he'll propose to his third-choice woman. And so on, until everyone is married off. Pretty simple, huh?

First question: why does this algorithm terminate in linear time?

Right: because each man proposes to a given woman at most once. So the total number of proposals is at most N^2, which is just the amount of memory we need to write down the preference lists in the first place.

Second question: when the algorithm does terminate, why is everyone married off?

Right: because if they weren't, then there'd be some woman who'd never been proposed to, and some man who'd never proposed to her. But this is impossible. Eventually, the man no one else wants will cave in, and propose to the woman no one else wants.

Third question: why is the pairing produced by this algorithm a *stable* one?

Right: because if it weren't, then there'd be one married couple (say, Bob and Alice), and another married couple (say, Charlie and Eve), such that Bob and Eve both prefer each other to their spouses. But in that case, Bob would've proposed to Eve *before* proposing to Alice. And if Charlie also proposed to Eve, then Eve would've made clear at the time that she preferred Bob. And this gives a contradiction.

In particular, we've shown, as promised, that there *exists* a stable pairing: namely, the pairing found by the Gale–Shapley algorithm.

PROBLEM SET

1. We saw that 3SAT is **NP**-complete. By contrast, it turns out that 2SAT – the version where we only allow two variables per clause – is solvable in polynomial time. Explain why.

2. Recall that **EXP** is the class of problems solvable in exponential time. One can also define **NEXP**: the class of problems for which a "yes" answer can be *verified* in exponential time. In other words, **NEXP** is to **EXP** as **NP** is to **P**. Now, we don't know if **P** = **NP**, and we also don't know if **EXP** = **NEXP**. But we *do* know that if **P** = **NP**, then **EXP** = **NEXP**. Why?

3. Show that **P** doesn't equal SPACE(n) (the set of problems solvable in linear space). *Hint:* You don't need to prove that **P** is not in SPACE(n), or that SPACE(n) is not in **P** – only that one or the other is true!

4. Show that, if **P** = **NP**, then there's a polynomial-time algorithm not only to decide whether a Boolean formula has a satisfying assignment but also to *find* such an assignment whenever one exists.

5. [**Extra credit**] Give an *explicit* algorithm that finds a satisfying assignment whenever are exists, and that runs in polynomial time assuming **P** = **NP**. (If there's no satisfying assignment, your algorithm can be have arbitrarily.) In other words, give an algorithm for problem 4 that you could implement and run right now – without involving any subroutine that you've assumed to exist but can't actually describe.

7 Randomness

In the last two chapters, we talked about computational complexity up till the early 1970s. Here, we'll add a new ingredient to our already simmering stew – something that was thrown in around the mid-1970s, and that now pervades complexity to such an extent that it's hard to imagine doing anything without it. This new ingredient is *randomness*.

Certainly, if you want to study quantum computing, then you first have to understand randomized computing. I mean, quantum amplitudes only become interesting when they exhibit some behavior that classical probabilities *don't*: contextuality, interference, entanglement (as opposed to correlation), etc. So we can't even begin to discuss quantum mechanics without first knowing what it is that we're comparing against.

Alright, so what is randomness? Well, that's a profound philosophical question, but I'm a simpleminded person. So, you've got some probability p, which is a real number in the unit interval $[0, 1]$. That's randomness.

But wasn't it a big achievement when Kolmogorov put probability on an axiomatic basis in the 1930s? Yes, it was! But in this chapter, we'll only care about probability distributions over finitely many events, so all the subtle questions of integrability, measurability, and so on won't arise. In my view, probability theory is yet another example where mathematicians immediately go to infinite-dimensional spaces, in order to solve the problem of having a nontrivial problem to solve! And that's fine – whatever floats your boat. I'm not *criticizing* that. But in theoretical computer science, we've already got our hands full with 2^n choices. We need $2^{\aleph_0}$ choices like we need a hole in the head.

Alright, so given some "event" A – say, the event that it will rain tomorrow – we can talk about a real number Pr[A] in [0, 1], which is the probability that A will happen. (Or rather, the probability we *think* A will happen – but I told you I'm a simpleminded person.) And the probabilities of different events satisfy some obvious relations, but it might be helpful to see them explicitly if you never have before.

First, the probability that A *doesn't* happen equals 1 minus the probability that it happens:

$$\Pr[\text{not}(A)] = 1 - \Pr[A].$$

Agree? I thought so.

Second, if we've got two events A and B, then

$$\Pr[A \text{ or } B] = \Pr[A] + \Pr[B] - \Pr[A \text{ and } B].$$

Third, an immediate consequence of the above, called the *union bound*:

$$\Pr[A \text{ or } B] \leq \Pr[A] + \Pr[B].$$

Or in English: if you're unlikely to drown and you're unlikely to get struck by lightning, then chances are you'll neither drown *nor* get struck by lightning, regardless of whether getting struck by lightning makes you more or less likely to drown. One of the few causes for optimism in this life.

Despite its triviality, the union bound is probably the most useful fact in all of theoretical computer science. I use it maybe 200 times in every paper I write.

What else? Given a numerical random variable X, the expectation of X, or $E[X]$, is defined to be $\sum_k \Pr[X = k]\, k$. Then given any two random variables X and Y, we have

$$E[X + Y] = E[X] + E[Y].$$

This is called *linearity of expectation*, and is probably the second most useful fact in all of theoretical computer science, after the union bound. Again, the key point is that any dependencies between X and Y are irrelevant.

Do we also have

$$E[XY] = E[X]\,E[Y]?$$

Right: we don't! Or rather, we do if X and Y are independent, but not in general.

Another important fact is Markov's inequality (or rather, one of his many inequalities): if $X \geq 0$ is a nonnegative random variable, then for all k,

$$\Pr[X \geq kE[X]] \leq 1/k.$$

Why? Well, if X were too many times larger than its expectation too often, then even if X were 0 the rest of the time, it still wouldn't be enough to balance the expectation out.

Markov's inequality leads immediately to the *third* most useful fact in theoretical computer science, called the Chernoff bound. The Chernoff bound says that if you flip a coin 1000 times, and you get heads 900 times, then chances are the coin was crooked. This is the theorem that casino managers implicitly use when they decide whether to send goons to break someone's legs.

Formally, let h be the number of times you get heads if you flip a fair coin n times. Then one way to state the Chernoff bound is

$$\Pr[|h - n/2| \geq \alpha] \leq 2e^{-c\alpha^2/n},$$

where c is a constant that you look up since you don't remember it. (Oh, alright: $c = 2$ will work.)

How can we prove the Chernoff bound? Well, there's a simple trick: let $x_i = 1$ if the ith coin flip comes up heads, and let $x_i = 0$ if tails. Then consider the expectation, not of $x_1 + \cdots + x_n$ itself, but of $\exp(x_1 + \cdots + x_n)$. Since the coin flips had better be uncorrelated with each other, we have

$$E[e^{x_1 + \cdots x_n}] = E[e^{x_1} \cdots e^{x_n}]$$
$$= E[e^{x_1}] \cdots E[e^{x_n}]$$
$$= \left(\frac{1 + e}{2}\right)^n.$$

Now we can just use Markov's inequality, and then take logs on both sides to get the Chernoff bound. I'll spare you the calculation (or rather, spare myself).

What do we need randomness for?

Even the ancients – Turing, Shannon, and von Neumann – understood that a random number source might be useful for writing programs. So, for example, back in the 1940s and 1950s, physicists invented a technique called *Monte Carlo simulation*, to study some weird question they were interested in at the time involving the implosion of hollow plutonium spheres. Monte Carlo simulation simply means gathering information about the *typical* or *average* behavior of a possibly complicated dynamical system, not by explicitly calculating the averages of various quantities that interest you, but simply by simulating the system a bunch of times with different random initial configurations and collecting statistics. Statistical sampling – say, of the different ways a hollow plutonium sphere might go *kaboom!* – is one perfectly legitimate use of randomness.

There are many, many reasons you might want randomness – for foiling an eavesdropper in cryptography, for avoiding deadlocks in communication protocols, and so on. But within complexity theory, the usual purpose of randomness is to "smear out error": that is, to take an algorithm that works on most inputs, and turn it into an algorithm that works on *all* inputs *most* of the time.

Let's see an example of a randomized algorithm. Suppose I describe a number to you by starting from 1, and then repeatedly adding, subtracting, or multiplying two numbers that were previously described (as in the card game "24"). Like so:

$$a = 1$$
$$b = a + a$$
$$c = b^2$$
$$d = c^2$$
$$e = d^2$$
$$f = e - a$$
$$g = d - a$$
$$h = d + a$$
$$i = gh$$
$$j = f - i$$

You can verify (if you're so inclined) that j, the "output" of the above program, equals zero. Now consider the following general problem: given such a program, does it output 0 or not? How could you tell?

Well, one way would just be to run the program, and see what it outputs! What's the problem with that?

Right: even if the program is very short, the numbers it produces at intermediate steps might be enormous – that is, you might need exponentially many digits even to write them down. This can happen, for example, if the program repeatedly generates a new number by squaring the previous one. So a straightforward simulation isn't going to be efficient.

What can you do instead? Well, suppose the program has n operations. Then here's the trick: first pick a random prime number p with n^2 digits. Then simulate the program, but *doing all the arithmetic modulo p*. Here there's a super-important point that often trips up beginners: the only place where our algorithm is allowed to use randomness is in its *own* choices – in this case, in its choice of the random prime number p. We're not allowed to consider any sort of average over possible *programs*, since the program is simply the input to the algorithm, and input is still worst case!

What can we say about the above algorithm? Well, it will certainly be efficient: that is, it will run in time polynomial in n. Also, if the output isn't zero modulo p, then you certainly conclude that isn't zero. However, this still leaves two questions unanswered.

1. Supposing the output *is* 0 modulo p, how confident can you be that it wasn't just a lucky fluke, and that the output is actually 0?
2. How do you pick a random prime number?

For the first question, let x be the program's output. Then $|x|$ can be at most 2^{2^n}, where n is the number of operations – since the fastest way to get big numbers is by repeated squaring. This immediately implies that x can have at most 2^n prime factors.

On the other hand, how many prime numbers are there with n^2 digits? The famous Prime Number Theorem tells us the answer: about $2^{n^2}/n^2$. Since $2^{n^2}/n^2$ is a lot bigger than 2^n, *most* of those primes can't possibly divide x (unless of course $x = 0$). So if we pick a random prime and it *does* divide x, then we can be very, very confident (but admittedly not certain) that $x = 0$.

So much for the first question. Now on to the second: how do you pick a random prime with n^2 digits? Well, our old friend the Prime Number Theorem tells us that, if you pick a random *number* with n^2 digits, then it has about a one in n^2 chance of being prime. So all you have to do is keep picking random numbers; after about n^2 tries you'll probably hit a prime! Instead of repeatedly picking a random number, why couldn't you just start at a fixed number, and then keep adding 1 until you hit a prime?

Sure, that would work – assuming a far-reaching extension of the Riemann Hypothesis! What you need is that the n^2-digit prime numbers are more-or-less evenly spaced, so that you can't get unlucky and hit some exponentially long stretch where everything's composite. Not even the Extended Riemann Hypothesis would give you that, but there *is* something called Cramér's Conjecture that would.

Of course, we've merely reduced the problem of picking a random prime to a different problem: namely, once you've picked a random number, how do you tell if it's prime? As I mentioned in the last chapter, figuring out if a number is prime or composite turns out to be much easier than actually factoring the number. Until recently, this primality-testing problem was *another* example where it seemed like you needed to use randomness – indeed, it was the granddaddy of all such examples.

The idea was this. Fermat's Little Theorem (not to be confused with his Last Theorem!) tells us that, if p is a prime, then $x^p = x(\text{mod } p)$ for every integer x. So if you found an x for which $x^p \neq x(\text{mod } p)$, that would immediately tell you that p was composite – even though you'd still know nothing about what its divisors were. The hope would be that, if you *couldn't* find an x for which $x^p \neq x(\text{mod } p)$, then you could say with high confidence that p was prime.

Alas, 'twas not to be. It turns out that there are composite numbers p that "pretend" to be prime, in the sense that $x^p = x(\text{mod } p)$ for every x. The first few of these pretenders (called the Carmichael numbers) are 561, 1105, 1729, 2465, and 2821. Of course, if there were only finitely many pretenders, and we knew what they were, everything would be fine. But Alford, Granville, and Pomerance[1] showed in 1994 that there are infinitely many pretenders.

But already in 1976, Miller and Rabin had figured out how to unmask the pretenders by tweaking the test a little bit. In other words, they found a modification of the Fermat test that always passes if p is prime, and that fails with high probability if p is composite. So, this gave a polynomial-time randomized algorithm for primality testing.

Then, in a breakthrough a decade ago that you've probably heard about, Agrawal, Kayal, and Saxena[2] found a *deterministic* polynomial-time algorithm to decide whether a number is prime. This breakthrough has no practical application whatsoever, since we've long known of randomized algorithms that are faster, and whose

---

[1] W. R. Alford, A. Granville and C. Pomerance, There are infinitely many Carmichael numbers, *Annals of Mathematics* 2:139 (1994), 703–722. http://www.math.dartmouth.edu/~carlp/PDF/paper95.pdf

[2] M. Agrawal, N. Kayal, and N. Saxena, PRIMES is in P, *Annals of Mathematics* **160**:2 (2004), 781–793. http://www.cse.iitk.ac.in/users/manindra/algebra/primality_v6.pdf

error probability can easily be made smaller than the probability of an asteroid hitting your computer in mid-calculation. But it's wonderful to know.

To summarize, we wanted an efficient algorithm that would examine a program consisting entirely of additions, subtractions, and multiplications, and decide whether or not it output 0. I gave you such an algorithm, but it needed randomness in two places: first, in picking a random number; and second, in testing whether the random number was prime. The second use of randomness turned out to be inessential – since we now have a deterministic polynomial-time algorithm for primality testing. But what about the *first* use of randomness? Was that use also inessential? As of 2013, no one knows! But large theoretical cruise missiles have been pummeling this very problem, and the situation on the ground is volatile. Consult your local theoretical computer science conference proceedings for more on this developing story.

Alright, it's time to define some complexity classes. (Then again, when *isn't* it time?)

When we talk about probabilistic computation, chances are we're talking about one of the following four complexity classes, which were defined in a 1977 paper of John Gill.[3]

- **PP (Probabilistic Polynomial-Time):** Yeah, apparently even Gill himself admitted that it's a lousy name. But this is a serious book, and I will *not* tolerate any seventh-grade humor. Basically, **PP** is the class of all decision problems for which there exists a polynomial-time randomized algorithm that accepts with probability greater than $1/2$ if the answer is yes, or less than $1/2$ if the answer is no. In other words, we imagine a Turing machine M that receives both an n-bit input string x, and an unlimited source of random bits. If x is a yes-input, then at least half of the random bit settings should cause M to accept; while

[3] J. Gill, Computational Complexity of Probabilistic Turing Machines, *SIAM Journal on Computing* 6:4 (1977), 675–695.

if x is a no-input, then at least half of the random bit settings should cause M to reject. Furthermore, M needs to halt after a number of steps bounded by a polynomial in n.

> Here's the standard example of a **PP** problem: given a Boolean formula ϕ with n variables, do at least *half* of the 2^n possible settings of the variables make the formula evaluate to TRUE? (Incidentally, just like deciding whether there *exists* a satisfying assignment is **NP**-complete, so this majority-vote variant can be shown to be **PP**-complete: that is, any other **PP** problem is efficiently reducible to it.)
>
> Now, why might **PP** *not* capture our intuitive notion of problems solvable by randomized algorithms?
>
> Right: because we want to avoid "Florida recount" situations! As far as **PP** is concerned, an algorithm is free to accept with probability $\frac{1}{2} + 2^{-n}$ if the answer is yes, and probability $\frac{1}{2} - 2^{-n}$ if the answer is no. But how would a mortal actually distinguish those two cases? If n was (say) 5000, then we'd have to gather statistics for longer than the age of the universe!
>
> And, indeed, **PP** is an extremely big class: for example, it certainly contains the **NP**-complete problems. Why? Well, given a Boolean formula ϕ with n variables, what you can do is accept right away with probability $\frac{1}{2} - 2^{-2n}$, and otherwise choose a random truth assignment and accept it if and only if it satisfies ϕ. Then your total acceptance probability will be more than $\frac{1}{2}$ if there's at least one satisfying assignment for ϕ, or less than $\frac{1}{2}$ if there isn't.
>
> Indeed, complexity theorists believe that **PP** is *strictly* larger than **NP** – although, as usual, we can't prove it.

The above considerations led Gill to define a more "reasonable" variant of **PP**, as follows.

- **BPP (Bounded-Error Probabilistic Polynomial-Time):** This is the class of decision problems for which there exists a polynomial-time randomized algorithm that accepts with probability greater than $\frac{2}{3}$ if the answer is yes, or less than $\frac{1}{3}$ if the answer is no. In other words: given any input, the algorithm can be wrong with probability at most $\frac{1}{3}$.

> What's important about $\frac{1}{3}$ is just that it's *some* positive constant smaller than $\frac{1}{2}$. Any such constant would be as good as any other. Why? Well, suppose we're given a **BPP** algorithm that errs with probability $\frac{1}{3}$. If we're so inclined, we can easily modify the algorithm to err with probability at most (say) 2^{-100}. How?

> Right: just rerun the algorithm a few hundred times; then output the majority answer! If we take the majority answer out of T independent trials, then our good friend the Chernoff bound tells us we'll be wrong with a probability that decreases exponentially in T.

> Indeed, not only could we replace $\frac{1}{3}$ by any constant smaller than $\frac{1}{2}$; we could even replace it by $\frac{1}{2} - 1/p(n)$, where p is any polynomial.

> So, that was **BPP**: if you like, the class of all problems that are feasibly solvable by computer in a universe governed by classical physics.

- **RP (Randomized Polynomial-Time):** As I said before, the error probability of a **BPP** algorithm can easily be made smaller than the probability of an asteroid hitting the computer. And that's good enough for *most* applications: say, administering radiation doses in a hospital, or encrypting multibillion-dollar bank transactions, or controlling the launch of nuclear missiles. But what about proving theorems? For certain applications, you really can't take chances.

> And that leads us to **RP**: the class of problems for which there exists a polynomial-time randomized algorithm that accepts with probability greater than $\frac{1}{2}$ if the answer is yes, or probability *zero* if the answer is no. To put it another way: if the

algorithm accepts even once, then you can be *certain* that the answer is yes. If the algorithm keeps rejecting, then you can be extremely confident (but never certain) that the answer is no.

RP has an obvious "complement," called **coRP**. This is just the class of problems for which there's a polynomial-time randomized algorithm that accepts with probability 1 if the answer is yes, or less than $\frac{1}{2}$ if the answer is no.

- **ZPP (Zero-Error Probabilistic Polynomial-Time):** This class can be defined as the intersection of **RP** and **coRP** – the class of problems in both of them. Equivalently, **ZPP** is the class of problems solvable by a polynomial-time randomized algorithm that has to be correct whenever it does output an answer, but can output "don't know" up to half the time. Again, equivalently, **ZPP** is the class of problems solvable by an algorithm that never errs, but that only runs *expected* polynomial time.

Sometimes you see **BPP** algorithms called "Monte Carlo algorithms," and **ZPP** algorithms called "Las Vegas algorithms." I've even seen **RP** algorithms called "Atlantic City algorithms." This always struck me as stupid terminology. (Are there also Indian reservation algorithms?)

Here are the known relationships among the basic complexity classes that we've seen so far in this book. The relationships I didn't discuss explicitly are left as exercises for the reader (i.e., you).

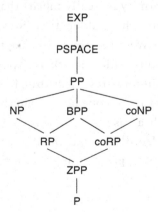

It might surprise you that we still don't know whether **BPP** is contained in **NP**. But think about it: even if a **BPP** machine accepted with probability close to 1, how would you prove that to a deterministic polynomial-time verifier who didn't believe you? Sure, you could show the verifier some random runs of the machine, but then she'd always suspect you of skewing your samples to get a favorable outcome.

Fortunately, the situation isn't *quite* as pathetic as it seems: we at least know that **BPP** is contained in **NP**$^{\textbf{NP}}$ (that is, **NP** with **NP** oracle), and hence in the second level of the polynomial hierarchy **PH**. Sipser, Gács, and Lautemann proved that in 1983. I'm actually going to skip it, because it's a bit technical. If you want it, here it is.[4]

Incidentally, while we know that **BPP** is contained in **NP**$^{\textbf{NP}}$, we don't know anything similar for **BQP**, the class of problems solvable in polynomial time on a quantum computer. **BQP** hasn't yet made its official entrance in this book – you'll have to wait a couple more chapters! – but I'm trying to foreshadow it by telling you what it apparently *isn't*. In other words, what do we know to be true of **BPP** that we *don't* know to be true of **BQP**? Containment in **PH** is only the first of three examples we'll see in this chapter.

In complexity theory, randomness turns out to be very closely related to another concept called *nonuniformity* – though we won't get to see the connection until later. Nonuniformity basically means that you get to choose a different algorithm for each input length n. Now, why would you want such a stupid thing? Well, remember in Chapter 5 I showed you the Blum Speedup Theorem – which says that it's possible to construct weird problems that admit no fastest algorithm, but only an infinite sequence of algorithms, with each one faster than the last on sufficiently large inputs? In such a case, nonuniformity would let you pick and choose from *all* algorithms, and thereby achieve the optimal performance. In other words, given an input of length n,

4 http://www.cs.berkeley.edu/~luca/cs278-01/notes/lecture9.ps

you could simply pick the algorithm that's fastest for inputs of that particular length!

But even in a world with nonuniformity, complexity theorists believe there would still be strong limits on what could efficiently be computed. When we want to talk about those limits, we use a terminology invented by Karp and Lipton in 1982.[5] Karp and Lipton defined the complexity class $P/f(n)$, or P with $f(n)$-size advice, to consist of all problems solvable in deterministic polynomial time on a Turing machine, with help from an $f(n)$-bit "advice string" a_n that depends only on the input length n.

You can think of the polynomial-time Turing machine as a grad student, and the advice string a_n as wisdom from the student's advisor. Like most advisors, this one is infinitely wise, benevolent, and trustworthy. He wants nothing more than to help his students solve their respective thesis problems: that is, to decide whether their respective inputs x in $\{0, 1\}^n$ are yes-inputs or no-inputs. But also like most advisors, he's too busy to find out what specific problems his students are working on. He therefore just doles out the same advice a_n to all of them, trusting them to apply it to their particular inputs x.

You *could* study advice that wasn't trustworthy, and in fact, I have. I defined some complexity classes based on untrustworthy advice, but in the usual definition of advice, we assume that it's trustworthy.

We'll be particularly interested in the class $P/poly$, which consists of all problems solvable in polynomial time using polynomial-size advice. In other words, $P/poly$ is the union of P/n^k over all positive integers k.

Now, is it possible that $P = P/poly$? As a first (trivial) observation, I claim the answer is no: P is *strictly* contained in $P/poly$, and indeed in $P/1$. In other words, even with a single bit of advice, you really can do more than with no advice. Why?

[5] R. M. Karp and R. J. Lipton, Turing machines that take advice, *L'Enseignement Mathématique* **28** (1982), 191–209.

Right! Consider the following problem:

Given an input of length n, decide whether the nth Turing machine halts.

Not only is this problem not in **P**, it's not even computable – for it's nothing other than a slow, "unary" encoding of the halting problem. On the other hand, it's easy to solve with a single advice bit that depends only on the input length n. For that advice bit could just tell you what the answer is!

Here's another way to understand the power of advice: while the number of problems in **P** is only countably infinite (why?), the number of problems in **P/1** is *uncountably* infinite. (Why?)

On the other hand, just because you can solve vastly more problems with advice than you can without, that doesn't mean advice will help you solve any *particular* problem you might be interested in. Indeed, a second easy observation is that advice doesn't let you do everything: there exist problems not in **P/poly**. Why?

Well, here's a simple diagonalization argument. I'll actually show a stronger result, that there exist problems not in **P/$n^{\log n}$**. Let $M_1, M_2, M_3, \ldots$ be a list of polynomial-time Turing machines. Also, fix an input length n. Then I claim that there exists a Boolean function $f:\{0, 1\}^n \to \{0, 1\}$ that the first n machines $(M_1, \ldots, M_n)$ all fail to compute, even given any $n^{\log n}$-bit advice string. Why? Just a counting argument: there are 2^{2^n} Boolean functions, but only n Turing machines and $2^{n^{\log n}}$ advice strings. So choose such a function f for every n; you'll then cause each machine M_i to fail on all but finitely many input lengths. Indeed, we didn't even need the assumption that the M_i run in polynomial time.

Why do I care about advice? First of all, it shows up again and again, even if, for example, all we want to know about is uniform computation. Even if all we want to know is if we can derandomize **BPP**, it turns out to be a question about advice. So it's very connected to the rest of complexity. Basically, you can think of an algorithm with advice as being no different than an infinite sequence of algorithms, just like what we saw with the Blum Speedup Theorem. It's just an algorithm, where as you go to larger and larger input lengths, you get to keep using new ideas and get more speedup. This is one way to think about advice.

I can give you another argument. You can think of advice as freeze-dried computation. There's some great, enormous sort of computational effort, that we then encapsulate in this convenient polynomially sized string over in the frozen foods section and that you can go and heat into the microwave to do work with.

Advice formalizes the possibility that such results of some uncomputable process have been hanging about the universe from the beginning of time. After all, we really don't know the initial conditions of the universe. The usual argument that it's a justified assumption is that, for whatever other state your computer might start in, there's some physical process that gave rise to that state. Presumably, this is only a polynomial-time physical process. So you could simulate the whole process that gave rise to that state, tracing it back to the Big Bang if needed. But is this really reasonable?

Of course, all this time we've been dancing around the real question: can advice help us solve problems that we actually care about, like the **NP**-complete problems? In particular, is **NP** $\subset$ **P/poly**? Intuitively, it seems unlikely: there are exponentially many Boolean formulas of size n, so even if you somehow received a polynomial-size advice string from God, how would that help you to decide satisfiability for more than a tiny fraction of those formulas?

But – and I'm sure this will come as a complete shock to you – we can't *prove it's impossible*. Well, at least in this case we have a good excuse for our ignorance, since if **P** $=$ **NP**, then obviously **NP** $\subset$ **P/poly** as well. But here's a question: if we *did* succeed in proving **P** $\neq$ **NP**, then would we also have proved that **NP** $\not\subset$ **P/poly**? In other words, would **NP** $\subset$ **P/poly** imply **P** $=$ **NP**? Alas, we don't even know the answer to *that*.

But as with **BPP** and **NP**, the situation isn't *quite* as pathetic as it seems. Karp and Lipton did manage to prove in 1982 that, if **NP** ⊂ **P/poly**, then the polynomial hierarchy **PH** would collapse to the second level (that is, to $\mathbf{NP^{NP}}$). In other words, if you believe the polynomial hierarchy is infinite, then you must also believe that **NP**-complete problems are not efficiently solvable by a nonuniform algorithm.

This "Karp–Lipton Theorem" is the most famous example of a very large class of complexity results, a class that's been characterized as "if donkeys could whistle, then pigs could fly." In other words, if one thing no one really believes is true were true, then another thing no one really believes is true would be true! Intellectual onanism, you say? Nonsense! What makes it interesting is that the two things that no one really believes are true would've previously seemed completely unrelated to each other.

It's a bit of a digression, but the proof of the Karp–Lipton Theorem is more fun than a barrel full of carp. So let's see the proof right now. We assume **NP** ⊂ **P/poly**; what we need to prove is that the polynomial hierarchy collapses to the second level – or equivalently, that $\mathbf{coNP^{NP}} = \mathbf{NP^{NP}}$. So let's consider an arbitrary problem in $\mathbf{coNP^{NP}}$, like so:

> *For all n-bit strings x, does there exist an n-bit string y such that* $\phi(x,y)$ *evaluates to TRUE?*

(Here ϕ is some arbitrary polynomial-size Boolean formula.)

We need to find an $\mathbf{NP^{NP}}$ question – that is, a question where the existential quantifier comes *before* the universal quantifier – that has the same answer as the question above. But what could such a question possibly be? Here's the trick: we'll first use the existential quantifier to guess a polynomial-size advice string a_n. We'll then use the universal quantifier to guess the string x. Finally, we'll use the advice string a_n – together with the assumption that **NP** ⊂ **P/poly** – to guess y on our own. Thus:

Does there exist an advice string a_n such that for all n-bit strings x, $\phi(x,M(x,a_n))$ evaluates to TRUE?

Here M is a polynomial-time Turing machine that, given x as input and a_n as advice, outputs an n-bit string y such that $\phi(x, y)$ evaluates to TRUE whenever such a y exists. By one of the problems from last chapter, we can easily construct such an M provided we can solve **NP**-complete problems in **P/poly**.

Alright, I told you before that nonuniformity was closely related to randomness – so much so that it's hard to talk about one without talking about the other. So, in the rest of this chapter, I want to tell you about two connections between randomness and nonuniformity: a simple one that was discovered by Adleman in the 1970s, and a deep one that was discovered by Impagliazzo, Nisan, and Wigderson in the 1990s.

The simple connection is that **BPP** $\subset$ **P/poly**: in other words, *nonuniformity is at least as powerful as randomness*. Why do you think that is?

Well, let's see why it is. Given a **BPP** computation, the first thing we'll do is amplify the computation to exponentially small error. In other words, we'll repeat the computation (say) n^2 times and then output the majority answer, so that the probability of making a mistake drops from $\frac{1}{3}$ to roughly 2^{-n^2}. (If you're trying to prove something about **BPP**, amplifying to exponentially small error is almost always a good first step!)

Now, how many inputs are there of length n? Right: 2^n. And for each input, only a 2^{-n^2} fraction of random strings cause us to err. By the union bound (the most useful fact in all of theoretical computer science), this implies that at most a 2^{n-n^2} fraction of random strings can *ever* cause us to err on inputs of length n. Since $2^{n-n^2} < 1$, this means there *exists* a random string, call it r, that never causes us to err on inputs of length n. So fix such an r, feed it as advice to the **P/poly** machine, and we're done!

So that was the simple connection between randomness and nonuniformity. Before moving on to the deep connection, let me make two remarks.

1. Even if $P \neq NP$, you might wonder whether **NP**-complete problems can be solved in *probabilistic* polynomial time. In other words, is **NP** in **BPP**? Well, we can already say something concrete about that question. If $NP \subseteq BPP$, then certainly $NP \subset P/poly$ (since $BPP \subset P/poly$). But that means **PH** collapses by the Karp–Lipton Theorem. So if you believe the polynomial hierarchy is infinite, then you also believe **NP**-complete problems are not efficiently solvable by randomized algorithms.

2. If nonuniformity can simulate randomness, then can it also simulate *quantumness*? In other words, is $BQP \subset P/poly$? Well, we don't know, but it isn't considered likely. Certainly Adleman's proof that **BPP** is in **P/poly** completely breaks down if we replace the **BPP** by **BQP**. But this raises an interesting question: *why* does it break down? What's the crucial difference between quantum theory and classical probability theory, which causes the proof to work in the one case but not the other? I'll leave the answer as an exercise for you.

Alright, now for the deep connection. Do you remember the primality-testing problem from earlier in the chapter? Over the years, this problem crept steadily down the complexity hierarchy, like a monkey from branch to branch.

- It's obvious that primality-testing is in **coNP**.
- In 1975, Pratt showed it was in **NP**.
- In 1977, Solovay, Strassen, and Rabin showed it was in **coRP**.
- In 1992, Adleman and Huang showed it was in **ZPP**.
- In 2002, Agrawal, Kayal, and Saxena showed it was in **P**.

The general project of taking randomized algorithms and converting them to deterministic ones is called *derandomization* (a name only a theoretical computer scientist could love). The history of the primality-testing problem can only be seen as a spectacular success of this project. But with such success comes an obvious question: can *every* randomized algorithm be derandomized? In other words, does **P** equal **BPP**?

Once again the answer is that we don't know. Usually, if we don't know if two complexity classes are equal, the "default conjecture" is that they're different. And so it was with **P** and **BPP** – *(ominous music)* – until now. Over the last decade and a half, mounting evidence has convinced almost all of us that in fact **P** = **BPP**. We won't be able to review this evidence in any depth. But let me quote one theorem, just to give you a flavor of it.

> **Theorem (Impagliazzo–Wigderson 1997):**[6] Suppose there
> exists a problem that's solvable in exponential time, and that's
> *not* solvable in subexponential time even with the help of a
> subexponential-size advice string. Then **P** = **BPP**.

Notice how this theorem relates derandomization to nonuniformity – and, in particular, to proving that certain problems are hard for nonuniform algorithms. The premise certainly seems plausible. From our current perspective, the conclusion (**P** = **BPP**) also seems plausible. And yet the two seem to have nothing to do with each other. So, this theorem might be characterized as "If donkeys can bray, then pigs can oink."

Where does this connection between randomness and nonuniformity come from? It comes from the theory of pseudorandom generators. We're gonna see a lot more about pseudorandom generators in the next chapter, when we talk about cryptography. But basically, a pseudorandom generator is just a function that takes as input a short string (called the *seed*), and produces as output a long string, in such a way that, if the seed is random, then the output *looks* random. Obviously, the output can't *be* random, since it doesn't have enough

---

[6] R. Impagliazzo and A. Wigderson, P = BPP if E requires exponential circuits: derandomizing the XOR lemma. In *Proceedings of ACM Symposium on Theory of Computing* (New York: ACM, 1997), pp. 220–9.

entropy: if the seed is k bits long, then there are only 2^k possible output strings, regardless of how long those output strings are. What we ask, instead, is that no polynomial-time algorithm can successfully distinguish the output of the pseudorandom generator from "true" randomness. Of course, we'd also like for the function mapping the seed to the output to be computable in polynomial time.

Already in 1982, Andy Yao realized that, if you could create a "good enough" pseudorandom generator, then you could prove **P = BPP**. Why? Well, suppose that, for any integer k, you had a way of stretching an $O(\log n)$-bit seed to an n-bit output in polynomial time, in such a way that no algorithm running in n^k time could successfully distinguish the output from true randomness. And suppose you had a **BPP** machine that ran in n^k time. In that case, you could simply loop over all possible seeds (of which there are only polynomially many), feed the corresponding outputs to the **BPP** machine, and then output the majority answer. The probability that the **BPP** machine accepts given a pseudorandom string *has* to be about the same as the probability that it accepts given a truly random string – since otherwise the machine would be distinguishing random strings from pseudorandom ones, contrary to assumption!

But what's the role of nonuniformity in all this? Well, here's the point: in addition to a random (or pseudorandom) string, a **BPP** machine also receives an input, x. And we need the derandomization to work for *every* x. But that means that, for the purposes of derandomization, we *must* think of x as an advice string provided by some superintelligent adversary for the sole purpose of foiling the pseudorandom generator. You see, this is why we had to assume a problem that was hard even in the presence of advice: because we need to construct a pseudorandom generator that's indistinguishable from random even in the presence of the "adversary," x.

To summarize: if we could prove that certain problems are sufficiently hard for nonuniform algorithms, then we would prove **P = BPP**.

This leads to my third difference between **BPP** and **BQP**: while most of us believe that **P** = **BPP**, most of us certainly *don't* believe that **P** = **BQP**. (Indeed, we *can't* believe that, if we believe factoring is hard for classical computers.) We don't have any "dequantization" program that's been remotely as successful as the derandomization program. Once again, it would seem there's a crucial difference between quantum theory and classical probability theory, which allows certain ideas (like those of Sipser–Gács–Lautemann, Adleman, and Impagliazzo–Wigderson) to work for the latter but not for the former.

Incidentally, Kabanets and Impagliazzo[7] (and others) managed to obtain a sort of converse to the derandomization theorems. What they showed is that, if we want to prove **P** = **BPP**, then we'll *have* to prove that certain problems are hard for nonuniform algorithms. This could be taken as providing some sort of explanation for why, assuming **P** = **BPP**, no one has yet managed to prove it. Namely, it's because if you want to prove **P** = **BPP**, then you'll have to prove certain problems are hard – and if you could prove those problems were hard, then you would be (at least indirectly) attacking questions like **P** vs. **NP**. In complexity theory, pretty much everything eventually comes back to **P** vs. **NP**.

PUZZLES

1. You and a friend want to flip a coin, but the only coin you have is crooked: it lands heads with some fixed but unknown probability p. Can you use this coin to *simulate* a fair coin flip? (I mean perfectly fair, not just approximately fair.)

2. There are n people standing in a circle. They're each wearing either a red hat or a blue hat, assigned uniformly and independently at random. They can each see everyone else's hats but not their own. They

[7] V. Kabanets and R. Impagliazzo, Derandomizing polynomial identity tests means proving circuit lower bounds. *Computational Complexity*, **13**:1/2 (2004), 1–46.

want to vote on whether the number of red hats is even or odd. Each person votes at the same time, so that no one's vote depends on anyone else's. What's the maximum probability with which the people can win this game? (By "win," I mean that their vote corresponds to the truth.) Assume for simplicity that n is odd.

8   Crypto

Puzzle 1. We are given a biased coin that comes up heads with probability p. Using this coin, construct an unbiased coin.

   Solution. The solution is the "von Neumann trick": flip the biased coin twice, interpreting HT as heads and TH as tails. If the flips come up HH or TT, then try again. Under this scheme, "heads" and "tails" are equiprobable, each occurring with probability $p(1 - p)$ in any given trial. Conditioned on either HT or TH occurring, it follows that the simulated coin is unbiased.

   Puzzle 2. n people sit in a circle. Each person wears either a red hat or a blue hat, chosen independently and uniformly at random. Each person can see the hats of all the other people, but not his/her own hat. Based only upon what they see, each person votes on whether or not the total number of red hats is odd. Is there a scheme by which the outcome of the vote is correct with probability greater than $\frac{1}{2}$?

   Solution. Each person decides his/her vote as follows: if the number of visible blue hats is larger than the number of visible red hats, then vote according to the parity of the number of visible red hats. Otherwise, vote the opposite of the parity of the number of visible red hats. If the number of red hats differs from the number of blue hats by at least 2, then this scheme succeeds with certainty. Otherwise, the scheme might fail. However, the probability that the number of red hats differs from the number of blue hats by less than 2 is small – $O(1/\sqrt{N})$.

CRYPTO

Cryptography has been a major force in human history for more than 3000 years. Numerous wars have been won or lost by the sophistication or stupidity of cryptosystems. If you think I'm exaggerating, read *The Codebreakers* by David Kahn,[1] and keep in mind that it was written before people knew about the biggest cryptographic story of all: the breaking of the Nazis' naval code in World War II, by a team that included Alan Turing.

And yet, even though cryptography has influenced human affairs for millennia, developments over the last 30 years have completely – yes, completely – changed our understanding of it. If you plotted when the basic mathematical discoveries in cryptography were made, you'd see a few in antiquity, maybe a few from the Middle Ages till the 1800s, one in the 1920s (the one-time pad), a few more around World War II, and then, after the birth of computational complexity theory in the 1970s, *boom boom boom boom boom boom boom...*

Our journey through the history of cryptography begins with the famous and pathetic "Caesar cipher" used by the Roman Empire. Here the plaintext message is converted into a ciphertext by simply adding 3 to each letter, wrapping around to A after you reach Z. Thus, D becomes G, Y becomes B, and DEMOCRITUS becomes GHPRFUL-WXV. More complex variants of the Caesar cipher have appeared, but given enough ciphertext they're all easy to crack, by using (for example) a frequency analysis of the letters appearing in the ciphertext. Not that that's stopped people from using these things! Indeed, as recently as 2006, the head of the Sicilian mafia[2] was finally caught after 40 years because he used the Caesar cipher – the *original* one – to send messages to his subordinates!

Could there be a cryptosystem that was *information-theoretically secure* – by which we mean, provably secure regardless of how much computation time an eavesdropper spent trying to crack it? Amazingly (if you've never seen this before), the answer turns out to be yes – but even more amazingly, it wasn't until the 1920s that such a system was discovered. For reasons we'll see shortly, the prototypical

[1] D. Kahn, *The Codebreakers* (New York: Scribner, 1996).
[2] See http://en.wikipedia.org/wiki/Pizzino.

information-theoretically secure system is called the *onetime pad*. The idea is simple: the plaintext message is represented by a binary string p, which is exclusive-ORed with a random binary key k of the same length. That is, the ciphertext c is equal to $p \oplus k$, where $\oplus$ denotes bitwise addition mod 2.

The recipient (who knows k) can decrypt the ciphertext with another XOR operation:

$$c \oplus k = p \oplus k \oplus k = p.$$

To an eavesdropper who doesn't know k, the ciphertext is just a string of random bits – since XORing any string of bits with a random string just produces another random string. The problem with the onetime pad, of course, is that the sender and recipient have to share a key that's as long as the message itself. Furthermore, if the same key is ever used to encrypt two or more messages, then the cryptosystem is no longer information-theoretically secure. (Hence, the name "onetime pad.") To see why, suppose two plaintexts p_1 and p_2 are both encrypted via the same key k to ciphertexts c_1 and c_2, respectively. Then we have

$$c_1 \oplus c_2 = p_1 \oplus k \oplus p_2 \oplus k = p_1 \oplus p_2,$$

and hence an eavesdropper can obtain the string $p_1 \oplus p_2$. By itself, this might or might not be useful, but it at least constitutes *some* information that an eavesdropper could learn about the plaintexts. But this is just a mathematical curiosity, right? Well, in the 1950s the Soviets got sloppy and reused some of their onetime pads. As a result, the NSA, through its VENONA project, was able to recover some (though not all) of the plaintext encrypted in this way. This seems to be how Julius and Ethel Rosenberg were caught.

In the 1940s, Claude Shannon proved that information-theoretically secure cryptography *requires* the sender and recipient to share a key at least as long as the message they want to communicate. Like pretty much all of Shannon's results, this one is trivial in retrospect. (It's good to be in on the ground floor!) Here's his proof: given the ciphertext and the key, the plaintext had better be uniquely recoverable. In other words, for any *fixed* key, the function that maps plaintexts to ciphertexts had better be an injective function. But this immediately implies that, for a given ciphertext *c*, *the number of plaintexts that could possibly have produced c is at most the number of keys*. In other words, if there are fewer possible keys than plaintexts, then an eavesdropper will be able to rule out some of the plaintexts – the ones that wouldn't encrypt to *c* for *any* value of the key. Therefore, our cryptosystem won't be perfectly secure. It follows that, if we want perfect security, then we need at least as many keys as plaintexts – or equivalently, the key needs to have at least as many bits as the plaintext.

I mentioned before that sharing huge keys is usually impractical – not even the KGB managed to do it perfectly! So we want a cryptosystem that lets us get away with smaller keys. Of course, Shannon's result implies that such a cryptosystem can't be information-theoretically secure. But what if we relax our requirements? In particular, what if we assume that the eavesdropper is restricted to running in polynomial time? This question leads naturally to our next topic...

PSEUDORANDOM GENERATORS

As I mentioned in the last chapter, a pseudorandom generator (PRG) is basically a function that takes as input a short, *truly* random string, and produces as output a long, *seemingly* random string. More formally, a pseudorandom generator is a function f with the following properties.

1. f maps an n-bit input string (called the *seed*) to a $p(n)$-bit output string, where $p(n)$ is some polynomial larger than n.
2. f is computable in time polynomial in n.
3. For every polynomial-time algorithm A (called the *adversary*), the difference

$$| \Pr_{n\text{-bit strings } x} [A \text{ accepts } f(x)] - \Pr_{p(n)\text{-bit strings } y} [A \text{ accepts } y] |$$

is *negligibly small* – by which I mean, it decreases faster than $1/q(n)$ for any polynomial q. (Of course, decreasing at an exponential rate is even better.) Or in English, no polynomial-time adversary can distinguish the output of f from a truly random string with any nonnegligible bias.

Now, you might wonder: how "stretchy" a PRG are we looking for? Do we want to stretch an n-bit seed to $2n$ bits? To n^2 bits? n^{100} bits? The answer turns out to be irrelevant!

Why? Because even if we only had a PRG f that stretched n bits to $n + 1$ bits, we could keep applying f *recursively* to its own output, and thereby stretch n bits to $p(n)$ bits for any polynomial p. Furthermore, if the output of this recursive process were efficiently distinguishable from a random $p(n)$-bit string, then the output of f itself would have been efficiently distinguishable from a random $(n + 1)$-bit string – contrary to assumption! Of course, there's something that needs to be proved here, but the something that needs to be proved *can* be proved, and I'll leave it at that.[3]

Now, I claim that *if* pseudorandom generators exist, *then* it's possible to build a computationally secure cryptosystem using only short encryption keys. Do you see why?

Right: first use the PRG to stretch a short encryption key to a long one – as long as the plaintext message itself. Then pretend that the long key is *truly* random, and use it exactly as you'd use a onetime pad!

Why is this scheme secure? As always in modern cryptography, what we do is to argue by reduction. Suppose that, given only the ciphertext message, an eavesdropper could learn something about the plaintext in polynomial time. We saw before that, if the encryption key were *truly* random (that is, were a onetime pad), then this would be impossible. It follows, then, that the eavesdropper would in effect

---

[3] Oh, OK. For those hungering for a proof, see (for example) Oded Goldreich, Foundations of Cryptography (Volume I: Basic Tools), Cambridge University Press, 2007.

be *distinguishing* the pseudorandom key from a random one. But this contradicts our assumption that no polynomial-time algorithm can distinguish the two!

Admittedly, this has all been pretty abstract and conceptual. Sure, we could do wonderful things if we had a PRG – but is there any reason to suppose PRGs actually exist?

A first, trivial observation is that PRGs can *only* exist if $P \neq NP$. Why?

Right: because if $P = NP$, then given a supposedly random string y, we can decide in polynomial time whether there's a short seed x such that $f(x) = y$. If y is random, then such a seed almost certainly won't exist – so if it *does* exist, we can be almost certain that y isn't random. We can therefore distinguish the output of f from true randomness.

Alright, but suppose we *do* assume $P \neq NP$. What are some concrete examples of functions that are believed to be pseudorandom generators?

One example is what's called the Blum–Blum–Shub[4] generator. Here's how it works: pick a large composite number N. Then the seed, x, will be a random element of Z_N. Given this seed, first compute x^2 mod N, $(x^2)^2$ mod N, $((x^2)^2)^2$ mod N, and so on. Then concatenate the least-significant bits in the binary representations of these numbers, and output that as your pseudorandom string $f(x)$.

Blum *et al.* were able to show that, if we had a polynomial-time algorithm to distinguish $f(x)$ from a random string, then (modulo some technicalities) we could use that algorithm to *factor N* in polynomial time. Or equivalently, if factoring is hard, then Blum–Blum–Shub is a PRG. This is yet another example where we "prove" something is

4 L. Blum, M. Blum and M. Shub, A Simple Unpredictable Pseudo-Random Number Generator, *SIAM Journal on Computing*, **15** (1996), 364–383.

hard by showing that, if it were easy, then something else that we think is hard would also be easy.

Alas, we *don't* think factoring is hard – at least, not in a world with quantum computers! So can we base the security of PRGs on a more quantum-safe assumption? Yes, we can. There are many, many ways to build a candidate PRG, and we have no reason to think that quantum computers will be able to break all of them. Indeed, you could even base a candidate PRG on the apparent unpredictability of (say) the "Rule 110" cellular automaton, as advocated by Stephen Wolfram in his groundbreaking, revolutionary, paradigm-smashing book.

Of course, our *dream* would be to base a PRG's security on the weakest possible assumption: $P \neq NP$ itself! But when people try to do that, they run into two interesting problems.

The first problem is that P versus NP deals only with the *worst* case. Imagine if you were a general or a bank president, and someone tried to sell you an encryption system with the sales pitch that there *exists* a message that's hard to decode. You see what the difficulty is: for both encryption systems and PRGs, we need NP problems that are hard *on average*, not just in the worst case. (Technically, we need problems that are hard on average with respect to *some* efficiently samplable distribution over the inputs – not necessarily the uniform distribution.) But no one has been able to prove that such problems exist, *even if we assume* $P \neq NP$.

That's not to say, though, that we know *nothing* about average-case hardness. As an example, consider the Shortest Vector Problem (SVP). Here, we're given a lattice L in R^n, consisting of all integer linear combinations of some given vectors $v_1, \ldots, v_n$ in R^n. Then the problem is to approximate the length of the shortest nonzero vector in L to within some multiplicative factor k.

SVP is one of the few problems for which we can prove a *worst-case/average-case equivalence* (that is, the average case is every bit as hard as the worst case), at least when the approximation ratio k is big enough. Based on that equivalence, Ajtai, Dwork,[5] Regev,[6] and others have constructed cryptosystems and pseudorandom generators whose security rests on the *worst-case* hardness of SVP. Unfortunately, the same properties that let us prove worst-case/average-case equivalence also make it unlikely that SVP is **NP**-complete for the relevant values of k. It seems more likely that SVP is *intermediate* between **P** and **NP**-complete, just like we think factoring is.

Alright, so suppose we just *assume* **NP**-complete problems are hard on average. Even then, there's a further difficulty in using **NP**-complete problems to build a PRG. This is that breaking PRGs just doesn't seem to have the right "shape" to be **NP**-complete. What do I mean by that? Well, think about how we prove a problem B is **NP**-complete: we take some problem A that's already known to be **NP**-complete, and we give a polynomial-time reduction that maps yes-instances of A to yes-instances of B, and no-instances of A to no-instances of B. In the case of breaking a PRG, presumably the yes-instances would be pseudorandom strings and the no-instances would be truly random strings (or maybe vice versa).

Do you see the problem here? If not, let me spell it out for you: *how do we describe a "truly random string" for the purpose of mapping to it in the reduction?* The whole point of a string being random is that we *can't* describe it by anything shorter than itself! Admittedly, this argument is full of loopholes, one of which is that the reduction might be randomized. Nevertheless, it *is* possible to conclude something from the argument: that if breaking PRGs is **NP**-complete, then the proof will have to be very different from the sort of **NP**-completeness proofs that we're used to.

[5] M. Ajtai and C. Dwork, A public-key cryptosystem with worst-case/average-case equivalence. In *Proceedings of 29th Annual ACM Symposium on Theory of Computing* (New York: ACM, 1997), pp. 284–93.

[6] O. Regev, On lattices, learning with errors, random linear codes, and cryptography. *Journal of the ACM*, **56**:6 (2009), 1–40.

ONE-WAY FUNCTIONS

One-way functions (OWFs) are the cousins of pseudorandom generators. Intuitively, an OWF is just a function that's easy to compute but hard to invert. More formally, a function f from n bits to $p(n)$ bits is a one way function if we have the following.

1. f is computable in time polynomial in n.
2. For every polynomial-time adversary A, the probability that A succeeds at inverting f,

$$\Pr_{n\text{-bit strings } x}[\, f(A(\,f(x))) = f(x)],$$

is negligibly small – that is, smaller than $1/q(n)$ for any polynomial q.

The event $f(A(f(x))) = f(x)$ appears in the definition instead of just $A(f(x)) = x$ in order to account for the fact that f might have multiple inverses. With this definition, we consider algorithms A that find *anything* in the preimage of $f(x)$, not just x itself.

I claim that the existence of PRGs implies the existence of OWFs. Can you tell me why?

Right: because a PRG *is* an OWF!

Alright then, can you prove that the existence of OWFs implies the existence of PRGs?

Yeah, this one's a little harder! The main reason is that the output of an OWF f doesn't have to appear random in order for f to be hard to invert. And indeed, it took more than a decade of work – culminating in a behemoth 1999 paper[7] of Håstad, Impagliazzo, Levin, and Luby – to figure out how to construct a pseudorandom generator from any one-way function. Because of the result of Håstad *et al.*, we now know that OWFs exist *if and only if* PRGs do. The proof, as you'd expect, is pretty complicated, and the reduction is not exactly

---

[7] J. Håstad, R. Impagliazzo, L. A. Levin and M. Luby, A pseudorandom generator from any one-way function. *SIAM Journal on Computing*, 28:4 (1999), 1364–96. http://citeseer.ist.psu.edu/hastad99pseudorandom.html

practical: the blowup is by about n^{10}! (Since improved to n^4.) This is the sort of thing that gives polynomial time a bad name – but it's the exception, not the rule! If we assume that the one-way function is a *permutation*, then the proof becomes much easier (it was already shown by Yao in 1982)[8] and the reduction becomes much faster. But of course that yields a less general result.

So far we've restricted ourselves to *private-key* cryptosystems, which take for granted that the sender and receiver share a secret key. But how would you share a secret key with (say) Amazon.com before sending them your credit card number? Would you *email* them the key? Oops – if you did that, then you'd better encrypt your email using *another* secret key, and so on ad infinitum! The solution, of course, is to meet with an Amazon employee in an abandoned garage at midnight.

No, wait... I meant that the solution is *public-key cryptography*.

PUBLIC-KEY CRYPTOGRAPHY

It's amazing, if you think about it, that so basic an idea had to wait until the 1970s to be discovered. Physicists were tidying up the Standard Model while cryptographers were still at the Copernicus stage!

So, how did public-key cryptography finally come to be? The first inventors – or rather discoverers – were Ellis, Cocks, and Williamson, working for GCHQ (the British NSA) in the early 1970s. Of course, they couldn't *publish* their work, so today they don't get much credit! Let that be a lesson to you.

The first *public* public-key cryptosystem was that of Diffie and Hellman, in 1976. A couple of years later, Rivest, Shamir, and Adleman discovered the famous RSA system that bears their initials.

---

[8] A. Chi-Chih Yao, Theory and applications of trapdoor functions [extended abstract]. In *Proceedings of 24th Annual IEEE Symposium on Foundations of Computer Science* (Silver Spring, MD: IEEE Computer Society Press, 1982), pp. 80–91.

Do any of you know how RSA was first revealed to the world? Right: as a puzzle in Martin Gardner's Mathematical Games column[9] for *Scientific American*!

RSA had several advantages over Diffie–Hellman: for example, it only required *one* party to generate a public key instead of both,[10] and it let users *authenticate* themselves in addition to communicating in private. But if you read Diffie and Hellman's paper,[11] pretty much all the main ideas are there.

Anyway, the core of any public-key cryptosystem is what's called a *trapdoor one-way function*.[12] This is a function that's

(1) easy to compute,
(2) hard to invert, and
(3) easy to invert *given some secret "trapdoor" information*.

The first two requirements are just the same as for ordinary OWFs. The third requirement – that the OWF should have a "trapdoor" that makes the inversion problem easy – is the new one. For comparison, notice that the existence of *ordinary* one-way functions implies the existence of secure *private-key* cryptosystems, whereas the existence of *trapdoor* one-way functions implies the existence of secure *public-key* cryptosystems.

So, what's an actual example of a public-key cryptosystem? Well, most of you have seen RSA at some point in your mathematical lives, so I'll go through it quickly.

---

[9] See Martin Gardner, *Penrose Tiles to Trapdoor Ciphers: And the Return of Dr. Matrix*, Mathematical Association of America, 1997.

[10] However, it was later realized that the original Diffie-Hellman idea can also be used to get an 'RSA-like' encryption scheme: the sender, Alice, can simply generate her Diffie-Hellman public key on the fly, rather than storing it in a file ahead of time. The result is called the ElGamal encryption scheme.

[11] http://citeseer.ist.psu.edu/340126.html

[12] Well, strictly speaking, the RSA cryptosystem is literally based on a trapdoor one-way function as I'm defining it here. But most other public-key cryptosystems—including lattice-based cryptosystems, and the ElGamal system from the previous footnote—are based on slight generalizations of the notion of trapdoor OWFs.

Let's say you want to send your credit card number to Amazon.com. What happens? First, Amazon randomly selects two large prime numbers p and q (which can be done in polynomial time), subject to the technical constraint that $p - 1$ and $q - 1$ should not be divisible by 3. (We'll see the reason for that later.) Then Amazon computes the product $N = pq$ and publishes it for all the world to see, while keeping p and q themselves a closely guarded secret.

Now, assume without loss of generality your credit card number is encoded by a positive integer x, smaller but not *too* much smaller than N. One note: it's extremely important that x contain, not just the credit card number itself, but also some random unrelated garbage. Why? Because otherwise you'd leave yourself open to attacks like doing a brute-force search through all possible credit card numbers (there aren't that many of them). Also, an attacker would see that two encryptions of your credit number looked exactly the same, which could be useful information. So, assume you've got this positive integer x, containing your credit card number padded art by random garbage. Now what do you do? Simple: you compute $x^3 \bmod N$ and send it over to Amazon! If a credit card thief intercepted your message en route, then she would have to recover x given only $x^3 \bmod N$. But computing cube roots modulo a composite number is believed to be an extremely hard problem, at least for classical computers! If p and q are both reasonably large (say, $10\,000$ digits each), then our hope would be that any classical eavesdropper would need millions of years to recover x.

This leaves an obvious question: how does Amazon *itself* recover x? Duh – by using its knowledge of p and q! We know from our friend Mr. Euler, way back in 1761, that the sequence

$$x \bmod N, \ x^2 \bmod N, \ x^3 \bmod N, \ldots$$

repeats with period $(p - 1)(q - 1)$. So provided Amazon can find an integer k such that

$$3k = 1 \bmod (p - 1)(q - 1),$$

it'll then have

$$(x^3)^k \bmod N = x^{3k} \bmod N = x \bmod N.$$

Now, we know that such a k exists, by the assumption that $p - 1$ and $q - 1$ are not divisible by 3. Furthermore, Amazon can *find* such a k in polynomial time, using Euclid's algorithm (from way *way* back, around 300 BC). Finally, given $x^3 \bmod N$, Amazon can compute $(x^3)^k$ in polynomial time by using a simple repeated squaring trick. So that's RSA.

To make everything as concrete and visceral as possible, I assumed that x always gets raised to the *third* power. The resulting cryptosystem is by no means a toy: as far as anyone knows, it's secure! In practice, though, people can and do raise x to arbitrary powers. As another remark, *squaring* x instead of cubing it would open a whole new can of worms, since any nonzero number that has a square root mod N has more than one of them.

Of course, if the credit card thief could *factor* N into pq, then she could run the exact same decoding algorithm that Amazon runs, and thereby recover the message x. So the whole scheme relies crucially on the assumption that factoring is hard! This immediately implies that RSA could be broken by a credit card thief with a quantum computer. Classically, however, the best-known factoring algorithm is the Number Field Sieve, which takes about $2^{n^{1/3}}$ steps.

As a side note, no one has yet proved that breaking RSA *requires* factoring: it's possible that there's a more direct way to recover the message x, one that doesn't entail learning p and q. On the other hand, in 1979 Rabin discovered a variant of RSA for which recovering the plaintext *is* provably as hard as factoring. Rabin's system, as it happens, was based on squaring rather than cubing. But as I mentioned before, the fact that square roots aren't unique opens a new can of worms. In particular, it means that ciphertexts in Rabin's system might have *two* possible decryptions (one of which hopefully looks like random garbage and can be discarded), rather than a single decryption.

Alright, but all this talk of cryptosystems based on factoring and modular arithmetic is so 1993! Today, we realize that as soon as we build a quantum computer, *Shor's algorithm* (to be discussed in Chapter 10) will break the whole lot of these things. Of course, this point hasn't been lost on complexity theorists, many of whom have since set to work looking for trapdoor OWFs that still seem safe against quantum computers. Currently, our best candidates for such trapdoor OWFs are based on lattice problems, like the Shortest Vector Problem (SVP) that I described earlier. Whereas factoring reduces to the *abelian* hidden subgroup problem, which is solvable in quantum polynomial time, SVP is only known to reduce to the *dihedral* hidden subgroup problem, which is *not* known to be solvable in quantum polynomial time despite more than a decade of effort.

Inspired by this observation, and building on earlier work by Ajtai and Dwork, Oded Regev proposed[13] public-key cryptosystems that are provably secure against quantum eavesdroppers, *assuming* SVP is hard for quantum computers. Note that his cryptosystems themselves are purely classical. On the other hand, even if you only wanted security against classical eavesdroppers, you'd *still* have to assume that SVP was hard for quantum computers, since the reduction from SVP to breaking the cryptosystem is a quantum reduction! Later, in 2009, Chris Peikert[14] discovered a way to "de-quantize" Regev's reduction, so that now one only needs to assume the *classical* hardness of SVP.

Even more dramatically, Craig Gentry showed[15] in 2009 that, by using the assumed hardness of certain lattice problems related to SVP, one can construct *fully homomorphic cryptosystems*: that

[13] http://www.cs.tau.ac.il/~odedr/papers/qcrypto.pdf

[14] C. Peikert, Public-key cryptosystems from the worst-case shortest vector problem [extended abstract]. In *Proceedings of Annual ACM Symposium on Theory of Computing* (New York: ACM, 2009), pp. 333–42.

[15] C. Gentry, Fully homomorphic encryption using ideal lattices. In *Proceedings of Annual ACM Symposium on Theory of Computing* (New York: ACM, 2009), pp. 169–78.

is, public-key cryptosystems that let you perform arbitrary computations on encrypted data without ever decrypting it. Why is this important? Well, for applications like "cloud computing," you might want (say) your mobile device to offload a long computation to some server somewhere, but in a way that *doesn't* let the server see any of your sensitive data. That is, you should be able to send the server encrypted inputs, and the server should be able to do the long computation you paid for and send you back an encrypted output, which you can then decrypt yourself (and maybe even verify) with the server remaining none the wiser. As a bonus, because our current fully homomorphic encryption systems are based on problems related to lattices, they share the property with Regev's systems that no one knows how to break them even with a quantum computer. Fully homomorphic cryptography is a possibility that was first raised in the 1970s, but no one knew how to do it until a few years ago. So, this is one of the biggest developments in theoretical cryptography for decades.

But is any of this *practical*? The conventional wisdom used to be no. A decade ago, the key lengths and message lengths of these lattice-based cryptosystems, though formally polynomial, were so large it was almost a joke: the blowup, in going from plaintext to ciphertext, could be by a factor of millions (depending on what kind of security you want). But lattice-based cryptography has been getting steadily closer to being practical – in part, frankly, because people realized that you can get big efficiency improvements if you're willing to be a little less rigorous about security. If scalable quantum computers able to break RSA ever look like a serious threat, I predict that the response will be to switch over to new public-key cryptosystems that look more like the lattice ones. And, again, the prospect of fully homomorphic encryption might provide a separate reason to make the switch.

What about elliptic-curve cryptosystems, another important class of public-key cryptosystems you might have heard about (and a class that, unlike lattice-based cryptosystems, is already deployed commercially today)? Unfortunately, elliptic-curve cryptosystems are

easily breakable by quantum computers, since the problem of breaking them can be expressed as an abelian hidden subgroup problem. (Elliptic-curve groups are abelian.) On the other hand, the best-
known *classical* algorithms for breaking elliptic-curve cryptosystems
apparently have higher running times than the Number Field Sieve
for breaking RSA – it's a question of $\sim 2^n$ versus $\sim 2^{n^{1/3}}$. That *could* be
fundamental, or it could just be because algorithms for elliptic-curve
groups haven't been studied as much.

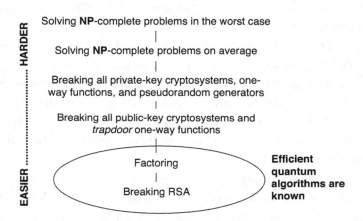

This completes our whirlwind tour of classical complexity and
cryptography; we are ready to talk about quantum mechanics.

9 Quantum

There are two ways to teach quantum mechanics. The first way – which for most physicists today is still the only way – follows the historical order in which the ideas were discovered. So, you start with classical mechanics and electrodynamics, solving lots of grueling differential equations at every step. Then, you learn about the "blackbody paradox" and various strange experimental results, and the great crisis these things posed for physics. Next, you learn a complicated patchwork of ideas that physicists invented between 1900 and 1926 to try to make the crisis go away. Then, if you're lucky, after years of study, you finally get around to the central conceptual point: that nature is described not by *probabilities* (which are always nonnegative), but by numbers called *amplitudes* that can be positive, negative, or even complex.

Look, obviously the physicists had their reasons for teaching quantum mechanics that way, and it works great for a certain kind of student. But the "historical" approach also has disadvantages, which in the quantum information age are becoming increasingly apparent. For example, I've had experts in quantum field theory – people who've spent years calculating path integrals of mind-boggling complexity – *ask me to explain the Bell inequality to them*, or other simple conceptual things like Grover's algorithm. I felt as if Andrew Wiles had asked me to explain the Pythagorean Theorem.

As a direct result of what I think of as the "QWERTY" approach to explaining quantum mechanics – which you can see reflected in almost every popular book and article, down to the present – the subject acquired an unnecessary reputation for being *complicated* and *hard*. Educated people memorized the slogans – "light is both a wave and a particle," "the cat is neither dead nor alive until you look," "you can ask about the position *or* the momentum, but not both," "one particle instantly learns the spin of the other through spooky action-at-a-distance," etc. But they also learned that they shouldn't even *try* to understand such things without years of painstaking work.

The second way to teach quantum mechanics eschews a blow-by-blow account of its discovery, and instead *starts directly from the conceptual core* – namely, a certain generalization of the laws of probability to allow minus signs (and more generally, complex numbers). Once you understand that core, you can *then* sprinkle in physics to taste, and calculate the spectrum of whatever atom you want. This second approach is the one I'll be following here.

So, what *is* quantum mechanics? Even though it was discovered by physicists, it's *not* a physical theory in the same sense as electromagnetism or general relativity. In the usual "hierarchy of sciences" – with biology at the top, then chemistry, then physics, then math – quantum mechanics sits at a level *between* math and physics that I don't know a good name for. Basically, *quantum mechanics is the operating system that other physical theories run on as application software* (with the exception of general relativity, which hasn't yet been successfully ported to this particular OS). There's even a word for taking a physical theory and porting it to this OS: "to quantize."

But if quantum mechanics isn't physics in the usual sense – if it's not about matter, or energy, or waves, or particles – then what *is* it about? From my perspective, it's about information and probabilities and observables, and how they relate to each other.

My contention in this chapter is the following: *Quantum mechanics is what you would inevitably come up with if you started from probability theory, and then said, let's try to generalize it so that the numbers we used to call "probabilities" can be negative numbers. As such, the theory could have been invented by mathematicians in the nineteenth century without any input from experiment. It wasn't, but it could have been.*

And yet, with all the structures mathematicians studied, none of them came up with quantum mechanics until experiment forced it on them. And that's a perfect illustration of why experiments are relevant in the first place! More often than not, the *only* reason we need experiments is that we're not smart enough. After the experiment has been done, if we've learned anything worth knowing at all, then *we hope* we've learned why the experiment wasn't necessary to begin with – why it wouldn't have made sense for the world to be any other way. But we're too dumb to figure it out ourselves!

Two other perfect examples of "obvious-in-retrospect" theories are evolution and special relativity. Admittedly, I don't know if the ancient Greeks, sitting around in their togas, could have figured out that these theories were *true*. But certainly – *certainly!* – they could've figured out that they were *possibly* true: that they're powerful principles that would've at least been on God's whiteboard when She was brainstorming the world.

In this chapter, I'm going to try to convince you – without any recourse to experiment – that quantum mechanics would *also* have been on the whiteboard. I'm going to show you why, if you want a universe with certain very generic properties, you seem forced to one of three choices: (1) determinism, (2) classical probabilities, or (3) quantum mechanics. Even if the "mystery" of quantum mechanics can never be banished entirely, you might be surprised by just how far people could've gotten without leaving their armchairs! That they *didn't* get far until atomic spectra and so on forced the theory down their throats is one of the strongest arguments I know for experiments being necessary.

A LESS THAN 0% CHANCE?

Alright, so what would it mean to have "probability theory" with negative numbers? Well, there's a reason you never hear the weather forecaster talk about a −20% chance of rain tomorrow – it really *does* make as little sense as it sounds. But I'd like you to set any qualms aside, and just think abstractly about an event with N possible outcomes. We can express the probabilities of those events by a vector of N real numbers:

$$(p_1, \ldots, p_N).$$

Mathematically, what can we say about this vector? Well, the probabilities had better be nonnegative, and they'd better sum to unity. We can express the latter fact by saying that the 1-norm of the probability vector has to be unity. (The 1-norm just means the sum of the absolute values of the entries.)

But the 1-norm is not the only norm in the world – it's not the only way we know to define the "size" of a vector. There are other ways, and one of the recurring favorites since the days of Pythagoras has been the *2-norm*, or *Euclidean norm*. Formally, the Euclidean norm means the square root of the sum of the squares of the entries. Informally, it means you're late for class, so instead of going this way and then that way, you cut across the grass.

Now, what happens if you try to come up with a theory that's *like* probability theory, but based on the 2-norm instead of the 1-norm? I'm going to try to convince you that quantum mechanics is what inevitably results.

Let's consider a single bit. In probability theory, we can describe a bit as having a probability p of being 0, and a probability $1 - p$ of being 1. But if we switch from the 1-norm to the 2-norm, now we no longer want two numbers that sum to unity, we want two numbers whose *squares* sum to unity. (I'm assuming we're still talking about real numbers.) In other words, we now want a vector (α, β) where $\alpha^2 + \beta^2 = 1$. Of course, the set of *all* such vectors forms a circle:

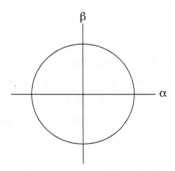

The theory we're inventing will *somehow* have to connect to observation. So, suppose we have a bit that's described by this vector (α, β). Then, we'll need to specify what happens if we *look* at the bit. Well, since it *is* a bit, we should see either 0 or 1! Furthermore, the probability of seeing 0 and the probability of seeing 1 had better add up to unity. Now, starting from the vector (α, β), how can we get two numbers that add up to unity? Simple: we can let α^2 be the probability of a 0 outcome, and let β^2 be the probability of a 1 outcome.

But in that case, why not forget about α and β, and just describe the bit *directly* in terms of probabilities? Ahh. The difference comes in how the vector changes when we apply an operation to it. In probability theory, if we have a bit that's represented by the vector $(p, 1 - p)$, then we can represent any operation on the bit by a *stochastic matrix*: that is, a matrix of nonnegative real numbers where every column adds up to unity. So, for example, the "bit flip" operation – which changes the probability of a 1 outcome from p to $1 - p$ – can be represented as follows:

$$\begin{pmatrix} 0 & 1 \\ 1 & 0 \end{pmatrix} \begin{pmatrix} p \\ 1 - p \end{pmatrix} = \begin{pmatrix} 1 - p \\ p \end{pmatrix}.$$

Indeed, it turns out that a stochastic matrix is the *most general* sort of matrix that always maps a probability vector to another probability vector.

Exercise 1 for the non-lazy reader: Prove this.

But now that we've switched from the 1-norm to the 2-norm, we have to ask: *what's the most general sort of matrix that always maps a unit vector in the 2-norm to another unit vector in the 2-norm?*

Well, we call such a matrix a *unitary matrix* – indeed, that's one way to define what a unitary matrix is! (Oh, alright. As long as we're only talking about real numbers, it's called an *orthogonal matrix*. But same difference.) Another way to define a unitary matrix, again in the case of real numbers, is as a matrix whose inverse equals its transpose.

Exercise 2 for the non-lazy reader: Prove that these two definitions are equivalent.

This "2-norm bit" that we've defined has a name, which as you might know is *qubit*. Physicists like to represent qubits using what they call "Dirac ket notation," in which the vector (α, β) becomes $\alpha|0\rangle + \beta|1\rangle$. Here, α is the *amplitude* of outcome $|0\rangle$, and β is the amplitude of outcome $|1\rangle$.

This notation usually drives computer scientists up a wall when they first see it – especially because of the asymmetric brackets! But if you stick with it, you see that it's really not so bad. As an example, instead of writing out a vector like $(0, 0, 3/5, 0, 0, 0, 4/5, 0)$, you can simply write $\frac{3}{5}|3\rangle + \frac{4}{5}|7\rangle$, omitting all of the 0 entries.

So given a qubit, we can transform it by applying any two-by-two unitary matrix – and that leads already to the famous effect of *quantum interference*. For example, consider the unitary matrix

$$\begin{pmatrix} \dfrac{1}{\sqrt{2}} & -\dfrac{1}{\sqrt{2}} \\ \dfrac{1}{\sqrt{2}} & \dfrac{1}{\sqrt{2}} \end{pmatrix}$$

which takes a vector in the plane and rotates it by 45 degrees counterclockwise. Now consider the state $|0\rangle$. If we apply U once to this state, we'll get $\frac{1}{\sqrt{2}}(|0\rangle + |1\rangle)$ – it's like taking a coin and flipping it. But then, if we apply the same operation U a second time, we'll get $|1\rangle$:

$$\begin{pmatrix} \dfrac{1}{\sqrt{2}} & -\dfrac{1}{\sqrt{2}} \\ \dfrac{1}{\sqrt{2}} & \dfrac{1}{\sqrt{2}} \end{pmatrix} \begin{pmatrix} \dfrac{1}{\sqrt{2}} \\ \dfrac{1}{\sqrt{2}} \end{pmatrix} = \begin{pmatrix} 0 \\ 1 \end{pmatrix}$$

So in other words, applying a "randomizing" operation to a "random" state produces a deterministic outcome! Intuitively, even though there are two "paths" that lead to the outcome $|0\rangle$, one of those paths has positive amplitude and the other has negative amplitude. As a result, the two paths *interfere destructively* and cancel each other out. By contrast, the two paths leading to the outcome $|1\rangle$ both have positive amplitude, and therefore interfere *constructively*.

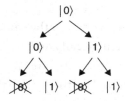

The reason you never see this sort of interference in the classical world is that probabilities can't be negative. So, cancellation between positive and negative amplitudes can be seen as the source of *all* "quantum weirdness" – the one thing that makes quantum mechanics different from classical probability theory. How I wish someone had told me that when I first heard the word "quantum"!

MIXED STATES

Once we have these quantum states, one thing we can always do is to take classical probability theory and "layer it on top." In other words, we can always ask, what if we don't *know* which quantum state we have? For example, what if we have a $\frac{1}{2}$ probability of $\frac{1}{\sqrt{2}}(|0\rangle + |1\rangle)$ and a $\frac{1}{2}$ probability of $\frac{1}{\sqrt{2}}(|0\rangle - |1\rangle)$? This gives us what's called a *mixed state*, which is the most general kind of state in quantum mechanics.

Mathematically, we represent a mixed state by an object called a *density matrix*. Here's how it works: say, you have this vector of N amplitudes, $(\alpha_1, \ldots, \alpha_N)$. Then, you compute the *outer product* of the vector with itself – that is, an $N \times N$ matrix whose (i, j) entry is $\alpha_i \alpha_j$ (again in the case of real numbers). Then, if you have a probability distribution over several such vectors, you just take a linear combination of the resulting matrices. So, for example, if you have probability p of some vector and probability $1 - p$ of a different vector, then it's p times the one matrix plus $1 - p$ times the other.

The density matrix encodes all the information that could ever be obtained from some probability distribution over quantum states, by first applying a unitary operation and then looking at the state (or, as we say in the business, *measuring* it).

Exercise 3 for the non-lazy reader: Prove this.

This implies that if two distributions give rise to the same density matrix, then those distributions are empirically indistinguishable, or in other words are *the same mixed state*. As an example, let's say you have the state $\frac{1}{\sqrt{2}}(|0\rangle + |1\rangle)$ with $\frac{1}{2}$ probability, and $\frac{1}{\sqrt{2}}(|0\rangle - |1\rangle)$ with $\frac{1}{2}$ probability. Then, the density matrix that describes your knowledge is

$$\frac{1}{2}\begin{pmatrix} \frac{1}{2} & \frac{1}{2} \\ \frac{1}{2} & \frac{1}{2} \end{pmatrix} + \frac{1}{2}\begin{pmatrix} \frac{1}{2} & -\frac{1}{2} \\ -\frac{1}{2} & \frac{1}{2} \end{pmatrix} = \begin{pmatrix} \frac{1}{2} & 0 \\ 0 & \frac{1}{2} \end{pmatrix}.$$

It follows, then, that no measurement you can ever perform will distinguish this mixture from a $\frac{1}{2}$ probability of $|0\rangle$ and a $\frac{1}{2}$ probability of $|1\rangle$.

THE SQUARING RULE

Why do we square the amplitudes instead of cubing them or raising them to the fourth power or whatever? Certainly, it agrees with experiment. But what we really want to know is: if *you* were designing the laws of physics, why would you do it this way and not some other way? Why not, for example, just use the absolute values of the amplitudes, or the absolute values cubed?

Alright, I can give you a couple of arguments for why to square the amplitudes.

The first argument is a famous result called Gleason's Theorem from the 1950s. Gleason's Theorem lets us assume *part* of quantum mechanics and then get out the rest of it! More concretely, suppose we have some procedure that takes as input a unit vector of real numbers, and that spits out the probability of an event. Formally, we have a function f that maps a unit vector $v \in \Re^N$ to the unit interval $[0, 1]$. And let's suppose $N = 3$ – the theorem actually works in any number of dimensions three or greater (but interestingly, *not* in two dimensions). Then, the key requirement we impose is that, whenever three vectors v_1, v_2, v_3 are all orthogonal to each other,

$$f(v_1) + f(v_2) + f(v_3) = 1.$$

Intuitively, if these three vectors represent "orthogonal ways" of measuring a quantum state, then they should correspond to mutually exclusive events. Crucially, we don't need *any* assumption other than that – no continuity, no differentiability, no nuthin'.

So, that's the setup. The amazing conclusion of the theorem is that, for *any* such f, there exists a mixed state such that f arises by measuring that state according to the standard measurement rule of quantum mechanics. I won't be able prove this theorem here, since it's pretty hard. But it's one way that you can "derive" the squaring rule without *exactly* having to put it in at the outset.

Exercise 4 for the non-lazy reader: Why does Gleason's Theorem *not* work in two dimensions?

If you like, I can give you a much more elementary argument. This is something I put it in one of my papers,[1] though I'm sure many others knew it before.

Let's say we want to invent a theory that's not based on the 1-norm-like classical probability theory, *or* on the 2-norm-like quantum mechanics, but instead on the p-norm for some $p \notin \{1, 2\}$. Call $(v_1, \ldots, v_N)$ a *unit vector in the p-norm* if

$$|v_1|^p + \cdots + |v_N|^p = 1.$$

Then, we'll need some "nice" set of linear transformations that map any unit vector in the p-norm to another unit vector in the p-norm.

It's clear that for any p we choose, there will be *some* linear transformations that preserve the p-norm. Which ones? Well, we can

---

[1] http://www.scottaaronson.com/papers/island.pdf

permute the basis elements, shuffle them around. That'll preserve the p-norm. And we can stick in minus signs if we want. That'll preserve the p-norm too. But here's the little observation I made: *if there are any linear transformations other than these trivial ones that preserve the p-norm, then either p = 1 or p = 2.* If $p = 1$, we get classical probability theory, while if $p = 2$, we get quantum mechanics. So if you don't want something boring, you have to set $p = 1$ or $p = 2$.

Exercise 5 for the non-lazy reader: Prove my little observation.

Alright, to get you started, let me give some intuition about why this observation *might* be true. Let's assume, for simplicity, that everything is real and that p is a positive even integer (though the observation also works with complex numbers and with any real $p \geq 0$). Then, for a linear transformation $A = (a_{ij})$ to *preserve the p-norm* means that

$$w_1^p + \cdots + w_N^p = v_1^p + \cdots + v_N^p$$

whenever

$$\begin{pmatrix} w_1 \\ \vdots \\ w_N \end{pmatrix} = \begin{pmatrix} a_{11} & \cdots & a_{1N} \\ \vdots & & \vdots \\ a_{N1} & \cdots & a_{NN} \end{pmatrix} \begin{pmatrix} v_1 \\ \vdots \\ v_N \end{pmatrix}.$$

Now we can ask: how many constraints are imposed on the matrix A by the requirement that this be true for every $v_1, \ldots, v_N$? If we work it out, in the case $p = 2$, we'll find that there are $N + \binom{N}{2}$ constraints. But since we're trying to pick an N-by-N matrix, that still leaves us $N(N-1)/2$ degrees of freedom to play with.

On the other hand, if (say) $p = 4$, then the number of constraints grows like $\binom{N}{4}$, which is *greater* than N^2 (the number of variables in the matrix). That suggests that it will be hard to find a nontrivial linear transformation that preserves the 4-norm. Of course, it doesn't *prove* that no such transformation exists – that's left as a puzzle for you.

Incidentally, this isn't the only case where we find that the 1-norm and 2-norm are "more special" than other p-norms. So, for example, have you ever seen the following equation?

$$x^n + y^n = z^n.$$

There's a cute little fact – unfortunately, I won't have time to prove it in this book – that the above equation has nontrivial integer solutions when $n = 1$ or $n = 2$, but not for any larger integers n. Clearly, then, if we use the 1-norm and the 2-norm more than other vector norms, it's not some arbitrary whim – these *really are* God's favorite norms! (And we didn't even need an experiment to tell us that.)

REAL VERSUS COMPLEX NUMBERS
Even after we've decided to base our theory on the 2-norm, we still have at least two choices: we could let our amplitudes be real numbers, *or* we could let them be complex. We know the solution Nature adopted: amplitudes in quantum mechanics are complex numbers. This means that you can't just square an amplitude to get a probability; first, you have to take the absolute value, and then you square *that*. So, why?

Years ago, at Berkeley, I was hanging out with some math grad students – I fell in with the wrong crowd – and I asked them that exact question. The mathematicians just snickered. "Give us a break – the complex numbers are algebraically closed!"[2] To them it wasn't a mystery at all.

---

[2] A field F of numbers is called "algebraically closed," if every algebraic equation involving numbers from F can also be solved using numbers from F (except for trivially unsolvable equations like $0 = 1$). To illustrate, the rational numbers aren't algebraically closed because the equation $x^2 = 2$ has only irrational solutions, and even the real numbers aren't algebraically closed because the equation $x^2 = -1$ has only imaginary solutions. But it was a big result in the early 1800s that the complex numbers *are* algebraically closed. A priori, you might have guessed that you'd need to invent an unending tower of more and more complicated numbers, in the course of trying to solve algebraic equations involving the previous numbers. But no, "the buck stops" at the complex numbers! For example, a solution to the equation $x^2 = i$ is $x = (1 + i)/\sqrt{2}$, which is still a complex number.

But to me it *is* sort of strange. I mean, complex numbers were seen for centuries as fictitious entities that human beings made up, in order that every quadratic equation should have a root. (That's why we talk about their "imaginary" parts.) So why should Nature, at its most fundamental level, run on something that *we* invented for our convenience?

Alright, yeah: suppose we require that, for every linear transformation U that we can apply to a state, there must be another transformation V such that $V^2 = U$. This is basically a *continuity* assumption: we're saying that, if it makes sense to apply an operation for one second, then it ought to make sense to apply that same operation for only half a second.

Can we get that with only real amplitudes? Well, consider the following linear transformation:

$$\begin{pmatrix} 1 & 0 \\ 0 & -1 \end{pmatrix}.$$

This transformation is just a *mirror reversal* of the plane. That is, it takes a two-dimensional Flatland creature and flips it over like a pancake, sending its heart to the other side of its two-dimensional body. But how do you apply *half* of a mirror reversal without leaving the plane? You can't! If you want to flip a pancake by a continuous motion, then you need to go into ... *dum dum dum* ... THE THIRD DIMENSION.

More generally, if you want to flip over an N-dimensional object by a continuous motion, then you need to go into the $(N+1)$st dimension.

Exercise 6 for the non-lazy reader: Prove that *any* norm-preserving linear transformation in N dimensions can be implemented by a continuous motion in $N + 1$ dimensions.

But what if you want every linear transformation to have a square root in the *same* number of dimensions? Well, in that case, you have to allow complex numbers. So that's one argument why you might want complex numbers at such a basic level of physics.

Alright, I can give you two other reasons why amplitudes should be complex.

The first comes from asking, how many independent real parameters are there in an N-dimensional mixed state? As it turns out, the answer is exactly N^2 – provided we assume, for convenience, that the state doesn't have to be normalized (i.e., that the probabilities can add up to less than unity). Why? Well, an N-dimensional mixed state is represented mathematically by an $N \times N$ hermitian[3] matrix with positive eigenvalues. Since we're not normalizing, we've got N independent real numbers along the main diagonal. *Below* the main diagonal, we've got $N(N-1)/2$ independent complex numbers, which means $N(N-1)$ real numbers. Since the matrix is hermitian, the complex numbers below the main diagonal *determine* the ones above the main diagonal. So the total number of independent real parameters is $N + N(N-1) = N^2$.

Now we bring in an aspect of quantum mechanics that I didn't mention before. If we know the states of *two* quantum systems individually, then how do we write their *combined* state? Well, we just form what's called the *tensor product*. So, for example, the tensor product of two qubits, $\alpha|0\rangle + \beta|1\rangle$ and $\gamma|0\rangle + \delta|1\rangle$, is given by

$$(\alpha|0\rangle + \beta|1\rangle) \otimes (\gamma|0\rangle + \delta|1\rangle) = \alpha\gamma|00\rangle + \alpha\delta|01\rangle$$
$$+ \beta\gamma|10\rangle + \beta\delta|11\rangle.$$

Here, I've used the notation $|00\rangle$ as a shorthand for $|0\rangle \otimes |0\rangle$, $|01\rangle$ as a shorthand for $|0\rangle \otimes |1\rangle$, and so on. (Sometimes I'll also use the notations $|0\rangle|0\rangle$ and $|0\rangle|1\rangle$. These again mean the same thing: one qubit in the first state, "tensored with" or "sitting next to" another qubit in the second state.) One important point about the tensor product is that it doesn't commute: $|0\rangle \otimes |1\rangle$ is a *different* state from $|1\rangle \otimes |0\rangle$! For the first corresponds to the binary string 01 (the first bit is 0 and the second bit is 1), while the second corresponds to the binary string 10 (the first bit is 1 and the second bit is 0).

---

[3] A matrix of complex numbers which is equal to its own conjugate transpose.

Again one can ask: did we *have* to use the tensor product? Could God have chosen some *other* way of combining quantum states into bigger ones? Well, as it happens, there *are* other ways to combine quantum systems – most notably, the so-called *symmetric* and *antisymmetric products* – and those other ways are actually used in physics, to describe the behavior of identical bosons and identical fermions, respectively. For me, though, saying we take the tensor product is almost what we *mean* when we say we're putting together two systems that are able to have an independent existence (as, I would say, identical bosons and identical fermions are not able).

As you probably know, there are two-qubit states that *can't* be written as the tensor product of one-qubit states. The most famous example is the EPR (Einstein–Podolsky–Rosen) pair:

$$\frac{|00\rangle + |11\rangle}{\sqrt{2}}.$$

Given a mixed state ρ on two subsystems A and B, if ρ can be written as a probability distribution over tensor product states $|\psi_A\rangle \otimes |\psi_B\rangle$, then we say ρ is *separable*. Otherwise, we say ρ is *entangled*.

Now let's come back to the question of how many real parameters are needed to describe a mixed state. Suppose we have a (possibly entangled) composite system AB. Then intuitively, it seems like the number of parameters needed to describe AB – which I'll call d_{AB} – should equal the *product* of the number of parameters needed to describe A and the number of parameters needed to describe B:

$$d_{AB} = d_A d_B.$$

If amplitudes are complex numbers, then happily this is true! Letting N_A and N_B be the number of dimensions of A and B, respectively, we have

$$d_{AB} = (N_A N_B)^2 = N_A^2 N_B^2 = d_A d_B.$$

But what if the amplitudes are real numbers? In that case, in an N-by-N density matrix, we'd only have $N(N + 1)/2$ independent real parameters. And it's *not* the case that, if $N = N_A N_B$, then

$$\frac{N(N+1)}{2} = \frac{N_A(N_A+1)}{2} \cdot \frac{N_B(N_B+1)}{2}.$$

Can this same argument be used to rule out quaternions? Yes! With real numbers, the left-hand side is too big, whereas with quaternions it's too small. Only with complex numbers is it juuuuust right!

There's actually another phenomenon with the same "Goldilocks" flavor, which was observed by Bill Wootters – and this leads to my third reason why amplitudes should be complex numbers. Let's say we choose a quantum state

$$\sum_{i=1}^{N} \alpha_i |i\rangle$$

uniformly at random (if you're a mathematician, under the Haar measure). And then we measure it, obtaining outcome $|i\rangle$ with probability $|\alpha_i|^2$. The question is, will the resulting probability vector *also* be distributed uniformly at random in the probability simplex? It turns out that, if the amplitudes are complex numbers, then the answer is yes. But if the amplitudes are real numbers or quaternions, then the answer is no!

LINEARITY

We've talked about why the amplitudes should be complex numbers, and why the rule for converting amplitudes to probabilities should be a squaring rule. But all this time, the elephant of *linearity* has been sitting there undisturbed. Why should quantum states evolve to other quantum states by means of linear transformations? One guess is that, if the transformations weren't linear, you could crunch vectors to be bigger or smaller. Close! Steven Weinberg[4] and others proposed nonlinear variants of quantum mechanics in which the state vectors do stay the same size. The trouble with these variants is that they'd let you take far-apart vectors and squash them together, *or* take extremely

---

[4] S. Weinberg, Precision tests of quantum mechanics, *Physical Review Letters* **62** (1989), 485.

close vectors and pry them apart! Indeed, that's essentially what it *means* for such theories to be nonlinear. So our configuration space no longer has this intuitive meaning of measuring the distinguishability of vectors. Two states that are exponentially close might in fact be perfectly distinguishable. And indeed, in 1998 Abrams and Lloyd[5] used exactly this observation to show that, *if* quantum mechanics were nonlinear, then one could build a computer to solve **NP**-complete problems in polynomial time. Of course, we don't know whether **NP**-complete problems are efficiently solvable in the physical world. But in a survey[6] I wrote years ago, I explained why the ability to solve **NP**-complete problems would give us "godlike" powers – arguably, even more so than the ability to transmit superluminal signals or reverse the Second Law of Thermodynamics. The basic point is that, when we talk about **NP**-complete problems, we're not just talking about scheduling airline flights (or for that matter, breaking the RSA cryptosystem). We're talking about *automating insight*: proving the Riemann Hypothesis, modeling the stock market, seeing whatever patterns or chains of logical deduction are there in the world to be seen.

So, suppose I maintain the working hypothesis that **NP**-complete problems are *not* efficiently solvable by physical means, and that if a theory suggests otherwise, more likely than not that indicates a problem with the theory. Then there are only two possibilities: either I'm right, or else I'm a god! And either one sounds pretty good to me . . .

Exercise 7 for the non-lazy reader: Prove that if quantum mechanics were nonlinear, then not only could you solve **NP**-complete problems in polynomial time, you could also use EPR pairs to transmit information faster than the speed of light.

Let me end this chapter by mentioning three central aspects of quantum mechanics that make some appearances in this book.

---

[5] http://www.arxiv.org/abs/quant-ph/9801041
[6] http://www.scottaaronson.com/papers/npcomplete.pdf

The first is the *No-Cloning Theorem*. This is simply the statement that there's no procedure, consistent with quantum mechanics, that takes as input an unknown quantum state $|\psi\rangle$, and produces as output two separate copies of $|\psi\rangle$ – that is, the tensor product $|\psi\rangle \otimes |\psi\rangle$. The proof of this thing is so trivial, it's debatable whether it even deserves the name "theorem" – but it's certainly important. Here's the proof: assume without loss of generality that $|\psi\rangle$ is just a single qubit, $|\psi\rangle = \alpha|0\rangle + \beta|1\rangle$. Then a "cloning map," which wrote a copy of $|\psi\rangle$ into another qubit initialized to (say) $|0\rangle$, would need to do the following:

$$(\alpha|0\rangle + \beta|1\rangle)|0\rangle \rightarrow (\alpha|0\rangle + \beta|1\rangle)(\alpha|0\rangle + \beta|1\rangle)$$
$$= \alpha^2|0\rangle|0\rangle + \alpha\beta|0\rangle|1\rangle + \alpha\beta|1\rangle|0\rangle + \beta^2|1\rangle|1\rangle.$$

Notice that α^2, $\alpha\beta$, and β^2 are quadratic functions of α and β. But unitary transformations can only ever produce *linear* combinations of amplitudes, and therefore can never produce an evolution of the above sort. And that's pretty much the No-Cloning Theorem! We see that, unlike classical information, which can be copied promiscuously across the universe, quantum information has a "privacy" to it – indeed in some ways, it's less like classical information than like gold, oil, or other "indivisible" resources.

A few comments about the No-Cloning Theorem are in order.

- The theorem is not just an artifact of an unphysical insistence on *perfect* copyability. Indeed, you might want to check that the linearity of quantum mechanics prohibits making even making a "pretty good" copy of a qubit, for a suitable definition of "pretty good."
- Of course we *can* map the state $(\alpha|0\rangle + \beta|1\rangle)|0\rangle$ to $\alpha|0\rangle|0\rangle + \beta|1\rangle|1\rangle$, by applying a Controlled-NOT gate from the first qubit to the second qubit. But that doesn't produce two copies of the original state $\alpha|0\rangle + \beta|1\rangle$; instead it produces an entangled state, where each *individual* qubit is in the mixed state $\begin{pmatrix} |\alpha|^2 & 0 \\ 0 & |\beta|^2 \end{pmatrix}$. Indeed, the only case where we can really regard that as "copying" is if $\alpha = 0$ or $\beta = 0$ – in which case we're talking about *classical* information, not quantum information.

- If the No-Cloning Theorem reminds you of Heisenberg's famous Uncertainty Principle – well, it should! The Uncertainty Principle says that there exist pairs of properties – most famously, the position and momentum of a particle – that can't both be measured to arbitrary accuracy, and it gives a quantitative tradeoff between the two. Now, to state the Uncertainty Principle as most physicists know it would require more physics than we've done in this book: I'd have to explain the relationship between position and momentum, and I'd even have to introduce Planck's constant ℏ (even if just to assert that, in my units, ℏ equals 1!). But even without that, let me show you how the No-Cloning Theorem implies a rough-and-ready, information-theoretic version of the Uncertainty Principle, and vice versa. On the one hand, if you could measure all properties of a quantum state to unlimited accuracy, then you *could* produce arbitrarily-accurate clones. On the other hand, if you could copy a state $|\psi\rangle$ an unlimited number of times, then you could learn all its properties to arbitrary accuracy – for example, by measuring the positions of some copies and the momenta of others.
- There's one sense in which the No-Cloning Theorem has almost nothing to do with quantum mechanics. Namely, a precisely analogous theorem can be proved for classical probability distributions. If you have a coin that's heads-up with some unknown probability p, there's no way to convert it into two coins that are *both* heads-up with probability p independently of each other. Sure, you can look at your coin, but the information about p you get that way is too limited to enable copying. What's new in quantum mechanics is simply that the no-cloning principle applies, not only to mixed states, but also to pure states—states such that, *if* you knew the right basis in which to measure, then you could learn which state you had with absolute certainty. But if you *don't* know the right basis, then you can neither learn these states nor copy them.

So, that was the No-Cloning Theorem. The second aspect of quantum mechanics that I should mention here is really a striking *application* of the No-Cloning Theorem. It's called *quantum key distribution* (QKD), and it's a protocol by which Alice and Bob can agree on a shared secret key, despite never having met in advance, and (unlike in public-key cryptography) without relying on any computational intractability assumption – indeed, the only real assumptions they need are the validity of quantum mechanics, and the availability of an authenticated *classical* channel. The possibility of this sort of cryptography was anticipated by Stephen Wiesner[7] in 1969, in a remarkable, decades-ahead-of-its-time paper that Wiesner didn't manage to get published for 15 years. (Recently, while visiting Jerusalem, I had the opportunity to meet Wiesner. He's now, by choice, a construction laborer there, and an extremely interesting guy.) The first *explicit* QKD proposal was by Bennett and Brassard[8] in 1984, and is creatively known as BB84. I won't present the full details of the protocol here: while they're not terribly complicated, they also won't be important for us; and in any case, there are plenty of good expositions of BB84 in textbooks and on the web.

Instead, let me just discuss the conceptual question of how quantum mechanics could *possibly* allow secret key agreement, without Alice and Bob ever meeting in person and without computational assumptions – something that's ruled out in the classical world by Shannon's arguments (see Chapter 8). The basic idea is that Alice and Bob send each other qubits prepared randomly in two or more non-orthogonal bases: for example, the four "BB84 states" $|0\rangle$, $|1\rangle$, $\frac{|0\rangle+|1\rangle}{\sqrt{2}}$, $\frac{|0\rangle-|1\rangle}{\sqrt{2}}$. They then measure *some* of the qubits they received in one of two random bases $\left(\{|0\rangle, |1\rangle\} \text{ or } \{\frac{|0\rangle+|1\rangle}{\sqrt{2}}, \frac{|0\rangle-|1\rangle}{\sqrt{2}}\}\right)$ and send each other the results over the authenticated classical channel, in order to check whether the transmission was successful. If it wasn't, then they can try again. If it *was*, then they can use other measurement outcomes – the ones they *haven't* communicated in the

---

[7] Stephen Wiesner, "Conjugate coding," *ACM SIGACT News* 15(1):78–88. 1983.

[8] Charles H. Bennett and Gilles Brassard, "Quantum Cryptography: Public key distribution and coin tossing," *Proceedings of the IEEE International Conference on Computers, Systems, and Signal Processing*, Bangalore, p. 175, 1984.

clear – to establish a shared secret key. Aha, but how do they know that an eavesdropper, Eve, wasn't secretly monitoring those qubits? The answer is the No-Cloning Theorem! Basically, one argues that if Eve had learned anything significant about those qubits, then she wouldn't *also* have been able to stick qubits back into the channel that would pass Alice's and Bob's verification tests with non-negligible probability. Because Eve wouldn't know the right basis in which to measure each qubit, Alice and Bob would be able to *detect* the very fact that Eve was monitoring the channel. The only thing Eve could do would be to commandeer the channel completely and impersonate either Alice or Bob, in a so-called "man-in-the-middle" (or woman-in-the-middle) attack. But that would require compromising not only the quantum channel, but also the classical channel that we assumed to be authenticated.

Incidentally, Wiesner's paper introduced another striking application of the No-Cloning Theorem, one that's interested me a great deal over the last few years: *quantum money*. The idea here is simple: if quantum states are really unclonable, why not exploit that to create cash that's physically impossible to counterfeit? As soon as you think about this, though, you notice a difficulty: money is only useful if someone can *verify* it as legitimate. So the question is: can you have quantum states $|\psi\rangle$ that legitimate users can measure in order to authenticate them, but that counterfeiters can't measure in order to copy them? Well, Wiesner gave a scheme for doing exactly that—interestingly, one that wasn't rigorously proved to be secure until quite recently.[9] His scheme involved exactly the four states $\left(|0\rangle, |1\rangle, \frac{|0\rangle+|1\rangle}{\sqrt{2}}, \frac{|0\rangle-|1\rangle}{\sqrt{2}}\right)$ that would later be known as the BB84 states.

However, the central drawback of Wiesner's scheme was that the only entity that knew how to verify a bill as legitimate was the bank that originally printed the bill. For only the bank knew in which bases $\left(\{|0\rangle, |1\rangle\} \text{ or } \{\frac{|0\rangle+|1\rangle}{\sqrt{2}}, \frac{|0\rangle-|1\rangle}{\sqrt{2}}\}\right)$ the qubits had been prepared, and it couldn't publish those bases without enabling

---

[9] Abel Molina, Thomas Vidick, and John Watrous, "Optimal counterfeiting attacks and generalizations for Wiesner's quantum money," 2012. http://arxiv.org/abs/1202/4010.

counterfeiters. Recently, there's been a spate of interest in what I call *public-key quantum money*: that is, quantum states that a bank can prepare, that no one can feasibly copy, and that *anyone* can authenticate. It's not hard to see that, if you want a public-key scheme, then you need computational assumptions: quantum mechanics by itself is not enough. (For a counterfeiter with unlimited computation time could always just do an exhaustive search, until it found a state that the publicly-known verification procedure accepted.) There have been lots of public-key quantum money schemes proposed over the past few years; unfortunately, the majority of them have been broken, and the rest tend to be *ad hoc*. Recently, though, Paul Christiano and I[10] proposed a new public-key quantum money scheme called the "hidden subspace scheme," which we can *prove* secure under a relatively "standard" cryptographic assumption. Our assumption – about the quantum hardness of solving a certain classical problem involving polynomials – is a strong one, but at least it's not "tautological"; it has nothing inherently to do with quantum money.

The third aspect of quantum mechanics, before we finish this chapter, is *quantum teleportation*. That name, of course, is catnip for journalists hungry to misunderstand, and to see quantum mechanics as making possible the world of *Star Trek*. Crucially, though, quantum teleportation solves a problem that wouldn't even exist, if not for quantum mechanics itself! Classically, you can always "teleport" information, for example by sending it over the Internet. (When I was five years old and watching my dad's fax machine, it was a great revelation to me that the paper wasn't being trans-materialized, but simply converted into information and reconstituted at the other end.) The quantum teleportation problem is: what if you want to send *qubits* over a classical channel? Naïvely, that seems totally impossible. Using a classical channel, the best you could ever do would be to send the results of *measuring* a state $|\psi\rangle$ in some basis – but unless the basis happened to contain $|\psi\rangle$, that clearly wouldn't be enough information to reconstruct $|\psi\rangle$ on the other end. That's why it was an amazing discovery – by Bennett *et al.*[11], in 1993 – that if Alice and Bob share an EPR pair $\frac{|00\rangle + |11\rangle}{\sqrt{2}}$, then Alice *can* transmit an arbitrary qubit to Bob, via a protocol wherein she sends Bob two classical bits, and Alice and Bob also measure their halves of the EPR pair ("using up" the EPR pair in the process).

How does the protocol work? Well, suppose Alice wants to teleport $|\psi\rangle = \alpha|0\rangle + \beta|1\rangle$ to Bob. Then the first thing she does is, she applies a Controlled-NOT from $|\psi\rangle$ onto her half of the EPR pair. This has the result

$$(\alpha|0\rangle + \beta|1\rangle) \otimes \frac{|00\rangle + |11\rangle}{\sqrt{2}}$$

$$\rightarrow \frac{\alpha}{\sqrt{2}}|000\rangle + \frac{\alpha}{\sqrt{2}}|011\rangle + \frac{\beta}{\sqrt{2}}|110\rangle + \frac{\beta}{\sqrt{2}}|101\rangle.$$

Next, Alice applies a Hadamard to her first qubit (the one that was originally $|\psi\rangle$). This produces the state

$$\frac{\alpha}{2}(|000\rangle + |100\rangle + |011\rangle + |111\rangle)$$

$$+ \frac{\beta}{2}(|010\rangle - |110\rangle + |001\rangle - |101\rangle).$$

Finally, Alice measures both of her qubits in the $\{|0\rangle, |1\rangle\}$ basis, and sends the result to Bob. Note that, *regardless* of what $|\psi\rangle$ was, Alice will see each of the four possible outcomes (00, 01, 10, and 11) with probability $\frac{1}{4}$. Furthermore, if she sees 00 then Bob's state is $\alpha|0\rangle + \beta|1\rangle$, if she sees 01 then Bob's state is $\beta|0\rangle + \alpha|1\rangle$, if she sees 10 then Bob's state is $\alpha|0\rangle - \beta|1\rangle$, and if she sees 11 then Bob's state is $\beta|0\rangle - \alpha|1\rangle$. Therefore, after receiving the two classical bits from Alice, Bob knows exactly which "corrections" to apply in order to recover the original state $\alpha|0\rangle + \beta|1\rangle$.

Two conceptual points: first, there's no instantaneous communication here. In order for $|\psi\rangle$ to be teleported, two classical bits need

---

[11] Charles H. Bennett, Gilles Brassard, Claude Crépeau, Richard Jozsa, Asher Peres, and William K. Wootters, "Teleporting an unknown quantum state via dual classical and Einstein-Podolsky-Rosen channels," *Physical Review Letters* 70:1895–1899, 1993.

to get from Alice to Bob, and those bits can only travel at the speed of light. Second, and even more interestingly, there's no violation of the No-Cloning Theorem. In order to teleport $|\psi\rangle$ to Bob, Alice had to *measure* her copy of $|\psi\rangle$ and thereby learn which classical bits to send him – and the measurement necessarily *destroyed* Alice's copy. Could there exist some cleverer teleportation protocol, which reproduced $|\psi\rangle$ on Bob's end but also left a copy of $|\psi\rangle$ intact on Alice's end? I claim that the answer is no. What makes me so certain? Why, the No-Cloning Theorem!

FURTHER READING
See this[12] now classic paper by Lucien Hardy for a "derivation" of quantum mechanics that's closely related to the arguments I gave, but much, much more serious and careful. Or, for a newer and different derivation, see this paper[13] by Chiribella *et al.*, which "derives" quantum mechanics as the unique theory that satisfies (1) various reasonable-sounding axioms that are also satisfied by classical probability theory, and (2) the axiom that every "mixed state" described by the theory must be obtainable by starting with a larger "pure state," then tracing out part of it. (Already in the 1930s, Schrödinger had called attention to the latter as a crucial and distinctive property of quantum mechanics. I confess, though, that I don't have any better intuition for why I would want to create a world that satisfied this particular "purification" axiom than for why I'd want to create a world that obeyed the 2-norm generalization of probability theory!) Finally, see pretty much anything Chris Fuchs[14] has written, especially this paper[15] by Caves, Fuchs, and Schack, which discusses why amplitudes should be complex numbers rather than reals or quaternions.

---

[12] http://www.arxiv.org/abs/quant-ph/0101012
[13] G. Chiribella, G. M. D'Ariano and P. Perinotti, Informational derivation of Quantum Theory. *Physical Review* A, **84** (2011), 012311. http://arxiv.org/abs/1011.6451
[14] http://www.perimeterinstitute.ca/personal/cfuchs/
[15] http://www.arxiv.org/abs/quant-ph/0104088

10 Quantum computing

Alright, so now we've got this beautiful theory of quantum mechanics, and the possibly-even-more-beautiful theory of computational complexity. Clearly, with two theories this beautiful, you can't just let them stay single – you have to set them up, see if they hit it off, etc.

And that brings us to the class **BQP**: Bounded-Error Quantum Polynomial-Time. We talked in Chapter 7 about **BPP**, or Bounded-Error *Probabilistic* Polynomial-Time. Informally, **BPP** is the class of computational problems that are efficiently solvable in the physical world if classical physics is true. Now we ask, what problems are efficiently solvable in the physical world if (as seems more likely) quantum physics is true?

To me it's sort of astounding that it took until the 1990s for anyone to really seriously ask this question, given that all the tools for asking it were in place by the 1960s or even earlier. It makes you wonder, what similarly obvious questions are there today that no one's asking?

So how do we define **BQP**? Well, there are four things we need to take care of.

1. **Initialization**. We say, we have a system consisting of n quantum bits (or *qubits*), and these are all initialized to some simple, easy-to-prepare state. For convenience, we usually take that to be a "computational basis state," though later in the book we'll consider relaxing that assumption. In particular, if the input string is x, then the initial state will have the form $|x\rangle|0\ldots0\rangle$: that is, $|x\rangle$ together

with as many "ancilla" qubits as we want initialized to the all-0 state.

2. **Transformations**. At any time, the state of our computer will be a superposition over all $2^{p(n)}$ $p(n)$-bit strings, where p is some polynomial in n:

$$|\psi\rangle = \sum_{z\in\{0,1\}^{p(n)}} \alpha_z |z\rangle.$$

But what operations can we use to transform one superposition state to another? Since this is quantum mechanics, the operations should be unitary transformations – but which ones? Given *any* Boolean function $f:\{0, 1\}^n \to \{0, 1\}$, there's *some* unitary transformation that will instantly compute the function for us. For example, we could take any unitary transformation that maps each basis state of the form $|x\rangle|0\rangle$ to $|x\rangle|f(x)\rangle$.

But, of course, for most functions f, we can't apply that transformation *efficiently*. Exactly by analogy to classical computing – where we're only interested in those circuits that can be built up by composing a small number of AND, OR, and NOT gates – here we're only interested in those unitary transformations that can be built up by composing a small number of quantum gates. By a "quantum gate," I just mean a unitary transformation that acts on a small number of qubits – say, one, two, or three.

Alright, let's see some examples of quantum gates. One famous example is the Hadamard gate, which acts as follows on a single qubit:

$$|0\rangle \to \frac{|0\rangle + |1\rangle}{\sqrt{2}}$$

$$|1\rangle \to \frac{|0\rangle - |1\rangle}{\sqrt{2}}$$

Another example is the Toffoli gate, which acts as follows on three qubits:

$$|000\rangle \rightarrow |000\rangle$$
$$|001\rangle \rightarrow |001\rangle$$
$$|010\rangle \rightarrow |010\rangle$$
$$|011\rangle \rightarrow |011\rangle$$
$$|100\rangle \rightarrow |100\rangle$$
$$|101\rangle \rightarrow |101\rangle$$
$$|110\rangle \rightarrow |111\rangle$$
$$|111\rangle \rightarrow |110\rangle$$

Or in words, the Toffoli gate flips the third qubit if and only if the first two qubits are both 1. Note that the Toffoli gate actually makes sense for classical computers as well.

Now, it was shown by Shi[1] that the Toffoli and Hadamard already constitute a *universal set of quantum gates*. This means, informally, that they're all you ever need for a quantum computer – since, if we wanted to, we could use them to approximate any other quantum gate arbitrarily closely. (Or technically, any gate whose unitary matrix has real numbers only, no complex numbers. But that turns out not to matter for computing purposes.) Furthermore, by a result called the Solovay-Kitaev Theorem,[2] any universal set of gates can simulate any other universal set efficiently – that is, with at most a polynomial increase in the number of gates. So as long as we're doing complexity theory, it really doesn't matter which universal gate set we choose.

This is exactly analogous to how, in the classical world, we could build our circuits out of AND, OR, and NOT gates, out of AND and NOT gates only, or even out of NAND gates only.

Now, you might ask: *which* quantum gate sets have this property of universality? Is it only very special ones? On the contrary, it turns out that in a certain precise sense, almost *any* set of one- and

[1] http://www.arxiv.org/abs/quant-ph/0205115
[2] See http://arxiv.org/abs/quant-ph/0505030

two-qubit gates (indeed, almost any *single* two-qubit gate) will be universal. But there are certainly exceptions to the rule. For example, suppose you had only the Hadamard gate (defined above) together with the following "controlled-NOT" gate, which flips the second qubit if the first qubit equals 1:

$$|00\rangle \rightarrow |00\rangle$$
$$|01\rangle \rightarrow |01\rangle$$
$$|10\rangle \rightarrow |11\rangle$$
$$|11\rangle \rightarrow |10\rangle$$

That seems like a natural universal set of quantum gates, but it isn't. The so-called Gottesman–Knill Theorem[3] shows that any quantum circuit consisting entirely of Hadamard and controlled-NOT gates can be simulated efficiently by a classical computer.

Now, once we fix a universal set (*any* universal set) of quantum gates, we'll be interested in those circuits consisting of at most $p(n)$ gates from our set, where p is a polynomial, and n is the number of bits of the problem instance we want to solve. We call these the *polynomial-size quantum circuits*.

3. **Measurement**. How do we read out the answer when the computation is all done? Simple: we measure some designated qubit, reject if we get outcome $|0\rangle$, and accept if we get outcome $|1\rangle$! Recall that for simplicity, we're only interested here in *decision problems* – that is, problems having a yes-or-no answer.

We further stipulate that, if the answer to our problem was "yes," then the final measurement should accept with probability at least $\frac{2}{3}$, whereas if the answer was "no," then it should accept with probability at most $\frac{1}{3}$. This is exactly the same requirement as for **BPP**. And as with **BPP**, we can replace the $\frac{2}{3}$ and $\frac{1}{3}$ by any other numbers we want (for example, $1 - 2^{-500}$ and 2^{-500}), by simply repeating the computation a suitable number of times and then outputting the majority answer.

---

[3] See http://www.arxiv.org/abs/quant-ph/9807006

Now, immediately there's a question: would we get a more powerful model of computation if we allowed not just one measurement, but many measurements throughout the computation!

It turns out that the answer is no – the reason being that you can always *simulate* a measurement (other than the final measurement, the one that "counts") using a unitary quantum gate. You can say, instead of measuring qubit A, let's apply a controlled-NOT gate from qubit A into qubit B, and then ignore qubit B for the rest of the computation. Then it's *as if* some third party measured qubit A – the two views are mathematically equivalent. (Is this a trivial technical point or a profound philosophical point? You be the judge . . .)

4. **Uniformity.** Before we can give the definition of **BQP**, there's one last technical issue we need to deal with. We talked about a "polynomial-size quantum circuit," but more correctly it's an infinite *family* of circuits, one for each input length n. Now, can the circuits in this family be chosen arbitrarily, completely independent of each other? If so, then we could use them to (for example) solve the halting problem, by just hardwiring into the nth circuit whether or not the nth Turing machine halts. If we want to rule that out, then we need to impose a requirement called *uniformity*. This means that there should exist a (classical) algorithm that, given n as input, outputs the nth quantum circuit in time polynomial in n.

Exercise. Show that letting a polynomial-time *quantum* algorithm output the nth circuit would lead to the same definition.

Alright, we're finally ready to put the pieces together and give a definition of **BQP**.

> **BQP** *is the class of languages $L \subseteq \{0, 1\}^*$ for which there exists a uniform family of polynomial-size quantum circuits, $\{C_n\}$, such that for all $x \in \{0, 1\}^n$:*
>
> - *if $x \in L$, then C_n accepts input $|x\rangle|0 \ldots 0\rangle$ with probability at least $2/3$.*
> - *if $x \notin L$, then C_n accepts input $|x\rangle|0 \ldots 0\rangle$ with probability at most $1/3$.*

UNCOMPUTING

So, what can we say about **BQP**?

Well, as a first question, let's say you have a **BQP** algorithm that calls another **BQP** algorithm as a subroutine. Could that be more powerful than **BQP** itself? Or in other words, could $\mathbf{BQP}^{\mathbf{BQP}}$ (that is, **BQP** with a **BQP** oracle) be more powerful than **BQP**?

It better not be! Incidentally, this is related to something I was once talking to Dave Bacon about. Why do physicists have so much trouble understanding the class **NP**? I suspect it's because **NP**, with its "magical" existential quantifier layered on top of a polynomial-time computation, is not the sort of thing they'd ever come up with. The classes that physicists would come up with – the *physicist complexity classes* – are hard to delineate precisely, but one property I think they'd definitely have is "closure under the obvious things," like one algorithm from the class calling another algorithm from the same class as a subroutine.

I claim that **BQP** is an acceptable "physicist complexity class" – and in particular, that $\mathbf{BQP}^{\mathbf{BQP}} = \mathbf{BQP}$. What's the problem in showing this?

Right, garbage! Recall that when a quantum algorithm is finished, you measure just a single qubit to obtain the yes-or-no answer. So, what to do with all the other qubits? Normally, you'd just throw them away. But now let's say you've got a superposition over different runs of an algorithm, and you want to bring the results of those runs together and interfere them. In that case, the garbage might prevent the different branches from interfering! So what do you do to fix this?

The solution, proposed by Charles Bennett in the 1980s, is to *uncompute*. Here's how it works.

1. You run the subroutine.
2. You copy the subroutine's answer qubit to a separate location.
3. You *run the entire subroutine backward*, thereby erasing everything except the answer qubit. (If the subroutine has some probability of error, this erasing step won't work perfectly, but it will still work pretty well.)

As you'd see if you visited my apartment, this is not the solution I generally adopt. But if you're a quantum computer, cleaning up the messes you make is a good idea.

RELATION TO CLASSICAL COMPLEXITY CLASSES

Alright, so how does **BQP** relate to the complexity classes we've already seen?

First, I claim that **BPP** $\subseteq$ **BQP**: in other words, anything you can do with a classical probabilistic computer, you can also do with a quantum computer. Why?

Right: because any time you were gonna flip a coin, you just apply a Hadamard gate to a fresh 0 qubit instead. In textbooks, this usually takes about a page to prove. We just proved it.

Can we get any *upper* bound on **BQP** in terms of classical complexity classes?

Sure we can! First of all, it's pretty easy to see that **BQP** $\subseteq$ **EXP**: anything you can compute in quantum polynomial time you can also compute in classical *exponential* time. Or to put it differently, quantum computers can provide *at most* an exponential advantage over classical computers. Why is that?

Right: because if you allow exponential slowdown, then a classical computer can just simulate the whole evolution of the state vector!

As it turns out, we can do a lot better than that. Recall that **PP** is the class of problems like the following.

- Given a sum of exponentially many real numbers, each of which can be evaluated in polynomial time, is the sum positive or negative (promised that one of these is the case)?
- Given a Boolean formula in n variables, do at least half of the 2^n possible variable settings make the formula evaluate to TRUE?
- Given a randomized polynomial-time Turing machine, does it accept with probability $\geq \frac{1}{2}$?

In other words, a **PP** problem involves summing up exponentially many terms, and then deciding whether the sum is greater or less than some threshold. Certainly, **PP** is contained in **PSPACE** is contained in **EXP**.

In their original paper on quantum complexity, Bernstein and Vazirani showed that **BQP** $\subseteq$ **PSPACE**. Shortly afterward, Adleman, DeMarrais, and Huang[4] improved their result to show that **BQP** $\subseteq$ **PP**. (This was also the first complexity result *I* proved. Had I known that Adleman *et al.* had proved it a year before, I might never have gotten started in this business! Occasionally, it's better to have a small academic light cone.)

So, why is **BQP** contained in **PP**? From a computer science perspective, the proof is maybe half a page. From a physics perspective, the proof is three words:

Feynman path integral!!!

Look, let's say you want to calculate the probability that a quantum computer accepts. The obvious way to do it is to multiply a bunch of $2^n \times 2^n$ unitary matrices, then take the sum of the squares of the absolute values of the amplitudes corresponding to accepting basis states (that is, basis states for which the output qubit is $|1\rangle$). What Feynman noticed in the 1940s is that there's a better way – a way that's vastly more efficient in terms of memory (or paper), though still exponential in terms of time.

The better way is to loop over accepting basis states, and for each one, loop over all *computational paths* that might contribute amplitude to that basis state. So, for example, let α_x be the final amplitude of basis state $|x\rangle$. Then we can write

$$\alpha_x = \sum_i \alpha_{x,i}$$

[4] L. M. Adleman, J. DeMarrais, and M.-D. A. Huang, Quantum Computability, *SIAM Journal on Computing*, **26**:5 (1997), 1524–1540.

where each $\alpha_{x,i}$ corresponds to a single leaf in an exponentially large "possibility tree," and is therefore computable in classical polynomial time. Typically, the $\alpha_{x,i}$ will be complex numbers with wildly differing phases, which will interfere destructively and cancel each other out; then α_x will be the tiny residue left over. *The reason quantum computing seems more powerful than classical computing is precisely that it seems hard to estimate that tiny residue using random sampling.* Random sampling would work fine for (say) a typical US election, but estimating α_x is more like the 2000 election.

Now, let S be the set of all accepting basis states. Then we can write the probability that our quantum computer accepts as

$$p_{accept} = \sum_{x \in S} \left| \sum_i \alpha_{x,i} \right|^2 = \sum_{x \in S} \sum_{i,j} \alpha_{x,i} \alpha_{x,j}^*$$

where $*$ denotes the complex conjugate. But this is just a sum of exponentially many terms, each of which is computable in **P**. We can therefore decide in **PP** whether $p_{accept} \leq \frac{1}{3}$ or $p_{accept} \geq \frac{2}{3}$.

From my perspective, Richard Feynman won the Nobel Prize in Physics essentially for showing **BQP** is contained in **PP**.

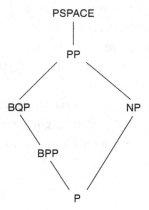

Of course, the question that really gets people hot under the collar is whether **BPP** $\neq$ **BQP**: that is, whether quantum computing is more powerful than classical. Today, we have evidence that this is indeed the case, most notably *Shor's algorithm* for factoring and discrete log. I'll assume you've heard of this algorithm, since it was one of the major scientific achievements of the late twentieth century, and is why we're talking about these things in the first place. If you *haven't* seen it, there are about 500 000 expositions on the Web.[5]

It's worth stressing that, even *before* Shor's algorithm, computer scientists had amassed formal evidence that quantum computers were more powerful than classical ones. Indeed, this evidence is what paved the way for Shor's algorithm.

One major piece of evidence was *Simon's algorithm*.[6] Suppose we have a function $f:\{0, 1\}^n \rightarrow \{0, 1\}^n$, which we can access only as a "black box," that is, by feeding it inputs and examining the outputs. We're promised that there exists a "secret XOR-mask" $s \in \{0, 1\}^n$, such that for all distinct (x, y) pairs $f(x) = f(y)$ if and only if $x \oplus y = s$. (Here $\oplus$ denotes bitwise XOR.) Our goal is to learn the identity of s. The question is, how many times do we need to query f to do that with high probability?

Classically, it's easy to see that $\sim 2^{n/2}$ queries are necessary and sufficient. As soon as we find a *collision* (a pair $x \neq y$ such that $f(x) = f(y)$), we know that $s = x \oplus y$, and hence we're done. But *until* we find a collision, the function looks essentially random. In particular, if we query the function on T inputs, then the probability of finding a collision is at most $\sim T^2/2^n$ by the union bound. Hence, we need $T \approx 2^{n/2}$ queries to find s with high probability.

On the other hand, Simon gave a quantum algorithm that finds s using only $\sim n$ queries. The basic idea is to query f in superposition, and thereby prepare quantum states of the form

$$\frac{|x\rangle + |y\rangle}{\sqrt{2}}$$

[5] For a gentle introduction, readers might want to try my own popular-level exposition of Shor's algorithm, entitled "Shor, I'll do it": http://www.scottaaronson.com/blog/?p=208.

[6] D. R. Simon, On the Power of Quantum Cryptography, *Proceedings of IEEE Symposium on Foundations of Computer Science*, (1994), 116–123.

for random (x, y) pairs such that $x \oplus y = s$. We then use the so-called *quantum Fourier transform* to extract information about s from these states. This use of the Fourier transform to extract "hidden periodicity information" provided a direct inspiration for Shor's algorithm, which does something similar over the abelian group $\mathbb{Z}_N$ instead of $\mathbb{Z}_2^n$. In a by-now famous story, Simon's paper got rejected the first time he submitted it to a conference – apparently Shor was one of the few people who got the point of it.

Again, I won't go through the details of Simon's algorithm; see here[7] if you want them.

So, the bottom line is that we get a problem – *Simon's problem* – that quantum computers can *provably* solve exponentially faster than classical computers. Admittedly, this problem is rather contrived, relying as it does on a mythical "black box" for computing a function f with a certain global symmetry. Because of its black-box formulation, Simon's problem certainly doesn't prove that **BPP** $\neq$ **BQP**. What it does prove is that there exists an *oracle* relative to which **BPP** $\neq$ **BQP**. This is what I meant by formal evidence that quantum computers are more powerful than classical ones.

As it happens, Simon's problem was *not* the first to yield an oracle separation between **BPP** and **BQP**. Just as Shor was begotten of Simon, so Simon was begotten of Bernstein–Vazirani. In the long-ago dark ages of 1993, Bernstein and Vazirani devised a black-box problem called *Recursive Fourier Sampling*. They were able to prove that any classical algorithm needs at least $\sim n^{\log n}$ queries to solve this problem, whereas there exists a quantum algorithm to solve it using only n queries.

Unfortunately, even to *define* the Recursive Fourier Sampling problem would take a longer digression than I feel is prudent. (If you think Simon's problem was artificial, you ain't seen nuthin'!) But the basic idea is this. Suppose we have black-box access to a Boolean

---

[7] http://www.cs.berkeley.edu/~vazirani/f04quantum/notes/lec7.ps

function $f:\{0, 1\}^n \rightarrow \{0, 1\}$. We're promised that there exists a "secret string" $s \in \{0, 1\}^n$, such that $f(x) = s \bullet x$ for all x (where $\bullet$ denotes the inner product mod 2). Our goal is to learn s, using as few queries to f as possible.

In other words: we know that $f(x)$ is just the XOR of some subset of input bits; our goal is to find out *which* subset.

Classically, it's obvious that n queries to f are necessary and sufficient: we're trying to learn n bits, and each query can only reveal one! But quantumly, Bernstein and Vazirani observed that you can learn s with just a single query. To do so, you simply prepare the state

$$\frac{1}{\sqrt{2^n}} \sum_{x \in \{0,1\}^n} (-1)^{f(x)} |x\rangle$$

then apply Hadamard gates to all n qubits. The result is easily checked to be $|s\rangle$.

What Bernstein and Vazirani did was to start from the problem described above – called *Fourier sampling* – and then compose it recursively. In other words, they created a Fourier sampling problem where, to learn one of the bits $f(x)$, you need to solve *another* Fourier sampling problem, and to learn one of the bits in *that* problem you need to solve a third problem, and so on. They then showed that, if the recursion is d levels deep, then any randomized algorithm to solve this Recursive Fourier Sampling problem must make at least $\sim n^d$ queries. By contrast, there exists a quantum algorithm that solves the problem using only 2^d queries.

Why 2^d queries, you ask, instead of just $1^d = 1$? Because at each level of recursion, the quantum algorithm needs to *uncompute* its garbage to get an interference effect – and that keeps adding an additional factor of 2. Like so:

```
Compute {
            Compute {
                        Compute
                        Uncompute
```

```
            }
            Uncompute {
                            Compute
                            Uncompute
                }
    }
Uncompute {
            Compute {
                            Compute
                            Uncompute
                }
            Uncompute {
                            Compute
                            Uncompute
                }
    }
```

Indeed, one of my results[8] shows that this sort of recursive uncomputation is an unavoidable feature of *any* quantum algorithm for Recursive Fourier Sampling.

So, once we have this gap of n^d versus 2^d, setting $d = \log n$ gives us $n^{\log n}$ queries on a classical computer versus $2^{\log n} = n$ queries on a quantum computer. Admittedly, this separation is not exponential versus polynomial – it's only "quasipolynomial" versus polynomial. But that's still enough to prove an oracle separation between **BPP** and **BQP**.

You might wonder: now that we have Simon's and Shor's algorithms – which *do* achieve an exponential separation between quantum and classical – why muck around with this recursive archeological relic? Well, one of the biggest open problems in quantum computing concerns the relationship between **BQP** and the polynomial hierarchy **PH** (defined in Chapter 6). Specifically, is **BQP** contained

[8] http://www.scottaaronson.com/papers/uncompute.pdf

in **PH**? Sure, it seems unlikely – but, as Bernstein and Vazirani asked back in 1993, can we actually find an *oracle* relative to which **BQP** $\not\subset$ **PH**? Alas, two decades and I don't know how many disillusioned grad students later, the answer is still no. Yet many of us still think a separation should be possible – and until recently Recursive Fourier Sampling was pretty much the only candidate problem we had for such a separation.

Finally, in 2009, I came up with a different candidate problem,[9] called "Fourier Checking," which ought to give not merely an oracle separation between **BQP** and **PH**, but (unlike Recursive Fourier Sampling) an *exponential* separation. Alas, *proving* this separation seems to require some new advances in classical complexity theory – specifically, in constant-depth circuit lower bounds – beyond what we know today. But, as a result of Fourier Checking, it's possible that Recursive Fourier Sampling has finally been superseded, retaining only its historical importance.

QUANTUM COMPUTING AND NP-COMPLETE PROBLEMS

From reading newspapers, magazines, and so on, one would think a quantum computer could "solve **NP**-complete problems in a heartbeat" by "trying every possible solution in parallel," and then instantly picking the correct one.

Well, arguably *that's* the central misconception about quantum computing among laypeople. Allow me to elaborate.

Obviously, we can't yet prove that quantum computers can't solve **NP**-complete problems efficiently – in other words, that **NP** $\not\subset$ **BQP** – since we can't even prove that **P** $\neq$ **NP**! Nor do we have any idea how to prove that *if* **P** $\neq$ **NP** *then* **NP** $\not\subset$ **BQP**.

What we do have is the early result of Bennett, Bernstein, Brassard, and Vazirani, that there exists an oracle relative to which

[9] S. Aaronson, BQP and the polynomial hierarchy. In *Proceedings of Annual ACM Symposium on Theory of Computing* (2010), pp. 141–50. http://www.scottaaronson.com/papers/bqpph.pdf

NP $\not\subset$ BQP. More concretely, suppose you're searching a space of 2^n possible solutions for a single valid one, and suppose that all you can do, given a candidate solution, is feed it to a "black box" that tells you whether that solution is correct or not. Then how many times do you need to query the black box to find the valid solution? Classically, it's clear that you need to query it $\sim 2^n$ times in the worst case (or $\sim 2^n/2$ times on average). On the other hand, Grover[10] famously gave a *quantum* search algorithm that queries the black box only $\sim 2^{n/2}$ times. But even before Grover's algorithm was discovered, Bennett *et al.* had proved that it was optimal! In other words, *any* quantum algorithm to find a needle in a size-2^n haystack needs at least $\sim 2^{n/2}$ steps. So the bottom line is that, for "generic" or "unstructured" search problems, quantum computers can give *some* speedup over classical computers – specifically, a quadratic speedup – but nothing like the exponential speedup of Shor's factoring algorithm.

You might wonder: why should the speedup be quadratic, rather than cubic or something else? Let me try to answer that question without getting into the specifics either of Grover's algorithm, *or* of the Bennett *et al.* optimality proof. Basically, *the reason we get a quadratic speedup is that quantum mechanics is based on the 2-norm rather than the 1-norm.* Classically, if there are N solutions, only one of which is right, then after one query we have a $1/N$ probability of having guessed the right solution, after two queries we have a $2/N$ probability, after three queries a $3/N$ probability, and so on. Thus, we need $\sim N$ queries to have a nonnegligible (i.e., close to unit) probability of having guessed the right solution. But quantumly, we get to apply linear transformations to vectors of *amplitudes*, which are the square roots of probabilities. So the way to think about it is this: after one query, we have a $1/\sqrt{N}$ amplitude of having guessed

---

[10] L. K. Grover, A Fast Quantum Mechanical Algorithm for Database Search, *Proceedings of ACM Symposium on Theory of Computing* (1996), 212–219. http://arxiv.org/abs/quant-ph/9605043

the right solution, after two queries we have a $2/\sqrt{N}$ amplitude, after three queries a $3/\sqrt{N}$ amplitude, and so on. So after T queries, the amplitude of having guessed a right solution is $T/\sqrt{N}$, and the probability is $|T/\sqrt{N}|^2 = T^2/N$. Hence, the probability will be close to unity after only $T \approx \sqrt{N}$ queries.

Alright, those of you who read my blog[11] must be tired of polemics about the limitations of quantum computers on unstructured search problems. So I'm going to take the liberty of ending this section now.

QUANTUM COMPUTING AND MANY-WORLDS

Since this book is titled *Quantum Computing since Democritus*, I guess I should end this chapter with a deep philosophical question. Alright, so how about this one: if we managed to build a nontrivial quantum computer, would that demonstrate the existence of parallel universes?

David Deutsch, one of the founders of quantum computing in the 1980s, certainly thinks that it would.[12] Though to be fair, Deutsch thinks the impact would "merely" be psychological – since for him, quantum mechanics has *already* proved the existence of parallel universes! Deutsch is fond of asking questions like the following: if Shor's algorithm succeeds in factoring a 3000-digit integer, then *where was the number factored?* Where did the computational resources needed to factor the number come from, if not from some sort of "multiverse" exponentially bigger than the universe we see? To my mind, Deutsch seems to be tacitly assuming here that factoring is not in **BPP** – but no matter; for purposes of argument, we can certainly grant him that assumption.

It should surprise no one that Deutsch's views about this are far from universally accepted. Many who agree about the possibility of building quantum computers, and the formalism needed to

---

[11] http://www.scottaaronson.com/blog/
[12] See for example David Deutsch, *The Fabric of Reality*, Penguin, 1997.

describe them, nevertheless disagree that the formalism is best interpreted in terms of "parallel universes." To Deutsch, these people are simply intellectual wusses – like the churchmen who agreed that the Copernican system was practically useful, so long as one remembers that obviously the Earth doesn't *really* go around the sun.

So, how do the intellectual wusses respond to the charges? For one thing, they point out that viewing a quantum computer in terms of "parallel universes" raises serious difficulties of its own. In particular, there's what those condemned to worry about such things call the "preferred basis problem." The problem is basically this: how do we define a "split" between one parallel universe and another? There are infinitely many ways you could imagine slicing up a quantum state, and it's not clear why one is better than another!

One can push the argument further. The key thing that quantum computers rely on for speedups – indeed, the thing that makes quantum mechanics different from classical probability theory in the first place – is *interference* between positive and negative amplitudes. But to whatever extent different "branches" of the multiverse can usefully interfere for quantum computing, to that extent they don't seem like separate branches at all! I mean, the whole *point* of interference is to mix branches together so that they lose their individual identities. If they retain their identities, then for exactly that reason we don't see interference.

Of course, a many-worlder could respond that, in order to lose their separate identities by interfering with each other, the branches had to *be there* in the first place! And the argument could go on (indeed, has gone on) for quite a while.[13]

Rather than take sides in this fraught, fascinating, but perhaps ultimately meaningless debate, I'd like to end with one observation

[13] For more about this argument, see Scott Aaronson, "Why Philosophers Should Care About Computational Complexity," in *Computability: Turing, Gödel, Church, and Beyond* (MIT Press, 2013; edited by Oron Shagrir), http://www.scottaaronson.com/papers/philos.pdf

that's *not* up for dispute. What the lower bound of Bennett *et al.* tells us is that, if quantum computing supports the existence of parallel universes, then it certainly doesn't do so in the way most people think! As we've seen, a quantum computer is *not* a device that could "try every possible solution in parallel" and then instantly pick the correct one. If we insist on seeing things in terms of parallel universes, then those universes all have to "collaborate" – more than that, have to *meld into one another* – to create an interference pattern that will lead to the correct answer being observed with high probability.

FURTHER READING

The definition and basic properties of **BQP** come from Ethan Bernstein and Umesh Vazirani, "Quantum Complexity Theory," *SIAM Journal on Computing* 26(5):1411–1473, 1997. The definitive introduction to quantum computing is Michael A. Nielsen and Isaac L. Chuang, *Quantum Computation and Quantum Information*, Cambridge University Press, 2011 (anniversary edition).

11 Penrose

This chapter is about Roger Penrose's arguments against the possibility of artificial intelligence, as famously set out in his books *The Emperor's New Mind*[1] and *Shadows of the Mind*.[2] It would be strange for a book like this one *not* to discuss these arguments, since, agree with them or not, they're some of the most prominent landmarks at the intersection of math, CS, physics, and philosophy. The reason we're discussing them *now* is that we finally have all the prerequisites (computability, complexity, quantum mechanics, and quantum computing).

Penrose's views are complicated: they involve speculations about an "objective collapse" of quantum states, which would arise from an as-yet-undiscovered quantum theory of gravity. Even more controversially, this hypothesized objective collapse would *play a role in human intelligence*, through its influence on cellular structures called microtubules in the brain.

But what is it that leads Penrose to make these exotic speculations in the first place? The core of Penrose's thesis is a certain argument purporting to show that human intelligence *can't* be algorithmic, for reasons related to Gödel's Incompleteness Theorem. And therefore, some nonalgorithmic element must be sought in human brain function, and the only plausible source of such an element is new physics (coming, for example, from quantum gravity). The "Gödel argument" itself didn't originate with Penrose: Gödel himself apparently believed some version of it (though he never published his views), and even in 1950 it was well enough known for Alan

[1] Oxford University Press, 2002 (reprint)
[2] Oxford University Press, 1996 (reprint)

Turing to rebut it in his famous paper "Computing machinery and intelligence." Probably the first detailed presentation of the Gödel argument in print came in 1961, from the philosopher John Lucas.[3] Penrose's main innovation is that he takes the argument seriously enough to explore, at length, what the universe and our brains would actually need to be like – or better, what they could *possibly* be like – if the argument were valid. Hence, all the stuff about quantum gravity and microtubules.

But let's start by summarizing, in a few sentences, the Gödel argument itself for why human thought can't be algorithmic. How about this: the First Incompleteness Theorem tells us that no computer, working within a fixed formal system F such as Zermelo–Fraenkel set theory, can prove the sentence

$G(F) =$ "This sentence cannot be proved in F."

But we humans can just "see" the truth of $G(F)$ – since if $G(F)$ were false, then it would be provable, which is absurd! Therefore, the human mind can do something that no present-day computer can do. Therefore, consciousness can't be reducible to computation.

Alright, what do people think of this argument?

Yeah, there are two rather immediate issues.

- Why does the computer have to work within a fixed formal system F?
- *Can* humans "see" the truth of $G(F)$?

Actually, the response I prefer encapsulates both of the above responses as "limiting cases." Recall from Chapter 3 that, by the *Second* Incompleteness Theorem, $G(F)$ is equivalent to $Con(F)$: the statement that F is consistent. Furthermore, this equivalence can be proved in F itself for any reasonable F. This has two important implications.

First, it means that, when Penrose claims that humans can "see" the truth of $G(F)$, really he's just claiming that humans can

[3] J. Lucas, Minds, Machines, and Gödel, *Philosophy* **XXXVI**: (1961), 112–127. http://users.ox.ac.uk/~jrlucas/Godel/mmg.html

see the consistency of F! When you put it that way, the problems become more apparent: *how* can humans see the consistency of F? Exactly which Fs are we talking about: Peano Arithmetic? ZF? ZFC? ZFC with large cardinal axioms? Can *all* humans see the consistency of *all* these systems, or do you have to be a Penrose-caliber mathematician to see the consistency of the stronger ones? What about the systems that people thought were consistent, but that turned out not to be? And even if you *did* see the consistency of (say) ZF, how would you convince someone else that you'd seen it? How would the other person know you weren't just pretending?

(Models of Zermelo–Fraenkel set theory are like those 3D dot pictures: sometimes you really have to squint . . .)

The second implication is that, if we grant a computer the same freedom that Penrose effectively grants to humans – namely, the freedom to *assume* the consistency of the underlying formal system – then the computer *can* prove G(F).

So the question boils down to this: can the human mind somehow peer into the Platonic heavens, in order to directly perceive (let's say) the consistency of ZF set theory? If the answer is no – if we can only approach mathematical truth with the same unreliable, savannah-optimized tools that we use for doing the laundry, ordering Chinese takeout, etc. – then it seems we ought to grant computers the same liberty of being fallible. But in that case, the claimed distinction between humans and machines would seem to evaporate.

Perhaps Turing himself said it best:[4] "If we want a machine to be intelligent, it can't also be infallible. There are theorems that say almost exactly that."

In my opinion, then, Penrose doesn't need to be talking about Gödel's theorem at all. The Gödel argument turns out to be just a mathematical restatement of a much older argument against reductionism: "sure a computer could *say* it perceives G(F), but it'd just be

---

[4] A. M. Turing, Computing machinery and intelligence, *Mind* **59** (1950), 433–460. http://www.loebner.net/Prizef/TuringArticle.html

shuffling symbols around! When *I* say I perceive G(F), I really *mean* it! There's something it *feels* like to be me!"

The obvious response is equally old: "what makes you sure that it doesn't feel like anything to be a computer?"

OPENING THE BLACK BOX

Alright, look: Roger Penrose is one of the greatest mathematical physicists on Earth. Is it possible that we've misconstrued his thinking?

To my mind, the most plausible versions of Penrose's argument are the ones based on an "asymmetry of understanding": namely, that, while we know the internal workings of a computer, we *don't* yet know the internal workings of the brain.

How can one exploit this asymmetry? Well, given any known Turing machine M, it's certainly possible to construct a sentence that stumps M:

S(M) = "Machine M will never output this sentence."

There are two cases: either M outputs S(M), in which case it utters a falsehood, or else M *doesn't* output S(M), in which case there's a mathematical truth to which it can never assent.

The obvious response is, why can't we play the same game with humans?

"Roger Penrose will never output this sentence."

Well, conceivably there's an answer: because we can formalize what it means for M to output something, by examining its inner workings. Indeed, "M" is really just shorthand for the appropriate Turing machine state diagram. But can we formalize what it means for *Penrose* to output something? The answer depends on what we believe about the internal workings of the brain, or more precisely, Penrose's brain! And this leads to Penrose's view of the brain as "noncomputational."

A common misconception is that Penrose thinks the brain is a quantum computer. In reality, a quantum computer would be *much* weaker than he wants! As we saw before, quantum computers don't even seem able to solve **NP**-complete problems in polynomial time. Penrose, by contrast, wants the brain to solve *uncomputable* problems, by exploiting hypothetical collapse effects from a yet-to-be-discovered quantum theory of gravity.

I once asked Penrose: why not go further, and conjecture that the brain can solve problems that are uncomputable even *given* an oracle for the halting problem, or an oracle for the halting problem for Turing machines with an oracle for the halting problem, etc.? His response was that yes, he'd conjecture that as well.

My own view has always been that, if Penrose really wants to speculate about the impossibility of simulating the brain on a computer, then he ought to talk not about computability but about *complexity*. The reason is simply that, in principle, we can always simulate a person by building a huge lookup table, which encodes the person's responses to every question that could ever be asked within (say) a million years. If we liked, we could also have the table encode the person's voice, gestures, facial expressions, etc. Clearly such a table will be finite. So there's always *some* computational simulation of a human being – the only question is whether or not it's an efficient one!

You might object that, if people could live for an infinite or even just an arbitrarily long time, then the lookup table *wouldn't* be finite. This is true but irrelevant. The fact is, people regularly *do* decide that other people have minds after interacting with them for just a few minutes! (Indeed, maybe just a few minutes of email or instant messaging.) So unless you want to retreat into Cartesian skepticism about everyone you've ever met on Facebook, Gmail chat, etc., there *must* be a relatively small integer n such that, by exchanging at most n bits, you can be reasonably sure that someone else has a mind.

In *Shadows of the Mind* (the "sequel" to *The Emperor's New Mind*), Penrose concedes that a human mathematician could always be simulated by a computer with a huge lookup table. He then argues that such a lookup table wouldn't constitute a "proper" simulation, since (for example) there'd be no reason to believe that any given statement in the table was true rather than false. The trouble with this argument is that it explicitly retreats from what one might have thought was Penrose's central claim: namely, that a machine can't even *simulate* human intelligence, let alone exhibit it!

In *Shadows*, Penrose offers the following classification of views on consciousness.

A. Consciousness is reducible to computation (the view of strong-AI proponents).
B. Sure, consciousness can be *simulated* by a computer, but the simulation couldn't produce "real understanding" (John Searle's view).
C. Consciousness can't even be simulated by computer, but nevertheless has a scientific explanation (Penrose's own view, according to *Shadows*).
D. Consciousness doesn't have a scientific explanation at all (the view of 99% of everyone who ever lived).

Now it seems to me that, in dismissing the lookup table as not a "real" simulation, Penrose is retreating from view C to view B. For as soon as we say that passing the Turing Test isn't good enough – that one needs to "pry open the box" and examine a machine's internal workings to know whether it thinks or not – what could possibly be the content of view C that would distinguish it from view B?

Again, though, I want to bend over backward to see if I can figure out what Penrose *might* be saying.

In science, you can always cook up a theory to "explain" the data you've seen so far: just list all the data you've got, and call that your "theory"! The obvious problem here is *overfitting*. Since your theory doesn't achieve any compression of the original data – i.e., since it takes as many bits to write down your theory as to write down the data itself – there's no reason to expect your theory to predict *future* data. In other words, your theory is worthless.

So, when Penrose says the lookup table isn't a "real" simulation, perhaps what he means is this. *Of course*, one could write a computer program to converse like Disraeli or Churchill, by simply storing every possible quip and counterquip. But that's the sort of overfitting up with which we must not put! The relevant question is not whether we can simulate Sir Winston by *any* computer program. Rather, it's whether we can simulate him by a program that can be *written down inside the observable universe* – one that, in particular, is dramatically shorter than a list of all possible conversations with him.

Now, here's the point I keep coming back to: if *this* is what Penrose means, then he's left the world of Gödel and Turing far behind, and entered *my* stomping grounds – the Kingdom of Computational Complexity. How does Penrose, or anyone else, *know* that there's no small Boolean circuit to simulate Winston Churchill? Presumably, we wouldn't be able to prove such a thing, even supposing (for the sake of argument) that we knew what a Churchill simulator *meant*! All ye who would claim the intractability of finite problems: that way lieth the **P** versus **NP** beast, from whose 2^n jaws no mortal hath yet escaped.[5]

AT RISK OF STATING THE OBVIOUS

Even if we supposed the brain *was* solving a hard computational problem, it's not clear why that would bring us any closer to understanding consciousness. If it doesn't feel like anything to be a Turing machine, then why does it feel like something to be a Turing machine with an oracle for the halting problem?

ALL ABOARD THE HOLISTIC QUANTUM GRAVY TRAIN

Let's set aside the specifics of Penrose's ideas, and ask a more general question. Should quantum mechanics have *any* effect on how we think about the brain?

---

[5] Again, for more, see Scott Aaronson, "Why Philosophers Should Care About Computational Complexity," in *Computability: Turing, Gödel, Church, and Beyond* (MIT Press, 2013; edited by Oron Shagrir), http://www.scottaaronson.com/papers/philos.pdf

The temptation is certainly a natural one: consciousness is mysterious, quantum mechanics is also mysterious; therefore, they *must* be related somehow! Well, maybe there's more to it than that, since the *source* of the mysteriousness seems the same in both cases: namely, how do we reconcile a third-person description of the world with a first-person experience of it?

When people try to make the question more concrete, they often end up asking "is the brain a quantum computer?" Well, it *might* be, but I can think of at least four good arguments against this possibility.

1. The problems for which quantum computers are believed to offer dramatic speedups – factoring integers, solving Pell's equation, simulating quark–gluon plasmas, approximating the Jones polynomial, etc. – just don't seem like the sorts of things that would have increased Oog the Caveman's reproductive success relative to his fellow cavemen.

2. Even if humans could benefit from quantum computing speedups, I don't see any evidence that they're actually doing so. It's said that Gauss could immediately factor large integers in his head – but if so, that only proves that *Gauss's* brain was a quantum computer, not that anyone else's is!

3. The brain is a hot, wet environment, and it's hard to understand how long-range coherence could be maintained there.[6] With today's understanding of quantum error correction, this is no longer a knock-down argument, but it's still an extremely strong one.

4. As I mentioned earlier, even if we suppose the brain *is* a quantum computer, it doesn't seem to get us anywhere in explaining consciousness, which is the usual problem that these sorts of speculation are invoked to solve!

Alright, look. So as not to come across as a total curmudgeon – for what could *possibly* be further from my personality? – let me at least tell you what sort of direction I *would* pursue if I *were* a quantum mystic.

---

[6] For more, see (for example) Max Tegmark, "The importance of quantum decoherence in brain processes," *Physical Review E*, 61:4194–4206, 1999. http://arxiv.org/abs/quant-ph/9907009

Near the beginning of *Emperor's New Mind*, Penrose brings up one of my all-time favorite thought experiments: the teleportation machine. This is a machine that whisks you around the galaxy at the speed of light, by simply scanning your whole body, encoding all the cellular structures as pure information, and then transmitting the information as radio waves. When the information arrives at its destination, nanobots (of the sort we'll have in a few decades, according to Ray Kurzweil *et al.*) use the information to reconstitute your physical body down to the smallest detail.

Oh, I forgot to mention: since obviously we don't want *two* copies of you running around, the original is destroyed by a quick, painless gunshot to the head. So, fellow scientific reductionists: which one of you wants to be the first to travel to Mars this way?

What, you feel squeamish about it? Are you going to tell me you're somehow attached to the *particular atoms* that currently reside in your brain? As I'm sure you're aware, those atoms are replaced every few weeks anyway. So it *can't* be the atoms themselves that make you you; it has to be the patterns of information they encode. And as long as the information is safely on its way to Mars, who cares about the original meat hard drive?

So, soul or bullet: take your pick!

Quantum mechanics *does* offer a third way out of this dilemma, one that wouldn't make sense in classical physics.

Suppose some of the information that made you you was actually *quantum* information. Then, even if you were a thoroughgoing materialist, you could still have an excellent reason not to use the teleportation machine: because, as a consequence of the No-Cloning Theorem, *no such machine could possibly work as claimed.*

This is not to say that you couldn't be teleported around at the speed of light. But the teleportation process would have to be very different from the one above: it could *not* involve copying you and then killing the original copy. Either you could be sent as quantum information, or else – if that wasn't practical – you could use the famous quantum teleportation protocol,[7] which sends only classical information, but also requires prior entanglement between the sender and the receiver. In either case, the original copy of you would disappear *unavoidably, as part of the teleportation process itself.* Philosophically, it would be just like flying from Newark to LAX: you wouldn't face any profound metaphysical dilemma about "whether to destroy the copy of you still at Newark."

Of course, this neat solution can only work if the brain stores quantum information. But crucially, in this case, we *don't* have to imagine that the brain is a quantum computer, or that it maintains entanglement across different neurons, or anything harebrained like that. As in quantum key distribution, all we need are individual coherent qubits.

Now, you might argue that in a hot, wet, decoherent place like the brain, not even a single qubit would survive for very long. And from what little I know of neuroscience, I'd tend to agree. In particular, it *does* seem that long-term memories are encoded as synaptic strengths, and that these strengths are purely classical information that a nanobot could in principle scan and duplicate without any damage to the original brain. On the other hand, what about (say) whether you're going to wiggle your left finger or your right finger one minute from now? Is *that* decision determined in part by quantum events?

Well, whatever else you might think about such a hypothesis, it's clear what it would take to falsify it. You'd simply have to build a machine that scanned a person's brain, and reliably predicted which finger that person would wiggle one minute from now. Today, as I'll discuss in Chapter 19, these are fMRI experiments that have made a start on this sort of prediction, but only a few seconds in advance and only somewhat better than chance.

[7] http://researcher.watson.ibm.com/researcher/view_project.php?id=2862

12 Decoherence and hidden variables

Why have so many great thinkers found quantum mechanics so hard to swallow? To hear some people tell it, the whole source of the trouble is that "God plays dice with the universe" – that, whereas classical mechanics could in principle predict the fall of every sparrow, quantum mechanics gives you only statistical predictions.

Well, you know what? Whoop-de-doo! If indeterminism were the only mystery about quantum mechanics, *quantum mechanics wouldn't be mysterious at all*. We could imagine, if we liked, that the universe *did* have a definite state at any time, but that some fundamental principle (besides the obvious practical difficulties) kept us from knowing the whole state. This wouldn't require any serious revision of our worldview. Sure, "God would be throwing dice," but in such a benign way that not even Einstein could have any real beef with it.

The real trouble in quantum mechanics is not that the future trajectory of a particle is indeterministic – it's that the *past* trajectory is also indeterministic! Or more accurately, the very notion of a "trajectory" is undefined, since until you measure, there's just an evolving wavefunction. And crucially, because of the defining feature of quantum mechanics – *interference* between positive and negative amplitudes – this wavefunction can't be seen as merely a product of our ignorance, in the same way that a probability distribution can.

Now, I want to tell you about *decoherence* and *hidden-variable theories*, which are two kinds of story that people tell themselves to feel better about these difficulties.

The hardheaded physicist will of course ask: given that quantum mechanics *works*, why should we waste our time trying to feel better about it? Look, if you teach an introductory course on quantum mechanics, and the students don't have nightmares for weeks, tear their hair out, wander around with bloodshot eyes, etc., then you probably didn't get the point across. So, rather than deny this aspect of quantum mechanics – rather than cede the field to the hucksters, the Deepak Chopras and *What the Bleep Do We Know?* people – shouldn't we map it out ourselves, even sell tickets to the tourists? I mean, if you're going to leap into the abyss, better you should go with someone who's already been there and back.

INTO THE ABYSS

Alright, so consider the following thought experiment. Let $|R\rangle$ be a state of all the particles in your brain, which corresponds to you looking at a red dot. Let $|B\rangle$ be a state that corresponds to you looking at a *blue* dot. Now imagine that, in the far future, it's possible to place your brain into a coherent superposition of these two states:

$$\frac{3}{5}|R\rangle + \frac{4}{5}|B\rangle.$$

At least to a believer in the Many-Worlds Interpretation, this experiment should be dull as dirt. We've got two parallel universes, one where you see a red dot and the other where you see a blue dot. According to quantum mechanics, you'll find yourself in the first universe with probability $|3/5|^2 = 9/25$, and in the second universe with probability $|4/5|^2 = 16/25$. What's the problem?

Well, now imagine that we apply some unitary operation to your brain, which changes its state to

$$\frac{4}{5}|R\rangle + \frac{3}{5}|B\rangle.$$

Still a cakewalk! Now you see the red dot with probability 16/25 and the blue dot with probability 9/25.

Aha! But **conditioned on seeing the red dot at the earlier time, what's the probability that you'll see the blue dot at the later time?**

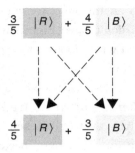

In ordinary quantum mechanics, this is a meaningless question! Quantum mechanics gives you the probability of getting a certain outcome if you make a measurement at a certain time, period. It doesn't give you *multiple-time* or *transition* probabilities – that is, the probability of an electron being found at point y at time $t + 1$, given that, had you measured the electron at time t (which you didn't), it "would have" been at point x. In the usual view, if you didn't actually measure the electron at time t, then it wasn't *anywhere* at time t: it was just in superposition. And if you *did* measure it at time t, then of course that would be a completely different experiment!

But why should we care about multiple-time probabilities? For me, it has to do with the reliability of memory. The issue is this: does the "past" have any objective meaning? Even if we don't know all the details, is there necessarily some fact-of-the-matter about what happened in history, about which trajectory the world followed to reach its present state? Or does the past only "exist" insofar as it's reflected in memories and records in the present?

The latter view is certainly the more natural one in quantum mechanics. But, as John Bell pointed out,[1] if we take it seriously, then it would seem difficult to do science! For what could it mean to make a *prediction* if there's no logical connection between past and future states – if by the time you finish reading this sentence, you might as well find yourself deep in the Amazon rainforest, with

---

[1] See John Bell, *Speakable and Unspeakable in Quantum Mechanics: Collected Papers on Quantum Philosophy* (second edition), Cambridge University Press, 2004.

all the memories of your trip there conveniently inserted, and all the memories of reading a quantum computing book conveniently erased?

Still here? Good!

Look, we all have fun ridiculing the creationists who think the world sprang into existence on October 23, 4004 BC at 9 a.m. (presumably Babylonian time), with the fossils already in the ground, light from distant stars heading toward us, etc. But if we accept the usual picture of quantum mechanics, then in a certain sense the situation is far worse: the world (as you experience it) might as well not have existed 10^{-43} seconds ago!

STORY 1. DECOHERENCE

The standard response to these difficulties appeals to a powerful idea called *decoherence*. Decoherence tries to explain why we don't notice "quantum weirdness" in everyday life – why the world of our experience is a more-or-less classical world. From the standpoint of decoherence, sure there might not be any objective fact about which slit an electron went through, but there *is* an objective fact about what you ate for breakfast this morning: the two situations are not the same!

The basic idea is that, as soon as the information encoded in a quantum state "leaks out" into the external world, that state will look locally like a classical state. In other words, as far as a local observer is concerned, there's no difference between a classical bit and a qubit that's become hopelessly entangled with the rest of the universe.

So, for example, suppose we have a qubit in the state

$$\frac{|0\rangle + |1\rangle}{\sqrt{2}}.$$

And suppose this qubit becomes *entangled* with a second qubit, to form the following joint state:

$$\frac{|00\rangle + |11\rangle}{\sqrt{2}}.$$

If we now ignore the second qubit and look only at the first qubit, the first qubit will be in what physicists call the *maximally mixed state*:

$$\rho = \begin{pmatrix} \frac{1}{2} & 0 \\ 0 & \frac{1}{2} \end{pmatrix}.$$

(Other people just call it a classical random bit.) In other words, no matter what measurement you make on the first qubit, you'll just get a random outcome. You're never going to see interference between the $|00\rangle$ and $|11\rangle$ "branches" of the wavefunction. Why? Because, according to quantum mechanics, two branches will only interfere if they become identical in *all* respects. But there's simply no way, by changing the first qubit alone, to make $|00\rangle$ identical to $|11\rangle$. The second qubit will always give away our would-be lovers' differing origins.

To see an interference pattern, you'd have to perform a *joint* measurement on the two qubits together. But what if the second qubit were a stray photon, which happened to pass through your experiment on its way to the Andromeda galaxy? Indeed, when you consider all the junk that might be entangling itself with your delicate experiment – air molecules, cosmic rays, geothermal radiation,... well, whatever, I'm not an experimentalist – it's as if the entire rest of the universe is constantly trying to "measure" your quantum state, and thereby force it to become classical! Sure, even if your quantum state *does* collapse (i.e., become entangled with the rest of the world), in principle you can still get the state back – by gathering together all the particles in the universe that your state has become entangled with, and then reversing everything that's happened since the moment of collapse. That would be sort of like Pamela Anderson trying to regain her privacy, by tracking down every computer on Earth that might contain photos of her!

If we accept this picture, then it explains two things.

1. Most obviously, it explains why in everyday life, we don't usually see objects quantumly interfering with their parallel-universe doppelgängers. (Unless we happen to live in a dark room with two

slits in the wall . . .) Basically, it's the same reason why we don't see eggs unscramble themselves.

2. As the flip side, the picture also explains why it's so hard to build quantum computers: because not only are we trying to keep errors from leaking into our computer, we're trying to keep the *computer* from leaking into the rest of the world! We're fighting against decoherence, one of the most pervasive processes in the universe. Indeed, it's precisely because decoherence is so powerful that the quantum fault-tolerance theorem[2] came as a shock to many physicists. (The fault-tolerance theorem says roughly that, if the rate of decoherence per qubit per gate operation is below a constant threshold, then it's possible in principle to correct errors faster than they occur, and thereby perform an arbitrarily long quantum computation.)

So, what about the thought experiment from before – the one where we place your brain into coherent superpositions of seeing a blue dot and seeing a red dot, and then ask about the probability that you see the dot change color? From a decoherence perspective, the resolution is that the thought experiment is completely ridiculous, since brains are big, bulky things that constantly leak electrical signals, and therefore, any quantum superposition of two neural firing patterns would collapse (i.e., become entangled with the rest of the universe) in a matter of nanoseconds.

Fine, a skeptic might retort. But what if in the far future, it were possible to upload your entire brain into a quantum computer, and then put the *quantum computer* into a superposition of seeing a blue dot and seeing a red dot? Huh? *Then* what's the probability that "you" (i.e., the quantum computer) would see the dot change color?

When I put this question to John Preskill years ago, he said that *decoherence itself* – in other words, an approximately classical universe – seemed to him like an important component of subjective experience as we understand it. And therefore, if you artificially removed decoherence, then it might no longer make sense to ask the same questions about subjective experience that we're used to asking.

---

[2] See http://www.arxiv.org/abs/quant-ph/9906129

I'm guessing that this would be a relatively popular response, among those physicists who are philosophical enough to say anything at all.

DECOHERENCE AND THE SECOND LAW

We *are* going to get to hidden variables. But first, I want to say one more thing about decoherence.

When I was talking before about the fragility of quantum states – how they're so easy to destroy, so hard to put back together – you might have been struck by a parallel with the Second Law of Thermodynamics. Obviously, that's just a coincidence, right? Duhhh, no. The way people think about it today, *decoherence is just one more manifestation of the Second Law.*

Let's see how this works. Given a probability distribution $D = (p_1, \ldots, p_N)$, there's this fundamental measure of the "amount of randomness" in D called the *entropy* of D, and denoted $H(D)$. Here's the formula for $H(D)$, if you've never seen it before:

$$H(D) = -\sum_i p_i \log p_i.$$

(Being a computer scientist, I'll stipulate that all the logarithms are base 2. Also, $p_i \log p_i$ is defined to be zero if $p_i = 0$.) Intuitively, $H(D)$ measures the minimum number of random bits that you'd need to generate a single sample from D – on average, if you were generating *lots* of independent samples. It also measures the minimum number of bits that you'd need to send your friend, if you wanted to tell her which element from D was chosen – again on average, if you were telling her about lots of independent draws. To illustrate, a distribution with *no* randomness has entropy zero, while an equal distribution over N possible outcomes has entropy $\log_2 N$ (thus, the entropy of a single fair coin flip is $\log_2 2 = 1$). Entropy was the central concept in Claude Shannon's *information theory* (which he announced, in nearly complete form, in a single paper in 1948).[3] But the roots of entropy

[3] C. E. Shannon. A Mathematical Theory of Communication, *Bell System Technical Journal* 27:3 (1948), 379–423. http://www.alcatel-lucent.com/bstj/vol27-1948/articles/bstj27-3-379.pdf

go all the way back to Boltzmann and those other thermodynamics dudes in the late 1800s.

Anyway, given a quantum mixed state ρ, the *von Neumann entropy* of ρ is defined to be the minimum, over all unitary transformations U, of the entropy of the probability distribution that results from measuring $U\rho U^{-1}$ in the standard basis. To illustrate, every pure state has an entropy of zero, whereas the one-qubit maximally mixed state has an entropy of unity.

Now, if we assume that the universe is always in a pure state, then the "entropy of the universe" starts out zero, and remains zero for all time! On the other hand, the entropy of the universe isn't really what we care about – we care about the entropy of this or that region. And we saw before that, as previously separate physical systems interact with each other, they tend to evolve from pure states into mixed states – and therefore their entropy goes up. In the decoherence perspective, this is simply the Second Law at work.

Another way to understand the relationship between decoherence and the Second Law is by taking a "God's-eye view" of the entire multiverse. Generically speaking, the different branches of the wavefunction could be constantly interfering with each other, splitting and merging in a tangled bush:

Universes

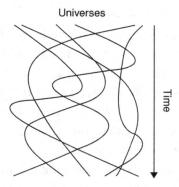

Time

What decoherence theory says is that in the real world, the branches look more like a nicely pruned tree:

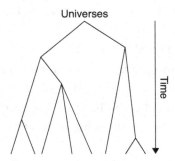

In principle, any two branches of this tree could collide with each other, thereby leading to "macroscopic interference effects," as in my story with the blue and red dots. But in practice, this is astronomically unlikely – since to collide, two branches would have to become identical in every respect.

Notice that if we accept this tree picture of the multiverse, then it immediately gives us a way to define the "arrow of time" – that is, to state noncircularly what the difference is between the future and the past. Namely, we can say that the past is the direction toward the root of the "multiverse tree," and the future is the direction toward the leaves. According to the decoherence picture, this is actually *equivalent* to saying that the future is the direction where entropy increases, and it's *also* equivalent to saying that the past is the direction we remember while the future is the direction we don't.

The tree picture also lets us answer the conundrums from before about the reliability of memory. According to the tree picture, even though in principle we need not have a unique "past," in practice we usually do: namely, the unique path that leads from the root of the multiverse tree to our current state. Likewise, even though in principle quantum mechanics need not provide multiple-time probabilities – that is, probabilities for what we're going to experience tomorrow, *conditioned* on what we're experiencing today – in practice such probabilities usually make perfect sense, for the same reason they make sense in the classical world. That is, when it comes to transitions between subjective experiences, in practice we're dealing not with unitary matrices but with *stochastic* matrices.

At this point, the sharp-eyed reader might notice a problem: won't the branches have to collide eventually, when the tree "runs out of room to expand"? The answer is yes. First, if the Hilbert space is finite dimensional, then obviously the parallel universes can only branch off a finite number times before they start bumping into one another. But even in an infinite-dimensional Hilbert space, we need to think of each universe as having some finite "width" (think of Gaussian wavepackets for example), so again we can only have a finite number of splittings.

The answer of decoherence theory is that yes, *eventually* the branches of the multiverse will start interfering with each other – just like eventually the universe will reach thermal equilibrium. But by that time, we'll presumably all be dead.

Incidentally, the fact that our universe is expanding exponentially – that there's this vacuum energy pushing the galaxies apart – seems like it might play an important role in "thinning out the multiverse tree," and thereby buying us more time until the branches start interfering with each other. This is something I'd like to understand better.

Oh, yes: I should also mention the "deep" question that I'm glossing over entirely here. Namely, *why did the universe start out in such a low-entropy, unentangled state to begin with?* Of course, one can try to give an anthropic answer to that question, but is there another answer?

STORY 2. HIDDEN VARIABLES

Despite how tidy the decoherence story seems, there are some people for whom it remains unsatisfying. One reason is that the decoherence story had to bring in a lot of assumptions seemingly extraneous to quantum mechanics itself: about the behavior of typical physical systems, the classicality of the brain, and even the nature of subjective experience. A second reason is that the decoherence story never *did* answer our question about the probability you see the dot change color – instead the story simply tried to convince us the question was meaningless.

So if the decoherence story doesn't make you sleep easier, then what else is on offer at the quantum bazaar? Well, now it's the hidden-variable theorists' turn to hawk their wares. (Most of the rest of this chapter will follow my paper "Quantum computing and hidden variables."[4])

The idea of hidden-variable theories is simple. If we think of quantum mechanics as describing this vast roiling ocean of parallel universes, constantly branching off, merging, and cancelling each other out, then we're now going to stick a little *boat* in that ocean. We'll think of the boat's position as representing the "real," "actual" state of the universe at a given point in time, and the ocean as just a "field of potentialities" whose role is to buffet the boat around. For historical reasons, the boat's position is called a *hidden variable* – even though, in some sense, it's the only part of this setup that's *not* hidden! Now, our goal will be to make up an evolution rule for the boat, such that, at any time, *the probability distribution over possible boat positions is exactly the $|\psi|^2$ distribution predicted by standard quantum mechanics.*

By construction, then, hidden-variable theories are experimentally indistinguishable from standard quantum mechanics. So presumably there can be no question of whether they're "true" or "false" – the only question is whether they're good or bad stories.

You might say, why should we worry about these unfalsifiable goblins hiding in quantum mechanics' closet? Well, I'll give you four reasons.

1. For me, part of what it means to understand quantum mechanics is to explore the space of possible stories that can be told about it. If we don't do so, then we risk making fools of ourselves by telling people that a certain sort of story can't be told when in fact it can, or vice versa. (There's plenty of historical precedent for this.)

---

[4] In *Physical Review A* 71:032325, 2005. http://www.scottaaronson.com/papers/qchvpra.pdf

2. As we'll see, hidden-variable theories lead to all sorts of meaty, non-trivial math problems, some of which are still open. And in the end, isn't that reason enough to study anything?

3. Thinking about hidden variables seems scientifically fruitful: it led Einstein, Podolsky, and Rosen to the EPR experiment, Bell to Bell's Inequality, Kochen and Specker to the Kochen–Specker Theorem, and me to the collision lower bound (to be discussed in Chapter 13).[5]

4. Hidden-variable theories will give me a perfect vehicle for discussing other issues in quantum foundations – like nonlocality, contextuality, and the role of time. In other words, you get *lots* of goblins for the price of one!

From my perspective, a hidden-variable theory is simply a rule for converting a unitary transformation into a classical probabilistic transformation. In other words, it's a function that takes as input an N-by-N unitary matrix $U = (u_{ij})$ together with a quantum state

$$|\psi\rangle = \sum_{i=1}^{N} \alpha_i |i\rangle,$$

and that produces as output an N-by-N stochastic matrix $S = (s_{ij})$. (Recall that a stochastic matrix is just a nonnegative matrix where every column sums to unity.) Given as input the probability vector obtained from measuring $|\psi\rangle$ in the standard basis, this S should produce as output the probability vector obtained from measuring $U|\psi\rangle$ in the standard basis. In other words, if

$$\begin{pmatrix} u_{11} & \cdots & u_{11} \\ \vdots & \ddots & \vdots \\ u_{11} & \cdots & u_{11} \end{pmatrix} \begin{pmatrix} \alpha_1 \\ \vdots \\ \alpha_N \end{pmatrix} = \begin{pmatrix} \beta_1 \\ \vdots \\ \beta_N \end{pmatrix},$$

then we must have

$$\begin{pmatrix} s_{11} & \cdots & s_{1N} \\ \vdots & \ddots & \vdots \\ s_{N1} & \cdots & s_{NN} \end{pmatrix} \begin{pmatrix} |\alpha_1|^2 \\ \vdots \\ |\alpha_N|^2 \end{pmatrix} = \begin{pmatrix} |\beta_1|^2 \\ \vdots \\ |\beta_N|^2 \end{pmatrix}.$$

[5] See http://www.scottaaronson.com/papers/collision.pdf

This is what it means for a hidden-variable theory to reproduce the predictions of quantum mechanics: it means that, whatever story we want to tell about correlations between boat positions at different times, certainly the *marginal* distribution over boat positions at any *individual* time had better be the usual quantum-mechanical one.

OK, obvious question: given a unitary matrix U and a state $|\psi\rangle$, does a stochastic matrix satisfying the above condition necessarily *exist*?

Sure it does! For we can always take the *product transformation*

$$S_{\text{prod}} = \begin{pmatrix} |\beta_1|^2 & \cdots & |\beta_1|^2 \\ \vdots & \ddots & \vdots \\ |\beta_N|^2 & \cdots & |\beta_N|^2 \end{pmatrix},$$

which just "picks the boat up and puts it back down at random," completely destroying any correlation between the initial and final positions.

NO-GO THEOREMS GALORE

So the question is not whether we can find a stochastic transformation $S(|\psi\rangle, U)$ that maps the initial distribution to the final one. Certainly we can! Rather, the question is whether we can find a stochastic transformation satisfying "nice" properties. But which "nice" properties might we want? I'm now going to suggest four possibilities – and then show that, alas, *not one of them can be satisfied*. The point of going through this exercise is that, along the way, we're going to learn an enormous amount about how quantum mechanics differs from classical probability theory. In particular, we'll learn about *Bell's Theorem*, the *Kochen–Specker Theorem*, and two other no-go theorems that as far as I know don't have names.

1. **Independence from the state:** Alright, so recall the problem at hand: we're given a unitary matrix U and quantum state $|\psi\rangle$, and want to cook up a stochastic matrix $S = S(|\psi\rangle, U)$ that maps the distribution

obtained by measuring $|\psi\rangle$ to the distribution obtained by measuring $U|\psi\rangle$.

The first property we might want is that S should depend only on the unitary U, and *not* on the state $|\psi\rangle$. However, this is easily seen to be impossible. For if we let

$$U = \begin{pmatrix} \dfrac{1}{\sqrt{2}} & -\dfrac{1}{\sqrt{2}} \\ \dfrac{1}{\sqrt{2}} & \dfrac{1}{\sqrt{2}} \end{pmatrix},$$

then

$$U \begin{pmatrix} \dfrac{1}{\sqrt{2}} \\ \dfrac{1}{\sqrt{2}} \end{pmatrix} = \begin{pmatrix} 0 \\ 1 \end{pmatrix}$$

implies

$$S = \begin{pmatrix} 0 & 0 \\ 1 & 1 \end{pmatrix},$$

whereas

$$U \begin{pmatrix} \dfrac{1}{\sqrt{2}} \\ -\dfrac{1}{\sqrt{2}} \end{pmatrix} = \begin{pmatrix} 1 \\ 0 \end{pmatrix}$$

implies

$$S = \begin{pmatrix} 1 & 1 \\ 0 & 0 \end{pmatrix}.$$

Therefore, S must be a function of U and $|\psi\rangle$ together.

2. **Invariance under time-slicings:** The second property we might want in our hidden-variable theory is *invariance under time-slicings*. This means that, if we perform two unitary transformations U and V in succession, we should get the same result if we apply the hidden-variable theory to VU as if we apply the theory to U and V separately

and then multiply the results. (Loosely speaking, the map from unitary to stochastic matrices should be "homomorphic.") Formally, what we want is that

$$S(|\psi\rangle, VU) = S(U|\psi\rangle, V)S(|\psi\rangle, U).$$

But again one can show that this is impossible – *except* in the "trivial" case that S is the product transformation S_{prod}, which destroys all correlations between the initial and final times.

To see this, observe that for all unitaries W and states $|\psi\rangle$, we can write W as a product $W = VU$, in such a way that $U|\psi\rangle$ equals a fixed basis state ($|1\rangle$, for example). Then applying U "erases" all the information about the hidden variable's initial value – so that if we later apply V, then the hidden variable's final value must be uncorrelated with its initial value. But this means that $S(|\psi\rangle, VU)$ equals $S_{\text{prod}}(|\psi\rangle, VU)$.

3. **Independence from the basis:** When I defined hidden-variable theories, some of you were probably wondering: why should we only care about measurement results in some *particular* basis, when we could've just as well picked any other basis? So, for example, if we're going to say that a particle has a "true, actual" location even before anyone measures that location, then shouldn't we say the same thing about the particle's momentum, and its spin, and its energy, and all the other observable properties of the particle? What singles out location as being more "real" than all the other properties?

Well, these are excellent questions! Alas, it turns out that *we can't assign definite values to all possible properties of a particle in any "consistent" way.* In other words, not only can we not define transition probabilities for all the particle's properties, we can't even handle all the properties simultaneously at any *individual* time!

This is the remarkable (if mathematically trivial) conclusion of the Kochen–Specker Theorem,[6] which was proved by Simon Kochen and Ernst Specker in 1967. Formally, the theorem says the following:

---

[6] See http://plato.stanford.edu/entries/kochen-specker/

suppose that for every orthonormal basis B in $\mathfrak{R}^3$, the universe wants to "precompute" what the outcome would be of making a measurement in that basis. In other words, the universe wants to pick one of the three vectors in B, designate that one as the "marked" vector, and return that vector later should anyone happen to measure in B. Naturally, the marked vectors ought to be "consistent" across different bases. That is, if two bases share a common vector, like so:

$$B_1 = \{|1\rangle, |2\rangle, |3\rangle\}$$

$$B_2 = \left\{|1\rangle, \frac{|2\rangle + |3\rangle}{\sqrt{2}}, \frac{|2\rangle - |3\rangle}{\sqrt{2}}\right\}$$

then the common vector should be the marked vector of one basis if and only if it's also the marked vector of the other.

Kochen and Specker prove that this is impossible. Indeed, they construct an explicit set of 117 bases (!) in $\mathfrak{R}^3$, such that marked vectors can't be chosen for those bases in any consistent way.

NerdNote: The constant 117 has since been improved to 31; see here[7] for example. Apparently, it's still an open problem whether that's optimal; the best lower bound I've seen mentioned is 18.

The upshot is that any hidden-variable theory will have to be what those in the business call *contextual*. That is, it will sometimes have to give you an answer that depends on which basis you measured in, with no pretense that the answer would've been the same had you measured in a different basis that also contained the same answer.

Exercise: Prove that the Kochen–Specker Theorem is false in two dimensions.

4. **Relativistic causality:** The final property we might want from a hidden-variable theory is adherence to the "spirit" of Einstein's special relativity. For our purposes, I'll define that to consist of two things.

[7] http://www.arxiv.org/abs/quant-ph/0304013

1. **Locality**. This means that, if we have a quantum state $|\psi_{AB}\rangle$ on two subsystems A and B, and we apply a unitary transformation U_A that acts only on the A system (i.e., is the identity on B), then the hidden-variable transformation $S(|\psi_{AB}\rangle, U_A)$ should also act only on the A system.

2. **Commutativity**. This means that, if we have a state $|\psi_{AB}\rangle$, and we apply a unitary transformation U_A to the A system only followed by another unitary transformation U_B to the B system only, then the resulting hidden-variable transformation should be the same as if we'd first applied U_B and then U_A. Formally, we want that

$$S(U_A|\psi_{AB}\rangle, U_B) \, S(|\psi_{AB}, U_A) = S(U_B|\psi_{AB}\rangle, U_A) \, S(|\psi_{AB}, U_B).$$

Now, you might've heard of a little thing called Bell's Inequality. As it turns out, Bell's Inequality doesn't *quite* rule out hidden-variable theories satisfying the two axioms above, but a slight strengthening of what Bell proved does the trick.

So what *is* Bell's Inequality? Well, if you look for an answer in almost any popular book or website, you'll find page after page about entangled photon sources, Stern–Gerlach apparatuses, etc., all of it helpfully illustrated with detailed experimental diagrams. This is necessary, of course, since if you took all the complications away, people *might* actually grasp the conceptual point!

However, since I'm not a member of the Physics Popularizers' Guild, I'm now going to break that profession's time-honored bylaws, and just tell you the conceptual point directly.

We've got two players, Alice and Bob, and they're playing the following game. Alice flips a fair coin; then, based on the result, she can either raise her hand or not. Bob flips another fair coin; then, based on the result, he can either raise *his* hand or not. What both players want is that **exactly one of them should raise their hand, if and only if both coins landed heads**. If that condition is satisfied, then they win the game; if it isn't then they lose. (This is a cooperative rather than competitive game.)

Now here's the catch: Alice and Bob are both in sealed rooms (possibly even on different planets), and *can't communicate with each other at all while the game is in progress.*

The question that interests us is: what is the maximum probability with which Alice and Bob can win the game?

Well, *certainly* they can win 75% of the time. Why?

Right: they can both just decide never to raise their hands, regardless of how the coins land! In that case, the only way they'll lose is if both of the coins land heads.

Exercise: Prove that this is optimal. In other words, *any* strategy of Alice and Bob will win at most 75% of the time.

Now for the punchline: suppose that Alice and Bob share the entangled state

$$|\Phi\rangle = \frac{|00\rangle + |11\rangle}{\sqrt{2}},$$

with Alice holding one half and Bob holding the other half. In that case, there exists a strategy[8] by which they can win the game with probability

$$\frac{2 + \sqrt{2}}{4} = 0.853\ldots.$$

To be clear, having the state $|\Phi\rangle$ does *not* let Alice and Bob send messages to each other faster than the speed of light – nothing does! What it lets them do is to win this particular game more than 75% of the time. Naïvely, we might have thought that would require Alice and Bob to "cheat" by sending each other messages, but that simply isn't true – they can also cheat by using entanglement!

So that was Bell's Inequality.

But what does this dumb little game have to do with hidden variables? Well, suppose we tried to model Alice's and Bob's measurements of the state $|\Phi\rangle$ using *two* hidden variables: one

---

[8] See http://www.cs.berkeley.edu/~vazirani/s07quantum/notes/lecture1.pdf

on Alice's side and the other on Bob's side. And, in keeping with relativistic causality, suppose we demanded that nothing that happened to Alice's hidden variable could affect Bob's hidden variable or vice versa. In that case, we'd predict that Alice and Bob could win the game at most 75% of the time. But this prediction would be wrong!

It follows that, if we want it to agree with quantum mechanics, then any hidden-variable theory has to allow "instantaneous communication" between any two points in the universe. Once again, this doesn't mean that quantum mechanics *itself* allows instantaneous communication (it doesn't), or that we can exploit hidden variables to send messages faster than light (we can't). It only means that, *if* we choose to describe quantum mechanics using hidden variables, then our *description* will have to involve instantaneous communication.

Exercise: Generalize Bell's argument to show that there's no hidden-variable theory satisfying the locality and commutativity axioms as given above.

So what we've learned, from Alice and Bob's coin-flipping game, is that any attempt to describe quantum mechanics with hidden variables will necessarily lead to tension with relativity. Again, none of this has any experimental consequences, since it's perfectly possible for hidden-variable theories to violate the "spirit" of relativity while still obeying the "letter." Indeed, hidden-variable fans like to argue that all we're doing is unearthing the repressed marital tensions between relativity and quantum mechanics themselves!

EXAMPLES OF HIDDEN-VARIABLE THEORIES
I know what you're thinking: after the pummeling we just gave them, the outlook for hidden-variable theories looks pretty bleak. But here's the amazing thing: *even in the teeth of four different no-go theorems, one can still construct interesting and mathematically nontrivial*

hidden-variable theories. I'd like to end this chapter by giving you three examples.

THE FLOW THEORY

Remember the goal of hidden-variable theories: we start out with a unitary matrix U and a state $|\psi\rangle$; from them we want to produce a stochastic matrix S that maps the initial distribution to the final distribution. Ideally, S should be derived from U in a "natural," "organic" way. So, for example, if the (i, j) entry of U is zero, then the (i, j) entry of S should also be zero. Likewise, making a small change to U or $|\psi\rangle$ should produce only a small change in S.

Now, it's not clear a priori that there even exists a hidden-variable theory satisfying the two requirements above. So what I want to do first is give you a simple, elegant theory that *does* satisfy those requirements.

The basic idea is to treat probability mass flowing through the multiverse just like oil flowing through pipes! We're going to imagine that initially we have $|\alpha_i|^2$ units of "oil" at each basis state $|i\rangle$, while by the end we want $|\beta_i|^2$ units of oil at each basis state $|i\rangle$. Here α_i and β_i are the initial and final amplitudes of $|i\rangle$, respectively. And we're also going to think of $|u_{ij}|$, the absolute value of the (i, j)th entry of the unitary matrix, as the capacity of an "oil pipe" leading from $|i\rangle$ to $|j\rangle$.

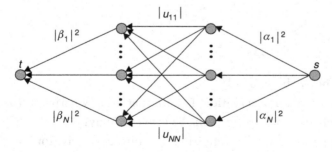

Then the first question is this: for any U and $|\psi\rangle$, *can* all of the 1 unit of oil be routed from s to t in the above network $G(U, |\psi\rangle)$, without exceeding the capacity of any of the pipes?

I proved[9] that the answer is yes. My proof uses a fundamental result from the 1960s called the *Max-Flow/Min-Cut Theorem*. Those of you who were/are computer science majors will vaguely remember this from your undergrad classes. For the rest of you, well, it's really worth seeing at least once in your life. (It's useful not *only* for the interpretation of quantum mechanics but also for stuff like internet routing!)

So what does the Max-Flow/Min-Cut Theorem say? Well, suppose we have a network of oil pipes like in the figure above, with a designated "source" called *s*, and a designated "sink" called *t*. Each pipe has a known "capacity," which is a nonnegative real number measuring how much oil can be sent through that pipe each second. Then the *max flow* is just the maximum amount of oil that can be sent from *s* to *t* every second, if we route the oil through the pipes in as clever a way as possible. Conversely, the *min cut* is the smallest real number *C* such that, by blowing up oil pipes whose total capacity is *C*, a terrorist could prevent *any* oil from being sent from *s* to *t*.

As an example, what's the max flow and min cut for the network below?

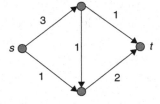

Right: they're both 3.

As a trivial observation, I claim that for any network, *the max flow can never be greater than the min cut*. Why?

Right: because by definition, the min cut is the total capacity of some "choke point" that all the oil has to pass through eventually! In other words, if blowing up pipes of total capacity *C* is enough to

---

[9] In http://www.scottaaronson.com/papers/qchvpra.pdf

cut the flow from s to t down to zero, then putting those same pipes back in can't increase the flow to more than C.

Now, the Max-Flow/Min-Cut Theorem says that the converse is also true: *for any network, the max flow and min cut are actually equal.*

Exercise (for those who've never seen it): Prove the Max-Flow/Min-Cut Theorem.

Exercise (hard): By using the Max-Flow/Min-Cut Theorem, prove that for any unitary U and any state $|\psi\rangle$ there exists a way to route all the probability mass from s to t in the network $G(U, |\psi\rangle)$ shown before.

So, we've now got our candidate hidden-variable theory! Namely: given U and $|\psi\rangle$, first find a "canonical" way to route all the probability mass from s to t in the network $G(U, |\psi\rangle)$. Then define the stochastic matrix S by $s_{ij} := p_{ij}/|\alpha_i|^2$, where p_{ij} is the amount of probability mass routed from $|i\rangle$ to $|j\rangle$. (For simplicity, I'll ignore what happens when $\alpha_i = 0$.)

By construction, this S maps the vector of the $|\alpha_i|^2$ to the vector of the $|\beta_i|^2$. It also has the nice property that, for all i, j, if $u_{ij} = 0$, then $s_{ij} = 0$ as well.

Why?

Right! Because if $u_{ij} = 0$, then no probability mass can get routed from $|i\rangle$ to $|j\rangle$.

Exercise (harder): Prove that it's possible to choose the "canonical" maximal flows in such a way that making a small change to U or $|\psi\rangle$ produces only a small change in the matrix (p_{ij}) of transition probabilities.

THE SCHRÖDINGER THEORY

So that was one cute example of a hidden-variable theory. I now want to show you an example that I think is even cuter. When I started thinking about hidden-variable theories, this was actually the first idea I came up with. Later I found out that Schrödinger had the same idea in a nearly forgotten 1931 paper.[10]

Specifically, Schrödinger's idea was to define transition probabilities in quantum mechanics by solving a system of coupled nonlinear equations. The trouble is that Schrödinger couldn't prove that his system *had* a solution (let alone a unique one); that had to wait for the work of Masao Nagasawa[11] in the 1980s. Luckily for me, I only cared about finite-dimensional quantum systems, where everything was much simpler, and where I could give a reasonably elementary proof that the equation system was solvable.

So what's the idea? Well, recall that, given a unitary matrix U, we want to "convert" it somehow into a stochastic matrix S that maps the initial distribution to the final one. This is basically equivalent to asking for a matrix P of *transition probabilities*: that is, a nonnegative matrix whose ith column sums to $|\alpha_i|^2$ and whose jth row sums to $|\beta_j|^2$. (This is just the requirement that the marginal probabilities should be the usual quantum-mechanical ones.)

Since we want to end up with a nonnegative matrix, a reasonable first step would be to replace every entry of U by its absolute value:

$$\begin{pmatrix} |u_{11}| & \cdots & |u_{1N}| \\ \vdots & \ddots & \vdots \\ |u_{N1}| & \cdots & |u_{NN}| \end{pmatrix}.$$

What next? Well, we want the ith column to sum to $|\alpha_i|^2$. So let's continue doing the crudest thing imaginable, and for every $1 \le i \le N$, just *normalize* the ith column to sum to $|\alpha_i|^2$!

Now, we also want the jth row to sum to $|\beta_j|^2$. How do we get that? Well, for every $1 \le j \le N$, we just normalize the jth row to sum to $|\beta_j|^2$.

[10] Erwin Schrödinger, "Über die Umkehrung der Naturgesetze," *Sitzungsber. Preuss. Akad. Wissen. Phys. Math. Kl.*, 1:144–153, 1931

[11] See, e.g., M. Nagasawa, *Schrödinger Equations and Diffusion Theory* (Basel: Birkhäuser, 1993).

Of course, after we normalize the rows, in general the ith column will no longer sum to $|\alpha_i|^2$. But that's no problem: we'll just normalize the columns again! Then we'll re-normalize the rows (which were messed up by normalizing the columns), then we'll re-normalize the columns (which were messed up by normalizing the rows), and so on *ad infinitum*.

Exercise (hard): Prove that this iterative process converges for any U and $|\psi\rangle$, and that the limit is a matrix $P = (p_{ij})$ of transition probabilities – that is, a nonnegative matrix whose ith column sums to $|\alpha_i|^2$ and whose jth row sums to $|\beta_j|^2$.

Open problem (if you get this, let me know): Prove that making a small change to U or $|\psi\rangle$ produces only a small change in the matrix $P = (p_{ij})$ of transition probabilities.

BOHMIAN MECHANICS

Some of you might be wondering why I haven't mentioned the most famous hidden-variable theory of all: Bohmian mechanics.[12] The answer is that, to discuss Bohmian mechanics, I'd have to bring in infinite-dimensional Hilbert spaces (blech!), particles with positions and momenta (double blech!), and other ideas that go against everything I stand for as a computer scientist.

Still, I should tell you a little about what Bohmian mechanics is and why it doesn't fit into my framework. In 1952, David Bohm proposed a *deterministic* hidden-variable theory: that is, a theory where not only do you get transition probabilities, but the probabilities are all either zero or unity! The way he did this was by taking as his hidden variable *the positions of particles in* $\Re^3$. He then stipulated that the probability mass for where the particles are should "flow" with the wavefunction, so that a region of configuration space with probability ε always gets mapped to another region with probability ε.

With one particle in one spatial dimension, it's easy to write down the (unique) differential equation for particle position that

---

[12] For an introduction, see for example http://plato.stanford.edu/entries/qm-bohm/

satisfies Bohm's probability constraint. Bohm showed how to generalize the equation to any number of particles in any number of dimensions.

To illustrate, here's what the Bohmian particle trajectories look like in the famous double-slit experiment:

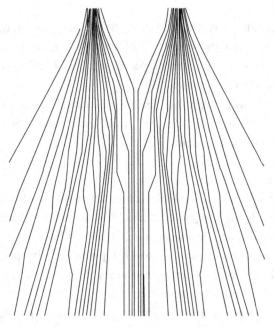

Again, the amazing thing about this theory is that it's *deterministic*: specify the "actual" positions of all the particles in the universe at any one time, and you've specified their "actual" positions at all earlier and later times. So, if you like, you can imagine that, at the moment of the Big Bang, God sprinkled particles across the universe according to the usual $|\psi|^2$ distribution; but after that He smashed His dice, and let the particles evolve deterministically forever after. And that assumption will lead you to exactly the same experimental predictions as the usual picture of quantum mechanics, the one where God's throwing dice up the wazoo.

The catch, from my point of view, is that this sort of determinism can *only* work in an infinite-dimensional Hilbert space, like the space of particle positions. I've almost never seen this observation discussed in print, but I can explain it in a couple of sentences.

Suppose we want a hidden-variable theory that's deterministic like Bohm's, but that works for quantum states in a *finite* number of dimensions. Then what happens if we apply a unitary transformation U that maps the state $|0\rangle$ to

$$\frac{|0\rangle + |1\rangle}{2}?$$

In this case, initially the hidden variable is $|0\rangle$ with certainty; afterward it's $|0\rangle$ with probability $\frac{1}{2}$ and $|1\rangle$ with probability $\frac{1}{2}$. In other words, applying U increases the entropy of the hidden variable from zero to unity. So to decide which way the hidden variable goes, clearly Nature needs to flip a coin!

A Bohmian would say that the reason determinism broke down here is that our wavefunction was "degenerate": that is, it didn't satisfy the continuity and differentiability requirements that are needed for Bohm's differential equation. But in a finite-dimensional Hilbert space, *every* wavefunction will be degenerate in that sense! And that's why, if our universe is discrete at the Planck scale, then it can't also be deterministic in the Bohmian sense.

I3 **Proofs**

We're going to start by beating a retreat from QuantumLand, back onto the safe territory of computational complexity. In particular, we're going to see how, in the 1980s and 1990s, computational complexity theory reinvented the millennia-old concept of *mathematical proof* – making it probabilistic, interactive, and cryptographic. But then, having fashioned our new pruning-hooks (proving-hooks?), we're going to return to QuantumLand and reap the harvest. In particular, I'll show you why, if you could see the entire trajectory of a hidden variable, then you could efficiently solve any problem that admits a "statistical zero-knowledge proof protocol," including problems like Graph Isomorphism for which no efficient quantum algorithm is yet known.

WHAT IS A PROOF?

Historically, mathematicians have had two very different notions of "proof."

The first is that a proof is something that induces in the audience (or at least the prover!) an intuitive sense of certainty that the result is correct. In this view, a proof is an inner transformative experience – a way for your soul to make contact with the eternal verities of Platonic heaven.

The second notion is that a proof is just a sequence of symbols obeying certain rules – or more generally, if we're going to take this view to what I see as its logical conclusion, *a proof is a computation*. In other words, a proof is a physical, mechanical process, such that, if the process terminates with a particular outcome, then you should accept that a given theorem is true. Naturally, you can never be more certain of the theorem than you are of the laws governing the machine. But as great logicians from Leibniz to Frege to Gödel understood, the weakness of this notion of proof is also its strength. If proof is purely mechanical, then in principle you can discover new mathematical truths by just turning a crank, without any understanding or insight. (As Leibniz imagined legal disputes one day being settled: "Gentlemen, let us calculate!")

The tension between the two notions of proof was thrown into sharper relief in 1976, when Kenneth Appel and Wolfgang Haken announced a proof of the famous Four-Color Map Theorem that every planar map can be colored with four colors, in such a way that no two adjacent countries are colored the same. The proof basically consisted of a brute-force enumeration of thousands of cases by computer; there's no feasible way for a human to apprehend it in its entirety.

If the Four-Color Theorem was basically proved by brute force, then how can they be sure they hit all the cases? The novel technical contribution that human mathematicians had to make was precisely that of reducing the problem to finitely many cases – specifically, about 2000 of them – which could then be checked by computer. Increasing our confidence is that the proof has since been redone by another group, which reduced the number of cases from about 2000 to about 1000.

Now, people will ask: how do you know that the computer didn't make a mistake? The obvious response is that human mathematicians *also* make mistakes. I mean, Roger Penrose likes to talk about making direct contact with Platonic reality, but it's a bit embarrassing when you *think* you've made such contact and it turns out the next morning that you were wrong!

We know the computer didn't make a mistake because we trust the laws of physics that the computer relies on, and that it wasn't hit by a cosmic ray during the computation. But in the last 20 years, there's been the question – why should we trust physics? We trust it in life-and-death situations every day, but should we really trust it with something as important as proving the Four-Color Theorem?

The truth is, we can play games with the definition of proof in order to expand it to unsettling levels, and we'll be doing this for the rest of the chapter.

PROBABILISTIC PROOFS

Recall that we can think of a proof as a computation – a purely mechanical process that spits out theorems. But what about a computation that errs with 2^{-1000} probability – is that a proof? That is, are **BPP** computations legitimate proofs? Well, if we can make the probability of error so small that it's more likely for a comet to suddenly smash our computer into pieces than for our proof to be wrong, it certainly seems plausible!

Now do you remember **NP**, the class of problems with polynomial-size certificates (for the "yes" answers) that can be checked in polynomial time? So, once we're thinking about randomized algorithms, the idea suggests itself of "combining" **NP** with **BPP**, to create a new complexity class where you get a polynomial-size certificate for the "yes" answers, and you also get to use a polynomial-time randomized algorithm to check the certificate. Well, that hybrid class has indeed been invented, by Laszlo Babai in the 1980s. But you probably won't guess what Babai *called* the class if you don't know already. Give up? It's called **MA**, for "Merlin-Arthur." Babai was imagining a game where "Merlin," an all-powerful but untrustworthy proving wizard, supplies a polynomial-size certificate, and then "Arthur," a skeptical, polynomial-time king, runs a randomized algorithm to check Merlin's certificate. More formally, **MA** can be defined as the class of languages L for which there exists a polynomial-time randomized algorithm V of Merlin such that for all x:

1. If $x \in L$, then there exists at least one certificate w such that $V(x,w)$ accepts with certainty.
2. If $x \notin L$, then regardless of w, $V(x,w)$ rejects with probability at least $\frac{1}{2}$.

It turns out that, if you replace "with certainty" by "with probability at least $\frac{2}{3}$" in point 1, then you get exactly the same class **MA**. (That takes a page or so to prove; we won't do it here.) One can also show that **NP** and **BPP** are contained in **MA**, and that **MA** is contained in **PP** and $\Sigma_2 P \cap \Pi_2 P$.

Now, once we have these characters Merlin and Arthur, we can also define more interesting games. In particular, suppose Arthur gets to submit a random *challenge* to Merlin, to which Merlin has to respond. Then you get a new class called **AM** (for "Arthur-Merlin"), which contains **MA** but is not known to equal it, and is contained in $\Pi_2 P$. Actually, I should tell you that most of us conjecture these days that **NP** = **MA** = **AM**; indeed, that's known to follow from circuit lower bound hypotheses similar to the ones that make **P** = **BPP** (see Chapter 7). But we're a long way from being able to prove that.

You might wonder, what happens if, after getting his answer from Merlin, Arthur gets to ask Merlin a followup question, or three or four followups? You'd think Merlin would be able to prove even more to Arthur, right? Wrong! Another surprising theorem says that **AM** = **AMAM** = **AMAMAM**... – that is, asking Merlin any constant number of questions gives Arthur exactly the same power as asking him just one question.

ZERO-KNOWLEDGE PROOFS

I was talking before about *stochastic* proofs, proofs that have an element of uncertainty about them. We can also generalize the notion of proof to include *zero-knowledge* proofs, proofs where the person seeing the proof doesn't learn anything about the statement in question except that it's true.

Intuitively, that sounds impossible, but I'll illustrate this with an example. Suppose we have two graphs. If they're isomorphic, that's easy to prove. But suppose they're not isomorphic. How could you prove that to someone, assuming you're an omniscient wizard?

Simple: have the person you're trying to convince pick one of the two graphs at random, randomly permute it, and send you the result. That person then asks: "which graph did I start with?" If the graphs are not isomorphic, then you should be able to answer this question with certainty. Otherwise, you'll only be able to answer it with probability $\frac{1}{2}$. And thus you'll almost surely make a mistake if the test is repeated a small number of times.

This is an example of an interactive proof system. Are we making any assumptions? We're assuming you don't know which graph the verifier started with, or that you can't access the state of his brain to figure it out. Or as theoretical computer scientists would say, we're assuming you can't access the verifier's "private random bits."

What's perhaps even more interesting about this proof system is that the verifier becomes convinced that the graphs are not isomorphic without learning anything else! In particular, the verifier becomes convinced of something, but *is not thereby enabled to convince anyone else* of the same statement.

A proof with this property – that the verifier doesn't learn anything besides the truth of the statement being proved – is called a *zero-knowledge proof*. Yeah, alright, you have to do some more work to define what it means for the verifier to "not learn anything." Basically, what it means is that, if the verifier were already convinced of the statement, he could've just *simulated the entire protocol on his own*, without any help from the prover.

Under a certain computational assumption – namely, that one-way functions exist – it can be shown that zero-knowledge proofs exist for every **NP**-complete problem. This was the remarkable discovery of Goldreich, Micali, and Wigderson in 1986.[1]

Because all **NP**-complete problems are reducible to each other (i.e., are "the same problem in different guises"), it's enough to give

---

[1] See Oded Goldreich, Silvio Micali, and Avi Wigderson, "Proofs that Yield Nothing but Their Validity, or All Languages in NP have Zero-Knowledge Proof Systems," *Journal of the ACM* **38**(3):691–729, 1991.

a zero-knowledge protocol for *one* **NP**-complete problem. And it turns out that a convenient choice is the problem of *three-coloring a graph*, meaning coloring every vertex red, blue, or green, so that no two neighboring vertices are colored the same. This book is black-and-white, but you can use your imagination and pretend that the graph below has two red vertices, two blue ones, and two green ones:

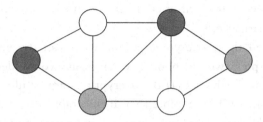

The question is: how can you convince someone that a graph is three-colorable, *without* revealing anything about the coloring?

Well, here's how. Given a three-coloring, first randomly permute the colors – for example, by changing every blue country to green, every green country to red, and every red country to blue. (There are $3! = 6$ possible permutations.) Next, send the verifier *encrypted messages* encoding all the colors, which have the effect of "digitally committing" you to those colors. In more detail, the messages should have the properties that

(1) the verifier can't read them (that is, breaking the encryption is computationally infeasible), but

(2) if you later decrypt the messages *for* the verifier, then the verifier can easily check for itself that you did so correctly – i.e., that you didn't cheat by substituting colors different from the ones that you previously committed to.

There's a relevant technical fact, which I'm simply going to assume without proof: that given any one-way function, it's *possible* to achieve this sort of commitment (though possibly in a way that needs lots of rounds of interaction). If you don't want to take that on faith, there are lots of easier ways to achieve digital commitment, provided you're willing to make a stronger cryptographic assumption. For example, if you assume factoring is hard, then the encrypted messages can be giant composite numbers, and the colors can be encoded by various properties of the prime factorizations of those numbers. Then you'd commit to the colors by sending the composite numbers, and you'd "decommit" (that is, reveal the colors) by sending the factorizations, whereupon the verifier could easily check for itself that those *were* the factorizations.

Anyway, given these encrypted colors, what can the verifier do? Simple: he can pick two neighboring vertices, ask you to decrypt the colors, and then check that (1) the decryptions are valid and (2) the colors are actually different. Note that, if the graph *wasn't* three-colorable, then either two adjacent countries must have gotten the same color, or else some country must not even have been colored red, blue, or green. In either case, the verifier will catch you cheating with probability at least $1/m$, where m is the number of edges.

Finally, if the verifier wants to increase his confidence, we can simply repeat the protocol a large (but still polynomial) number of times. Note that each time you choose a fresh permutation of the colors as well as fresh encryptions. If after (say) m^3 repetitions, the verifier still hasn't caught you cheating, he can be sure that the probability you *were* cheating is vanishingly small.

But why is this protocol zero-knowledge? Intuitively, it's "obvious": when you decrypt two colors, all the verifier learns is that two neighboring vertices were colored differently – but then, they *would* be colored differently if it's a valid three-coloring, wouldn't they? Alright, to make this more formal, you need to prove that the verifier "doesn't learn anything," by which we mean that *by himself*, in polynomial time, the verifier could've produced a probability distribution over sequences of messages that was indistinguishable, by any polynomial-time algorithm, from the actual sequence of messages that the verifier exchanged with you. As you can imagine, it gets a bit technical.

Is there any difference between the two zero-knowledge examples I just showed you? Sure: the zero-knowledge proof for three-coloring a map depended crucially on the assumption that the verifier can't, in polynomial time, decrypt the map by himself. (If he could, he would be able to learn the three-coloring!) This is called a *computational* zero-knowledge proof, and the class of all problems admitting such a proof is called **CZK**. By contrast, in the proof for Graph Non-Isomorphism, the verifier couldn't cheat even with unlimited computational power. This is called a *statistical* zero-knowledge proof, a proof in which the distributions given by an honest prover and a cheating prover need to be close in the statistical sense. The class of all problems admitting this kind of proof is called **SZK**.

Clearly **SZK** $\subseteq$ **CZK**, but is the containment strict? Intuitively, we'd guess that **CZK** is a larger class, since we only require a protocol to be zero-knowledge against *polynomial-time* verifiers, not verifiers with unlimited computation. And indeed, it's known that if one-way functions exist, then **CZK** = **IP** = **PSPACE** – in other words, **CZK** is "as big as it could possibly be." On the other hand, it's also known that **SZK** is contained in the polynomial hierarchy. (In fact, under a derandomization assumption, **SZK** is even in **NP** $\cap$ **coNP**).

PCP

A PCP (Probabilistically Checkable Proof) is yet another impossible-seeming game one can play with the concept of "proof." It's a proof that's written down in such a way that you, the lazy grader, only need to flip it open to *a few random places* to check (in a statistical sense) that it's correct. Indeed, if you want very high confidence (say, to one part in a thousand) that the proof is correct, you never need to examine more than about 30 bits. Of course, the hard part is encoding the proof so that this is possible.

It's probably easier to see this with an example. Do you remember the Graph Nonisomorphism problem? We'll show that there is a proof that two graphs are nonisomorphic, such that anyone verifying the proof only needs to look at a constant number of bits (though admittedly, the proof itself will be exponentially long).

First, given any pair of graphs G_0 and G_1 with n nodes each, the prover sends the verifier a specially encoded string proving that G_0 and G_1 are nonisomorphic. What's in this string? Well, we can choose some ordering of all possible graphs with n nodes, so call the ith graph H_i. Then for the ith bit of the string, the prover puts a 0 there if H_i is isomorphic to G_0, a 1 if H_i is isomorphic to G_1, and otherwise (if H_i is isomorphic to neither) he arbitrarily places a 0 or a 1. How does this string prove to the verifier that G_0 and G_1 are nonisomorphic? Easy: the verifier flips a coin to get G_0 or G_1, and randomly permutes it to get a new graph H. Then, she queries for the bit of the proof corresponding to H, and accepts if and only if the queried bit matches her original graph. If indeed G_0 and G_1 are nonisomorphic, then the verifier will always accept, and if not, then the probability of acceptance is at most $\frac{1}{2}$.

In this example, though, the proof was exponentially long and only worked for Graph Nonisomorphism. What kinds of result do we have in general? The famous PCP Theorem[2] says that *every* **NP** problem admits PCPs – and furthermore, PCPs with polynomially long proofs! This means that *every* mathematical proof can be encoded in such a way that any error in the original proof translates into errors *almost everywhere* in the new proof.

One way of understanding this is through 3SAT. The PCP theorem is equivalent to the **NP**-completeness of the problem of solving 3SAT with the promise that either the formula is satisfiable, or else there's no truth assignment that satisfies more than (say) 90% of the clauses. Why? Because you can encode the question of whether some mathematical statement has a proof with at most n symbols as a 3SAT instance – in such a way that if there's a valid proof, then the formula is satisfiable, and if not, then no assignment satisfies more than 90%

---

[2] The literature on the PCP Theorem is large; at least a dozen people made major contributions to discovering and refining its proof. For a recent popular-level overview, see Dana Moshkovitz, "The Tale of the PCP Theorem," *ACM Crossroads* **18**(3):23–26, 2012. http://people.csail.mit.edu/dmoshkov/XRDS.pdf

of the clauses. So given a truth assignment, you only need to distinguish the case that it satisfies all the clauses from the case that it satisfies at most 90% of them – and this can be done by examining a few dozen random clauses, completely independently of the length of the proof.

COMPLEXITY OF SIMULATING
HIDDEN-VARIABLE THEORIES

We talked last chapter about the path of a particle's hidden variable in a hidden-variable theory, but what is the complexity of *finding* such a path? This problem is certainly at least as hard as quantum computing – since even to sample a hidden variable's value at any *single* time would in general require a full-scale quantum computation. Is sampling a whole trajectory an even harder problem?

Here's another way to ask this question. Suppose that at the moment of your death, your whole life flashes before you in an instant – and suppose you can then perform a polynomial-time computation on your life history. What does that let you compute? Assuming, of course, that a hidden-variable theory is true, and that while you were alive, you somehow managed to place your own brain in various nontrivial superpositions.

To study this question, we can define a new complexity class called **DQP**, or Dynamical Quantum Polynomial-Time. The formal definition of this class is a bit hairy (see my paper[3] for details). Intuitively, though, **DQP** is the class of problems that are efficiently solvable in the "model" where you get to sample the whole trajectory of a hidden variable, under some hidden-variable theory that satisfies "reasonable" assumptions.

Now, you remember the class **SZK**, of problems that have statistical zero-knowledge proof protocols? The main result from my paper was that **SZK** ⊆ **DQP**. In other words, if only we could measure the whole trajectory of a hidden variable, we could use a quantum computer to solve every **SZK** problem – including Graph Isomorphism

---

[3] http://www.scottaaronson.com/papers/qchvpra.pdf

and many other problems not yet known to have efficient quantum algorithms!

To explain why that is, I need to tell you that in 1997 Sahai and Vadhan discovered an extremely nice "complete promise problem" for **SZK**. That problem is the following.

> Given two efficiently-samplable probability distributions D_1 and D_2, are they close or far in statistical distance (promised that one of those is the case)?

This means that when thinking about **SZK**, we can forget about zero-knowledge proofs, and just assume we have two probability distributions and we want to know whether they're close or far.

But let me make it even more concrete. Let's say that you have a function $f:\{1, 2, \ldots, N\} \to \{1, 2, \ldots, N\}$, and you want to decide whether f is one-to-one or two-to-one, promised that one of these is the case. This problem – which is called the *collision problem* – doesn't *quite* capture the difficulty of all **SZK** problems, but it's close enough for our purposes.

Now, how many queries to f do you need to solve the collision problem? If you use a classical probabilistic algorithm, then it's not hard to see that $\sqrt{N}$ queries are necessary and sufficient. As in the famous "birthday paradox" (where if you put 23 people in a room, there's at least even odds that two of the people share a birthday), you get a square-root savings over the naïve bound, since what matters is the number of *pairs* for which a collision could occur. But unfortunately, if N is exponentially large, as it is in the situations we're thinking about, then $\sqrt{N}$ is still completely prohibitive: the square root of an exponential is still an exponential.

So what about *quantum* algorithms? In 1997, Brassard, Høyer, and Tapp[4] showed how to *combine* the $\sqrt{N}$ savings from the

---

[4] G. Brassard, P. Høyer, and A. Tapp, Quantum cryptanalysis of hash and claw-free functions, *SIGACT News* **28**:2 (1997), 14–19. http://arxiv.org/abs/quant-ph/9705002

birthday paradox with the unrelated $\sqrt{N}$ savings from Grover's algorithm, to obtain a quantum algorithm that solves the collision problem in (this is going to sound like a joke) $\sim N^{1/3}$ queries. So, yes, quantum computers *do* give at least a slight advantage for this problem. But is that the best one can do? Or could there be a better quantum algorithm, that solves the collision problem in (say) $\log(N)$ queries, or maybe even less?

In 2002, I proved the first nontrivial lower bound[5] on the quantum query complexity of the collision problem, showing that any quantum algorithm needs at least $\sim N^{1/5}$ queries. This was later improved to $\sim N^{1/3}$ by Yaoyun Shi,[6] thereby showing that the algorithm of Brassard, Høyer, and Tapp was indeed optimal.

On the other hand – to get back to our topic – suppose you could see the whole trajectory of a hidden variable. In that case, I claim that you could solve the collision problem with only a *constant* number of queries (independent of N)! How? The first step is to prepare the state

$$\frac{1}{\sqrt{N}} \sum_{i=1}^{N} |i\rangle |f(i)\rangle.$$

Now measure the second register (which we won't need from this point onward), and think only about the resulting state of the first register. If f is one-to-one, then in the first register, you'll get a classical state of the form $|i\rangle$, for some random i. If f is two-to-one, on the other hand, then you'll get a state of the form $\frac{|i\rangle + |j\rangle}{\sqrt{2}}$, where i and j are two values with $f(i) = f(j)$. If only you could perform a further measurement to tell these states apart! But alas, as soon as you measure, you destroy the quantum coherence, and the two types of state look completely identical to you.

[5] S. Aaronson, Quantum Lower Bound for the Collision Problem, *Proceedings of ACM Symposium on Theory of Computing*, (2002), 635–642. http://www.scottaaronson. com/papers/collision.pdf

[6] Y. Shi, Quantum Lower Bounds for the Collision and the Element Distinctness Problems, *Proceedings of IEEE Symposium on Foundations of Computer Science*, (2002), 513–519. http://arxiv.org/abs/quant-ph/0112086

Aha, but remember we get to see an entire hidden-variable trajectory! Here's how we can exploit that. Starting from the state $\frac{|i\rangle+|j\rangle}{\sqrt{2}}$, first apply a Hadamard gate to every qubit. This produces a "soup" of exponentially many basis vectors – but if we then Hadamard every qubit a *second* time, we get back to the original state $\frac{|i\rangle+|j\rangle}{\sqrt{2}}$. Now, the idea is that when we Hadamard everything, the particle "forgets" whether it was at i or j. (This can be proved under some weak assumptions on the hidden-variable theory.) Then, when we observe the history of the particle, we'll learn something about whether the state had the form $|i\rangle$ or $\frac{|i\rangle+|j\rangle}{\sqrt{2}}$. For in the former case, the particle will always return to i, but in the latter case, it will "forget," and will need to pick randomly between i and j. As usual, by repeating the "juggling" process polynomially many times one can make the probability of failure exponentially small. (Note that this does not require observing more than one hidden-variable trajectory: the repetitions can all happen within a single trajectory.)

What are the assumptions on the hidden-variable theory that are needed for this to work? The first is basically that if you have a bunch of qubits and you apply a Hadamard to one of them, then you should only get to transfer between hidden-variable basis states that differ in the first qubit.

Note that this assumption is very different from (and weaker than) requiring the hidden-variable theory to be "local," in the sense physicists usually mean by that. *No* hidden-variable theory can be local. Some guy named Bell proved that.

And the second assumption is that the hidden-variable theory is "robust" to small errors in the unitaries and quantum states. This assumption is needed to define the complexity class **DQP** in a reasonable way.

As we've seen, **DQP** contains both **BQP** and the Graph Isomorphism problem. But interestingly, it seems unlikely that **DQP** contains the **NP**-complete problems, just as it seems unlikely that **BQP** contains them! It turns out that in the hidden-variable model, you can search an unordered list of size N using $\sim N^{1/3}$ queries instead of the $\sim\sqrt{N}$ you'd get from Grover's algorithm: all you need to do, basically, is run Grover's algorithm for $\sim N^{1/3}$ steps in order to amplify the probability of the marked item to $1/N^{1/3}$, then use your history-sampling capability to simulate "measuring the same state $\sim N^{1/3}$ times" until the marked item comes out. But of course, if $N{=}2^n$ then $N^{1/3}$ is still exponential.

Here I should make a *mea culpa*: in my 2005 paper (and indeed, in earlier printings of this book), I claimed to have a proof that $N^{1/3}$ queries are needed for unordered search even in the hidden-variable model (and hence, that exists an oracle A relative to which $\mathbf{NP}^A \not\subset \mathbf{DQP}^A$). This would've been significant evidence against the solvability of $\mathbf{NP}$-complete problems in the hidden-variable model. In 2013, however, my students Mitchell Lee and Adam Bouland discovered a flaw in my proof. We're working on fixing the flaw: I still think it's unlikely that sampling histories would let you search an N-element list in log N steps or solve $\mathbf{NP}$-complete problems in polynomial time, but a new proof will be needed. In any case, I find $\mathbf{DQP}$ interesting as almost the only example I've seen of a "modest" generalization of quantum computing: something that really does give you more power than $\mathbf{BQP}$ (e.g., the ability to find collisions and solve Graph Isomorphism), but probably not "absurdly" more power (e.g., the ability to solve $\mathbf{NP}$-complete problems).

14　How big are quantum states?

I'm going to talk about the title question, but first, a little digression. In science, there's this traditional hierarchy where you have biology on top, and chemistry underlies it, and then physics underlies chemistry. If the physicists are in a generous mood, they'll say that math underlies physics. Then, computer science is over somewhere with soil engineering or some other nonscience.

Now, my point of view is a bit different: computer science is what mediates between the physical world and the Platonic world. With that in mind, "computer science" is a bit of a misnomer; maybe it should be called "quantitative epistemology." It's sort of the study of the capacity of finite beings such as us to learn mathematical truths. I hope I've been showing you some of that.

How do we reconcile this with the notion that any actual implementation of a computer must be based on physics? Wouldn't the order of physics and CS be reversed?

Well, by similar logic one could say that any mathematical proof has to be written on paper, and therefore physics should go below math in the hierarchy. Or one could say that math is basically a field that studies whether particular kinds of Turing machine will halt or not, and so CS is the ground that everything else sits on. Math is then just the special case where the Turing machines enumerate topological spaces or do something else that mathematicians care about. But then, the strange thing is that physics, especially in the form of quantum probability, has lately been seeping down the intellectual hierarchy, contaminating the "lower" levels of math and CS. This is how I've always thought about quantum computing: as a case of physics not staying where it's supposed to in the intellectual hierarchy! If you like, I'm professionally interested in physics precisely to the *extent* that it seeps down into the "lower" levels, which are supposed to be the least arbitrary ones, and forces me to rethink what I thought I understood about those levels.

Anyway, on to the subject of this chapter, I think that it's helpful to classify interpretations of quantum mechanics, or at least to reframe debates about them, by asking where they come down on the question of the exponentiality of quantum states. To describe the state of a hundred or a thousand atoms, do you really need more classical bits of information than you could write down in the observable universe?

Roughly speaking, the Many-Worlds interpretation would say "absolutely." This is a view that David Deutsch defends very explicitly; if the different universes (or components of the wavefunction) used in Shor's algorithm are not physically there, then where was the number factored?

We also talked about Bohmian mechanics, which says "yes," but that one component of the vector is "more real" than the rest. Then, there is the view that used to be called the Copenhagen view, but is now called the Bayesian view, the information-theoretic view, or one of a host of other names.

In the Bayesian view, a quantum state is an exponentially long vector of amplitudes in more-or-less the same sense that a classical probability distribution is an exponentially long vector of probabilities. If you were to take a coin and flip it 1000 times, you would have some set of 2^{1000} possible outcomes, but we don't because of that decide to regard all of those outcomes as physically real.

At this point, I should clarify that I'm not talking about the *formalism* of quantum mechanics; that's something that (almost) everyone agrees about. What I'm asking is whether quantum mechanics describes an actual, real "exponential-sized object" existing in the physical world. So, the move that you make when you take the Copenhagen view is to say that the exponentially long vector is "just in our heads."

The Bohmian view is this strange kind of intermediate position. In the Bohmian view, you do sort of see these exponential numbers of possibilities as somehow real; they're the guiding field, but there's this one "more real" thing that they're guiding. In the Copenhagen interpretation, these exponentially many possibilities really are just in your head. Presumably, they correspond to something in the external world, but what that something is, we either don't know or aren't allowed to ask. Chris Fuchs says that there's some physical context to quantum mechanics – something outside of our heads – but that we don't know what that context is. Niels Bohr tended to make the move toward "you aren't allowed to ask."

Now that we have quantum computing, can we bring the intellectual arsenal of computational complexity theory to bear on this sort of question? I hate to disappoint you, but we can't resolve this debate using computational complexity. It's not well defined enough. Although we can't declare one of these views to be the ultimate victor, what we *can* do is to put them into various "staged battles" with each other and see which one comes out the winner. To me, this is the real motivation for studying questions about quantum proofs, advice, and communication, like the ones we're going to see in this chapter. Namely, we want to understand: if you have a quantum state of n qubits, does it act more like n or like 2^n classical bits? Of course there's always a sort of exponentiality in our formal description of a quantum state, but we want to know to what extent we can actually get at it, or root it out.

Before we embark on this quest, we'll need to arm ourselves with some complexity classes. I know, I know: we have all these complexity classes, and they seem kind of esoteric. Maybe it's just a bad historical accident that we use all of these acronyms to express our ideas, rather than coming up with sexy names like "black hole," "quark," or "supersymmetry" as a physicist would. It's like the joke about the prisoners where one of them calls out "37" and all of them will fall on the floor laughing, then another calls out "22" but no one laughs because it's all in the telling. There are these staggering, mind-bending mysteries about truth, proof, computers, physics, and the very limits of the knowable, and for ease of reference, we bottle the mysteries up using inscrutable sequences of three or four capital letters. Maybe we shouldn't do that.

But we're going to do it anyway, starting with **QMA** (Quantum Merlin-Arthur), the quantum generalization of **MA**. You can think of **QMA** as the set of truths such that, if you had a quantum computer, you could be convinced of the answer by being given a quantum state. More formally, it's the set of problems that admit a polynomial-time quantum algorithm Q such that for every input x the following holds.

- If, on the input x, the answer to the problem is "yes," then there exists some quantum state $|\phi\rangle$ of a polynomial number of qubits such that Q accepts $|x\rangle|\phi\rangle$ with probability greater than $\frac{2}{3}$.
- If, on the input x, the answer to the problem is "no," then there does not exist any polynomial-sized quantum state $|\phi\rangle$ such that Q accepts $|x\rangle|\phi\rangle$ with probability greater than $\frac{1}{3}$.

What I mean is that the number of qubits of $|\phi\rangle$ should be bounded by a polynomial in the length n of x. You can't be given some state of 2^n qubits. If you could, then that would sort of trivialize the problem.

We want there to be a quantum state of reasonable size that convinces you of a "yes" answer. So when the answer is "yes," there's a state that convinces you, and when the answer is "no," there's no such state. **QMA** is sort of the quantum analog of **NP**. Recall that we have the Cook–Levin Theorem, which gives us that the Boolean satisfiability problem (SAT) is **NP**-complete. There is also a Quantum Cook–Levin Theorem – which is a great name, since both Cook and Levin are quantum computing skeptics (though Levin much more so than Cook). The Quantum Cook–Levin theorem tells us that we can define a quantum version of the 3SAT problem, which turns out to be **QMA**-complete as a promise problem.

A promise problem is some problem you only have to get the right answer if there's some promise on the input. If you, as the algorithm, have been screwed over by crappy input, then any court is going to rule in your favor and you can do whatever you want. It may even be a very difficult computation to decide if the promise holds or not, but that's not your department. There are certain complexity classes for which we don't really believe that there are complete problems, but for which there are complete promise problems. **QMA** is one such class. The basic reason we need a promise is because of the gap between $\frac{1}{3}$ and $\frac{2}{3}$. Maybe you would be given some input, but you'd accept with some probability that is neither greater than $\frac{2}{3}$ nor less than $\frac{1}{3}$. In that case, you've done something illegal, and so we assume that you aren't given such an input.

So what is this quantum 3SAT problem?[1] Basically, think of n qubits stuck in an ion trap (hey, I'm trying to bring in some physics), and now we describe a bunch of measurements, each of which involves at most three of the qubits. Each measurement i accepts with probability equal to P_i. These measurements are not hard to describe, since they involve at most three qubits. Let's say that we add up n of the measurements. Then, the promise will be either there is a state such that this sum is very large, or that for all states, the sum is much smaller. Then, the problem is to decide which of the two conditions holds. This problem is complete in **QMA** in the same sense that the classical analog, 3SAT, is complete in **NP**. This was first proved by Kitaev, and was later improved by many others.[2]

The real interest comes with the question of how *powerful* the **QMA** class is. Are there truths that you can verify in a reasonable amount of time with quantum computers, but which you can't verify with a classical computer? This is an example of what we talked about earlier, where we're trying to put realistic and subjective views of quantum states into "staged battle" with each other and see which one comes out the winner.

---

[1] Actually, what I'm calling "quantum 3SAT" is normally called the "Local Hamiltonian problem" in the literature, and it's more analogous to MAX3SAT than to ordinary 3SAT. But those distinctions aren't important for us here.

[2] See for example J. Kempe, A. Kitaev, and O. Regev, The Complexity of the local Hamiltonian problem. *SIAM Journal on Computing* 35:5 (2006), 1070–1097. http://arxiv.org/abs/quant-ph/0406180.

There's a result of John Watrous[3] which gives an example where it seems that being given an exponentially long vector really does give you some sort of power. The problem is called *group non-membership*. You're given a finite group G. We think of this as being exponentially large, so that you can't be given it explicitly by a giant multiplication table. You're given it in some more subtle way. We will think of it as a *black-box group*, which means that we have some sort of black box which will perform the group operations for you. That is, it will multiply and invert group elements for you. You're also given a polynomially long list of generators of the group.

Each element of the group is encoded in some way by some n-bit string, though you have no idea *how* it's encoded. The point is that there are exponentially many group elements, but only polynomially many generators.

So now we're given a subgroup $H \leq G$, which can also be given to us as a list of generators. Now the problem is an extremely simple one: we're given an element x of the group, and want to know whether or not it's in the subgroup. I've specified this problem abstractly in terms of these black boxes, but you can instantiate it, if you have a specific example of a group. For example, these generators could be matrices over some finite field, and you're given some other matrix and are asked whether you can get to it from your generators. It's a very natural question.

Let's say the answer is "yes." Then, could that be proved to you?

You can show how x was generated. There's one thing you need to say (not a very hard thing), which is that if $x \in H$, then there is some "short" way of getting to it. Not necessarily by multiplying the generators you started with, but by recursively generating new elements and adding those to your list, and using those to generate new elements, and so on.

[3] J. Watrous, Succinct quantum proofs for properties of finite groups. In *Proceedings of IEEE Symposium on Foundations of Computer Science* (2000), pp. 537–46. http://arxiv.org/abs/cs.CC/0009002

For example, if we started with the group $\mathbb{Z}_n$, the additive group modulo n, and if we have some single starting element 1, we can we just keep adding 1 to itself, but it will take us a while to get to 2^{5000}. But if we recursively build $2 = 1 + 1$, $4 = 2 + 2$ and so on by repeatedly applying the group operation to our new elements, we'll get to whatever element we want quickly.

Is it always possible to do it in polynomial time? It turns out the answer is yes, for any group. The way to see that is to construct a chain of subgroups from the one you started with. It takes a little work to show, but it's a theorem of Babai and Szemerédi, which holds whether or not the group is solvable.

Now here's the question: what if $x \notin H$? Could you demonstrate that to someone? Sure, you could give them an exponentially long proof, and if you had an exponentially long time, you could demonstrate it, but this isn't feasible. We still don't know quite how to do with this, even if you were given a classical proof and allowed to check it via quantum computation, though we do have some conjectures about that case.

Watrous showed that you can prove non-membership if you're given a certain quantum state, which is a superposition over all the elements of the subgroup. Now this state might be very hard to prepare. Why?

It's exponentially large, but there are other exponentially large superposition states which are easy to prepare, so that can't be the whole answer. The problem turns out to be one of uncomputing garbage.

So we know how to take a random walk on a group, and so we know how to sample a random element of a group. But here, we're asked for something more. We're asked for a coherent superposition of the group's elements. It's not hard to prepare a state of the form $\sum |g\rangle |\text{garbage}_g\rangle$. Then how do you get rid of that garbage? That's the question. Basically, this garbage will be the random walk or whatever process you use to get to g, but how do you forget how you got to that element?

But what Watrous said is to suppose we had an omniscient prover, and suppose that prover was able to prepare that state and give it to you. Well then, you could verify that an element is not in the subgroup H. We can do this in two steps.

1. Verify that we really were given the state we needed (we'll just assume this part for now).
2. Use the state $|H\rangle$ to prove that $x \notin H$ by using controlled left-multiplication:

$$\frac{1}{\sqrt{2}} \left(|0\rangle |H\rangle + |1\rangle |xH\rangle \right).$$

Then, do a Hadamard and measure the first qubit. In more detail, you have the left qubit act as the control qubit. If $x \in H$, then xH is a permutation of H, and so we get interference fringes (the light went both through the x slit and the xH slit). If $x \notin H$, then we have that xH is a coset, and thus shares no elements in common with H. Hence, $\langle H|xH \rangle = 0$, and so we measure random bits. We can tell these two cases apart.

You also have to verify that this state $|H\rangle$ really was what we were given. To do this, we will do a test like what we just did. Here, we pick the element x by taking a classical random walk on the subgroup H. Then, if $|H\rangle$ were really the superposition over the subgroup, $|xH\rangle$ would just be shifted around by x, whereas if $x \notin H$, we get something else. You have to prove that this is not only a necessary test, but a sufficient one as well. That's basically what Watrous proved.

This gives us one example where it seems like having a quantum state actually helps you, as if you could really get at the exponentiality of the state. Maybe this isn't a staggering example, but it's something.

An obvious question is whether, in all of those cases where a quantum proof seems to helps you, you could do just as well if you were given a classical proof that you then verified via quantum computation. Are we really getting mileage from having the quantum state, or is our mileage coming from the fact that we have a quantum computer to do the checking? We can phrase the question by asking if **QMA** = **QCMA**, where **QCMA** is like **QMA** except that the proof now has to be a classical proof. Greg Kuperberg and I wrote a paper[4] where we tried to look at this question directly. One thing we showed looks kind of bad for the realistic view of quantum states (at least in this particular battle): if the Normal Hidden Subgroup problem (what the problem is isn't important right now) can be solved in quantum-polynomial time, and it seems like it can, and if we make some other group-theoretic assumptions that seem plausible according to all the group theorists that we asked, then the Group Non-membership Problem is actually in **QCMA**. That is, you can dequantize the proof and replace it with a classical one.

On the other hand, we showed that there exists a quantum oracle A relative to which **QMA**$^A \neq$ **QCMA**^{A}. This is a really simple thing to describe. To start with, what is a quantum oracle? Quantum oracles are just quantum subroutines to which we imagine that both a **QMA** and a **QCMA** machine have access. While classical oracles act on the computational basis (possibly in superposition within a quantum state), quantum oracles can act on an arbitrary basis. To see the idea behind the oracle that we used, let's say that you're given some n-qubit unitary operation U. Moreover, let's say that you're promised that either U is the identity matrix I or that there exists some secret "marked state" $|\psi\rangle$ such that $U|\psi\rangle = -|\psi\rangle$; that is, that U has some secret eigenvector corresponding to an eigenvalue of -1. The problem is then to decide which of these conditions holds.

It's not hard to see that this problem, as an oracle problem, is in **QMA**. Why is it in **QMA**? Because the prover would just have to give the verifier $|\psi\rangle$, and the verifier would apply $U|\psi\rangle$ to verify that, yes, $U|\psi\rangle = -|\psi\rangle$. So that's not saying a whole lot.

What we proved is that this problem, as an oracle problem, is *not* in **QCMA**. So even if you had both of the resources of this

4 S. Aaronson and G. Kuperberg, Quantum Versus Classical Proofs and Advice, *Theory of Computing* 3:7 (2007), 129–157. http://arxiv.org/abs/quant-ph/0604056

unitary operation U and some polynomial-sized classical string to kind of guide you to this secret negative eigenvector, you'd still need exponentially many queries to find $|\psi\rangle$.

This gives some evidence in the other direction, that maybe **QMA** is more powerful than **QCMA**. If they were equivalent in power, then that would have to be shown using a quantumly nonrelativizing technique: that is, a technique that is sensitive to the presence of quantum oracles. We don't really know of such a technique right now, besides techniques that are also *classically* nonrelativizing and don't seem applicable to this problem.

So there's really another sort of metaquestion here, which is if there's some kind of separation between quantum and classical oracles. That is, if there's some kind of question that we can only answer with quantum oracles. Could we get a classical oracle separation between **QMA** and **QCMA**? Greg Kuperberg and I tried for a while and couldn't do it. Very recently, Andy Lutomirski[5] proposed a candidate problem that he (and I) conjecture *should* give such a separation, but no one has been able to prove it yet. If you can, that'd be great.

OK. So that was quantum proofs. There are other ways we can try and get at the question of how much stuff is there to be extracted from a quantum state. Holevo's Theorem deals with the following question: if Alice wants to send some classical information to Bob, and if she has access to a quantum channel, can she use this to her advantage? If quantum states are these exponentially long vectors, then intuitively, we might expect that, if Alice could send some n-qubit state, maybe she could use this to send Bob 2^n classical bits. We can arrive at this from a simple counting argument. The number of quantum states of n qubits, any pair of which are of almost zero inner product with each other, is doubly exponential in n. All we're saying is that, in order to specify such a state, you need exponentially many bits. Thus, we might hope that we could get some kind of

---

[5] http://arxiv.org/abs/1107.0321

exponential compression of information. Alas, Holevo's Theorem tells us that it is not to be. You need n qubits to reliably transmit n classical bits, with just some constant factor representing that you're willing to tolerate some probability of error, but really nothing better than you would get with a classical probabilistic encoding.

Here's a handwaving intuition: You can only measure it once. Each bit of information you extract cuts in half the dimensionality of the Hilbert space. Sure, in some sense, you can encode more than n bits, but then you can't reliably retrieve them.

This theorem was actually known in the 1970s, and was ahead of its time.

It was only recently that anyone asked a very natural and closely related question: what if Bob doesn't want to retrieve the whole string? We know from Holevo's theorem that getting the whole string is impossible, but what if Bob only wants to retrieve one bit and Alice doesn't know which one ahead of time? Can Alice create a quantum state $|\psi_x\rangle$ such that, for whichever bit x_i Bob wants to know, he can just measure $|\psi_x\rangle$ in the appropriate basis and would then learn that particular bit? After he's learned x_i, then he's destroyed the state and can't learn any more, but that's OK. Alice wants to send Bob a quantum phonebook, and Bob only wants to look up one number. It turns out that, via a proof from Ambainis, Nayak et al.,[6] this is still not possible. What they proved is that to encode n bits in this manner, so that any one can be read out, you need at least $\frac{n}{\log n}$ qubits.

Maybe you could get some small savings, but certainly not an exponential saving. Shortly after, Nayak proved that actually, if you want to encode n bits, you need n qubits. If we're willing to lose a logarithmic factor or two, I can show rather easily how this is a consequence of Holevo's theorem. The reason that it's true illustrates a technique that I've gotten a lot of mileage out of, and there might be more mileage that can still be gotten out of it.

---

[6] See A. Ambainis, A. Nayak, A. Ta-Shma, and U. V. Vazirani, Dense quantum coding and quantum finite automata, *Journal of the ACM*, **49**:4 (2002), 496–511. This paper also contains Nayak's later improvement.

Suppose, by way of contradiction, that we had such a protocol that would reliably encode n bits into no more than $\log n$ qubits in such a way that any one bit could then be retrieved with high probability – we'll say with error at most one-third. Then, what we could do is to take a bunch of copies of the state. We just want to push down the error probability, so we take a tensor product of, say, $\log n$ copies. Given this state, what Bob can do is to run the original protocol on each copy to get x_i and then take the majority vote. For some sufficiently large constant times $\log n$, this will push down the error rate to at most n^{-2}. So for any particular bit i, Bob will be able to output a bit y_i such that $\Pr[y_i = x_i] \geq 1 - n^{-2}$. Now, since Bob can do that, what else can he do? He can keep repeating this, and get greedy. I'm going to run this process and get x_1, but now, because the outcome of this measurement could be predicted almost with certainty given this state, you can prove, because of that, that you aren't getting a lot of information, and so the state is only slightly disturbed by the measurement. This is just a general fact about quantum measurements. If you could predict the outcome with certainty, then the state wouldn't be disturbed *at all* by the measurement.[7]

So this is what we do. We've learned x_1 and the state has been damaged only slightly. When we run the protocol again, we learn what x_2 is with only small damage. Since small damage plus small damage is still small damage, we can find what x_3 is and so on. So, we can recover all of the bits of the original string using fewer qubits then the bound shown by Holevo. Based on this, we can say that we can't have such a protocol.

Why do we care about any of this? Well, maybe we don't, but I can tell you how this stuff entered my radar screen. Now, we're not going to ask about quantum proofs, but about a closely related concept called quantum advice. So we'll bring in a class called **BQP/qpoly**: the set of problems efficiently solvable by a quantum computer, given a

---

[7] What we're talking about here is simply what, in another language, the physicist Yakir Aharonov and his collaborators call the concept of "weak measurement."

polynomially sized quantum advice state. What's the difference between advice and proof? As we discussed in Chapter 7, advice depends only on the input length n but is absolutely trustworthy, whereas a proof depends on the actual input but needs to be checked.

So the advantage of advice is that you can trust it, but the disadvantage is that it might not be as useful as it isn't tailored to the particular problem instance that you're trying to solve. So we can imagine that maybe it's hard for quantum computers to solve **NP**-complete problems, but only if the quantum computer has to start in some all-zero initial state. Maybe there are some very special states that were created in the Big Bang and that have been sitting around in some nebula ever since (somehow not decohering), and if we get on a spaceship and find these states, they obviously can't anticipate what particular instance of SAT we wanted to solve, but they sort of anticipated that we would want to solve *some* instance of SAT. Could there be this one generic SAT-solving state $|\psi_n\rangle$, such that, given any Boolean formula P of size n, we could, by performing some quantum computation on $|\psi_n\rangle$, figure out whether P is satisfiable? What we're really asking here is whether **NP** $\subset$ **BQP/qpoly**.

What can we say about the power of **BQP/qpoly**? We can adapt Watrous's result about quantum proofs to this setting of quantum advice. Returning to the Group Nonmembership Problem, if the Big Bang anticipated what subgroup we wanted to test membership in, but not what element we wanted to test, then it could provide us with the state $|H\rangle$ that's a superposition over all the elements of H, and then whatever element we wanted to test for membership in H, we could do it. This shows that a version of the Group Non-membership Problem is in **BQP/qpoly**.

I didn't mention this earlier, but we can prove that **QMA** $\subseteq$ **PP**,[8] so there's evidently some limit on the power of **QMA**. You can see that, in the worst case, all you would have to do is search through all possible quantum proofs (all possible states of n qubits), and see if

---

[8] For a nice proof due to Vyalyi, see eccc.hpi-web.de/eccc-reports/2003/TR03-021/

there's one that causes our machine to accept. You can do better than that, and that's where the bound of **PP** comes from.

What about **BQP/qpoly**? Can you see any upper bound on the power of this class? That is, can you see any way of arguing what it *can't* do?

Do we even know that **BQP/qpoly** isn't equal to **ALL**, the set of all languages whatsoever (including uncomputable languages)? Let's say you were given an exponentially long classical advice string. Well, then, it's not hard to see that you could then solve any kind of problem whatsoever. Why? Because say that $f : \{0, 1\}^n \to \{0, 1\}$ is the Boolean function we want to compute. Then, we just let the advice be the entire truth table for the function, and then we just need to look up the appropriate entry in the truth table, and we've solved any problem of size n we want to solve. The halting problem, you name it.

For another example, consider the famous constant Ω defined by Gregory Chaitin.[9] Informally, Ω is the probability that a "randomly generated computer program" halts on a blank input, in some fixed Turing-universal programming language. (Technically, in order for the probability to be well defined, the programming language needs to be "self-delimiting," which means that you can never produce a valid program by adding more bits to the end of another valid program.) The bits in the binary expansion of Ω are almost of like the wisdom of God: they encode the answers to a huge number of mathematical questions (Goldbach's Conjecture, the Riemann Hypothesis, etc.) in what one could call a maximally efficient way. It would be wild to be given such a thing as "advice"! (Though note that, as a practical matter, extracting interesting information from the advice – the truth or falsehood of Goldbach's Conjecture and so forth – would require immense computations and would almost certainly be completely impractical. *In practice* Ω would probably look no different to you than a uniformly random string. But still: *dude!*)

---

[9] See, for example, Chaitin's article http://www.cs.auckland.ac.nz/CDMTCS/chaitin/sciamer3.html for a nice popular account of Ω.

Intuitively, it seems a bit implausible that **BQP/qpoly = ALL**, because being given a polynomial number of qubits really isn't like being given an exponentially long string of classical bits. The question is, how much can this "sea" of exponentially many classical bits that are needed to describe a quantum state determine what we get out?

I guess I'll cut to the chase and tell you that, at a workshop years ago, Harry Buhrman asked me this question, and it was obvious to me that **BQP/qpoly** wasn't everything, and he told me to prove it. And eventually I realized that anything you could do with polynomially sized quantum advice, you could do with polynomially sized classical advice, provided that you can make a measurement and then postselect on its outcome. That is, I proved that **BQP/qpoly ⊆ PostBQP/poly**. In particular, this implies that **BQP/qpoly ⊆ PSPACE/poly**.[10] (Later on, in 2010, Andrew Drucker and I[11] improved this result still further, to show that in fact **BQP/qpoly ⊆ QMA/poly**, which in some sense gives the "optimal" upper bound on **BQP/qpoly** in terms of a classical advice class, assuming **BQP/qpoly** isn't just flat-out equal to **BQP/poly**. But I won't say more about that here.) The upshot is that anything you can be told by quantum advice, you can also be told by classical advice of a comparable size, provided that you're willing to spend exponentially more computational effort to extract what that advice is trying to tell you.

It's again a two-minute endeavor to give a handwaving proof that **BQP/qpoly ⊆ PSPACE/poly**. I like the way that Greg Kuperberg described the proof. What he said is that what we do if we have some quantum advice and we want to simulate it using classical advice by postselection is to use a "Darwinian training set" of inputs. We'll say that we've got this machine that takes classical advice, and then

[10] S. Aaronson, Limitations of Quantum Advice and One-Way Communication, *Theory of Computing* 1 (2005), 1–28. http://theoryofcomputing.org/articles/v001a001/v001a001.pdf

[11] S. Aaronson and A. Drucker, A full characterization of quantum advice. In *Proceedings of Annual ACM Symposium on Theory of Computing* (2010), pp. 131–40. http://arxiv.org/abs/1004.0377

we want to describe to this machine some set of quantum advice using only classical advice. To do so, we consider some test inputs X_1, $X_2, \ldots, X_T$. Note, by the way, that our classical advice machine doesn't know the true quantum advice state $|\psi\rangle$. The classical advice machine starts by guessing that the quantum advice is the maximally mixed state, since without a-priori knowledge any quantum state is equally likely to be the advice state. Then, X_1 is an input to the algorithm such that if the maximally mixed state is used in place of the quantum advice, the algorithm produces the wrong answer with a probability of greater than one-third. If the algorithm still guesses the right answer, then making a measurement changes the advice state to some new state ρ_1. So why is this process described as "Darwinian?" The next part of the classical advice, X_2 describes some input to the algorithm such that the wrong answer will be produced with probability greater than one-third if the state ρ_1 is used in place of the actual quantum advice. If, despite the high chance of getting the wrong answer when run with X_1 and X_2 as input, the algorithm still produces two correct answers, then we use the resultant estimate of the advice state ρ_2 to produce the next part of the classical advice X_3. Basically, we're trying to teach our classical advice machine the quantum state by repeatedly telling it, "supposing you got all the previous lessons right, here's a new test you're still going to fail. Go and learn, my child."

The point is that, if we let $|\psi_n\rangle$ be the true quantum advice, then since we can decompose the maximally mixed state into whatever basis we want, we can imagine it as a mixture of the true advice state that we're trying to learn, and a bunch of things that are all orthogonal to it. Each time that we give a wrong answer with a probability greater than one-third, it's like we're lopping off another third of this space. We then postselect on succeeding. We also know that if we were to start with the true advice state, then we would succeed, and so this process has to bottom out somewhere; we eventually winnow away all the chaff and run out of examples where the algorithm fails.

So, in this setting, quantum states are not acting like exponentially long vectors. They're acting like they only encode some polynomial amount of information, although *extracting* what you want to know might be exponentially more efficient than if the same information were presented to you classically. Again, we're getting ambiguous answers, but that's what we expected. We knew that quantum states occupy this weird kind of middle realm between probability distributions and exponentially long strings. It's nice to see exactly how this intuition plays out, though, in each of these concrete scenarios. I guess this is what attracts me to quantum complexity theory. In some sense, this is same stuff that Bohr and Heisenberg argued about, but we're now able to ask the questions in a much more concrete way – and sometimes even answer them.

15 Skepticism of quantum computing

Last chapter, we talked about whether quantum states should be thought of as exponentially long vectors, and I brought up class **BQP/qpoly** and concepts like quantum advice. Actually, I'd say that the main reason why I care is something I didn't mention last time, which is that it relates to whether we should expect quantum computing to be fundamentally possible or not. There are people, like Leonid Levin and Oded Goldreich, who just take it as obvious that quantum computing *must* be impossible.[1] Part of their argument is that it's extravagant to imagine a world where describing the state of 200 particles takes more bits then there are particles in the universe. To them, this is a clear indication something is going to break down. So part of the reason that I like to study the power of quantum proofs and quantum advice is that it helps us answer the question of whether we really should think of a quantum state as encoding an exponential amount of information.

So, on to the Eleven Objections.

1. Works on paper, not in practice.
2. Violates Extended Church–Turing Thesis.
3. Not enough "real physics."
4. Small amplitudes are unphysical.
5. Exponentially large states are unphysical.
6. Quantum computers are just souped-up analog computers.
7. Quantum computers aren't like anything we've ever seen before.
8. Quantum mechanics is just an approximation to some deeper theory.

---

[1] See, for example, http://www.cs.bu.edu/fac/lnd/expo/qc.htm and http://www.wisdom. weizmann.ac.il/~oded/on-qc.html

9. Decoherence will always be worse than the fault-tolerance threshold.
10. We don't need fault-tolerance for classical computers.
11. Errors aren't independent.

What I did is to write out every skeptical argument against the possibility of quantum computing that I could think of. We'll just go through them, and make commentary along the way. Let me just start by saying that my point of view has always been rather simple: it's entirely conceivable that quantum computing is impossible for some fundamental reason. If so, then that's by far the most exciting thing that could happen for us. That would be much more interesting than if quantum computing were possible, because it changes our understanding of physics. To have a quantum computer capable of factoring 10 000-digit integers is the relatively *boring* outcome – the outcome that we'd expect based on the theories we already have.

I like to engage skeptics for several reasons. First of all, because I like arguing. Second, often I find that the best way to come up with new results is to find someone who's saying something that seems clearly, manifestly wrong to me, and then try to think of counterarguments. Wrong claims are a fertile source of research ideas.

So what are some of the skeptical arguments that I've heard? The one I hear more than any other argument is "well, it works formally, on paper, but it's not gonna work in the *real* world." People actually say this, and they actually treat it like it was an argument. For me, the fallacy here is not that people can have ideas that don't work in the real world, but rather that if they don't work in the real world, they can still "work on paper." Of course, there could be assumptions such that an idea only works if the assumptions are satisfied. Thus, the question becomes if the assumptions are stated clearly or not.

I was happy to find out that I wasn't the first person to point out this particular fallacy. Immanuel Kant wrote an entire treatise demolishing it: *On the Common Saying: "That may be right in theory but does not work in practice."*

The second argument is that quantum computing must be impossible because it violates the Extended Church–Turing Thesis: "Anything that is efficiently computable in the physical world is computable in polynomial time on a standard Turing machine." That is, we know that quantum computing can't be possible (assuming **BPP** $\neq$ **BQP**), because we know that **BPP** defines the limit of the efficiently computable.

So, we have this thesis, and quantum computing violates the thesis, so (if you have faith in the thesis) it must be impossible. On the other hand, if you replaced factoring with **NP**-complete problems, then this argument would actually become more plausible to me, because I would think that any world in which we could solve **NP**-complete problems efficiently would not look much like our world. For **NP**-intermediate problems like factoring and Graph Isomorphism, I'm not willing to take some sort of a-priori theological position. But the diagram below shows how I think things most likely stand.

So that was the second argument. On to the third: "I'm suspicious of all these quantum computing papers because there isn't enough of the *real* physics that I learned in school. There's too many unitaries and not enough Hamiltonians. There's all this entanglement, but my professor told me not to even think about entanglement, because it's all just kind of weird and philosophical, and has nothing to do with the structure of the helium atom." What can one say to this? Certainly, this argument succeeds in establishing that we have a different way of talking about quantum mechanics now, in addition to the ways people have had for many years. Those making this argument are advancing an additional claim, though, which is that the new way of talking about quantum mechanics is *wrong*. And that claim, of course, requires a separate argument. I don't know if any further response is needed.

The fourth argument is that "these exponentially small amplitudes are clearly unphysical." This is another argument that Leonid Levin has made. Consider some state of 1000 qubits, such that each component has an amplitude of 2^{-500}. We don't know of any physical law that holds to more than about a dozen decimal places, and you're asking for accuracy to *hundreds* of decimal points. Why should someone even imagine that makes any sense whatsoever?

The obvious repudiation of argument 4, then, is that I can take a classical coin and flip it a thousand times. Then, the probability of any particular sequence is 2^{-1000}, which is far smaller than any constant we could ever measure in nature. Does this mean that probability theory is some "mere" approximation of a deeper theory, or that it's going to break down if I start flipping the coin too many times?

For me the key point is that amplitudes evolve linearly, and in that respect are similar to probabilities. We've got minus signs, and so we've got interference, but maybe if we really thought about why probabilities are okay, we could argue that it's not just that we're always in a deterministic state and just don't know what it is, but that this property of linearity is something more general. Linearity is the thing that prevents small errors from creeping up on us. If we have a bunch of small errors, the errors add rather than multiplying. That's linearity.

Argument 5 gets back to what we were talking about in the previous chapter: "it's obvious that quantum states are these extravagant objects; you can't just take 2^n bits and pack them into n qubits." Actually, I was arguing with Paul Davies, and he was making this argument, appealing to the holographic principle and saying that we have a finite upper bound on the number of bits that can be stored in a finite region of spacetime. If you have some 1000-qubit quantum state, it requires 2^{1000} bits, and according to Davies, we've just violated the holographic bound.[2]

So how should one respond to that? First of all, this information, whether or not we think it's "there," can't generally be read out. This is the content of results like Holevo's theorem. In some sense, you might be able to pack 2^n bits into a state, but the number of bits that you can reliably get out is only n.

The holographic bound says, informally, that the maximum number of bits that can be stored in any finite region is proportional to the region's surface area, at roughly the rate of one bit per Planck area, or 1.4×10^{69} bits per meter squared. Why should the maximum number of bits grow like the *surface area*, rather than the volume? That's a very profound question that people like Ed Witten and Juan Maldacena probably stay up at night worrying about. The doofus answer is that if you try to take lots and lots of bits and pack them into some volume (such as a cubical hard disk), then at some point, your cubical hard disk will collapse and form a black hole. A flat drive will also collapse, but a one-dimensional drive *won't* collapse.

Here's the thing: there seem to be all these bits near the event horizon of the black hole. Why the event horizon? Because if you're standing outside a black hole, then you never actually see anything fall *through* the event horizon. Instead, because of time dilation, all

---

[2] Davies subsequently published this argument; see http://arxiv.org/abs/quant-ph/0703041

the infalling objects will seem to get eerily frozen just outside the event horizon – approaching it, Zeno-like, but never reaching it.

Then, if you want to preserve unitarity, and not have pure states evolve into mixed states when something gets dropped into a black hole, you say that when the black hole evaporates via Hawking radiation, then the bits get peeled off like scales, and go flying out into space. Again, this is not something that people really understand. People treat the holographic bound (rightfully) as the one of the few clues we have for a quantum theory of gravity, but they don't yet have the detailed theory that implements the bound, except for some special model systems.

The other funny thing about this is that, in classical general relativity, the event horizon doesn't play a particularly special role. You could pass through it and you wouldn't even notice. Eventually, you'll know you passed through it, because you'll be sucked into the singularity, but while you're passing through it, it doesn't feel special. On the other hand, this information point of view says that as you pass through, you'll pass a lot of bits near the event horizon. What is it that singles out the event horizon as being special in terms of information storage? It's very strange, and I wish I understood it (see Chapter 22 for further discussion).

There actually is an interesting question here. The holographic principle says that you can store only so much information within a region of space, but what does it *mean* to have stored that information? Do you have to have random access to the information? Do you have to be able to access whatever bit you want and get the answer in a reasonable amount of time? In the case that these bits are stored in a black hole, apparently if there are n bits on the surface, then it takes on the order of $n^{3/2}$ time for the bits to evaporate via Hawking radiation. So, the time-order of retrieval is polynomial in the number of bits, but it still isn't particularly efficient. A black hole should not be one's first choice for a hard disk.

Argument 6: "a quantum computer would merely be a souped-up analog computer." This I've heard again and again, from people like

Robert Laughlin, Nobel laureate, who espoused this argument in his popular book *A Different Universe*.[3] This is a popular view among physicists. We know that analog computers are not that reliable, and can go haywire because of small errors. The argument proceeds to ask why a quantum computer should be any different, since you have these amplitudes which are continuously varying quantities.

But the response to this argument has been known since 1996 or so. It's called the *Threshold Theorem*.[4] Informally, the Threshold Theorem says that, if you can just make the probability of error per qubit per time step sufficiently small – less than some constant, which was traditionally estimated at 10^{-6}, but might be as high as 0.1 or 0.2 – then you can do something called *quantum fault-tolerance*, which stops the errors from ever building up and destroying the computation. An analogous fault-tolerance theorem for classical computing was proved by John von Neumann in the 1950s, but in some sense, it ultimately ended up not being needed, since once transistors came along they were so reliable that people almost *never* had to worry about them failing. In the mid-1990s, some physicists conjectured that the "analog" nature of quantum computers would make *quantum* fault-tolerance impossible. In more detail, the intuition was that, since measurement in quantum mechanics is a destructive process, the very act of making a measurement to see whether an error had occurred, or to copy quantum information as a safeguard against future errors, would already destroy the information you were trying to protect. But this intuition turned out to be mistaken: there are clever ways to measure only the "error syndrome," which tells you whether an error has occurred and how to fix it, *without* measuring and destroying the "legitimate" quantum information. What ultimately makes such measurements possible is the *linearity* of quantum mechanics: the spear on which a thousand wrong intuitions about how quantum mechanics works have died!

---

[3] Basic Books, 2006.

[4] For gentle introductions to the Threshold Theorem, see for example http://arxiv.org/abs/quant-ph/9705031 by John Preskill or http://arxiv.org/abs/quant-ph/9812037 by Dorit Aharonov.

Is there a similar Threshold Theorem for analog computers? No, and there *can't* be. The point is, there's a crucial property that is shared by discrete theories, probabilistic theories and quantum theories, but that is *not* shared by analog or continuous theories. That property is insensitivity to small errors. Once again, that's really a consequence of linearity.

Note that, if we want a weaker Threshold Theorem, we could consider a computation taking t time steps, where the amount of error per time step could be $1/t$. Then, the Threshold Theorem would be trivial to prove. If we had a product of unitaries $U_1 U_1 \ldots U_{100}$, and each one were to be corrupted by $1/t$ $(1/100$ in this case$)$, then we'd have a product like

$$(U_1 + U_1'/t)(U_2 + U_2'/t)\cdots(U_{100} + U_{100}'/t).$$

The product of all these errors still won't be much, again because of linearity. An observation made by Bernstein and Vazirani[5] was that quantum computation is sort of naturally robust against one-over-polynomial errors. "In principle," that could be taken to answer the question already; what remains is "merely" showing how to tolerate larger and more realistic amounts of error, rather than just one-over-polynomial error.

On to argument 7. This is an argument raised, for example, by Michel Dyakonov.[6] The argument goes that all the systems we have experience with involve very rapid decoherence, and thus that it isn't plausible to think that we could "just" engineer some system which is not like any of the systems in nature that we have any experience with.

A nuclear fission reactor is also unlike any naturally occurring system in many ways. What about a spacecraft? Things don't normally use propulsion to escape the earth. We haven't seen anything doing that in nature. Or a classical computer.

---

[5] www.cs.berkeley.edu/~vazirani/pubs/bv.ps

[6] See http://arxiv.org/abs/quant-ph/0610117, or the more recent http://arxiv.org/abs/1212.3562

Next, there are the people who just take it for granted that quantum mechanics must be an approximate theory that only works for a small number of particles. When you go to a larger number of particles, something else must take over. The trouble is, there *have* been experiments that have tested quantum mechanics with fairly large numbers of particles, like the Zeilinger group's experiment with buckyballs. There have also been SQUID experiments that have prepared the "Schrödinger cat state" $|0 \ldots 0\rangle + |1 \ldots 1\rangle$ on n qubits, where, depending on what you want to count as a degree of freedom, n is as large as several billion.

Again, though, the fundamental point is that discovering a breakdown of QM would be the most exciting possible outcome of trying to build a quantum computer. And, how else are you going to discover that, but by investigating these things experimentally and seeing what happens? Astonishingly, I meet people (especially computer scientists) who ask me "what, you're going to expect a Nobel Prize if your quantum computer *doesn't* work?" To them, it's just so obvious that a quantum computer isn't going to work that it isn't even interesting.

Some people will say, "no, no, I want to make a separate argument. I don't believe that quantum mechanics is going to break down, but even if it doesn't, quantum computing could still be fundamentally impossible, because there's just too much decoherence in the world." These people are claiming that decoherence is a *fundamental* problem. That is, that the error will always be worse than the fault-tolerance threshold, or that some nasty little particle will always pass through and decohere your quantum computer.

The next argument is a little more subtle: for a classical computer, we don't have to go through all this effort. You just get fault-tolerance naturally. You have some voltage that either is less than a lower threshold or is greater than an upper threshold, and that gives us two easily distinguishable states that we can identify as 0 and 1. We don't have to go through the same amount of work to get fault-tolerance. In modern microprocessors, for example, they don't even bother to build in much redundancy and fault-tolerance, because the components are just so reliable that such safeguards aren't needed. The argument then proceeds by noting that you can, in principle, do universal quantum computing by exploiting this fault-tolerant machinery, but that this should raise a red flag – why do you *need* all that error correction machinery? Shouldn't this make you suspicious?

Here's my response. The only reason we don't need fault-tolerance machinery for classical computers is that the components are so reliable, but we haven't been able to build reliable quantum computer components yet. In the early days of classical computing, it wasn't clear at all that reliable classical components would exist. Von Neumann actually proved a classical analog of the Threshold Theorem, then later, it was found that we didn't need it. He did this to answer skeptics who said there was always going to be something making a nest in your JOHNNIAC, insects would always fly into the machine, and that these things would impose a physical limit on classical computation. Sort of feels like history's repeating itself.

We can already see hints of how things *might* eventually turn out. People have been looking at proposals such as non-abelian anyons where your quantum computer is "naturally fault tolerant," since the only processes that can cause errors have to go around the quantum computer with a nontrivial topology. These proposals show that it's conceivable we'll someday be able to build quantum computers that have the same kind of "natural" error correction that we have in classical computers.

I wanted to have a round number of arguments, but I wound up with 11. So, Argument 11 comes from people who understand the Fault-Tolerance Theorem, but who take issue with the assumption that the errors are independent. This argument posits that it's ridiculous to suppose that errors are uncorrelated, or even that they're only weakly correlated, from one qubit to the next. Instead, the claim is that such errors *are* correlated, albeit in some very complicated way. In order to understand this argument, you have to work from the skeptics' mind-set: to them, this isn't an engineering issue, it's given a priori that quantum computing is not going to work. The question is how to correlate the errors such that quantum computing won't work.

My favorite response to this argument comes from Daniel Gottesman, who was arguing about this against Levin, who believes that the errors will be correlated in some conspiracy that defies the imagination. Gottesman said, supposing the errors were correlated in such a diabolical fashion and that Nature should undergo so much work to kill off quantum computation, why couldn't you turn that around and use whatever diabolical process Nature employs to get access to even *more* computational power? Maybe you could even solve **NP**-complete problems. It seems like Nature would have to expend enormous amounts of effort just to correlate qubits so as to kill quantum computation.

In other words, not only would your errors have to be correlated in some diabolical way, they'd have to be correlated in some *unpredictable* diabolical way. Otherwise, you could deal with the problem in general.

To summarize, I think that arguing with skeptics is not only fun but extremely useful. It could be that quantum computing is impossible for some fundamental reason. But I'm still waiting for an argument that really engages my imagination nontrivially. People are objecting to this or to that, but they aren't coming up with some fully imagined, alternative picture of the world in which quantum computing *wouldn't* be possible. That's what's missing for me, what I keep looking for and not finding.[7]

I'll close with a question that you should think about before the next chapter. If we see 500 ravens, which are all black, should we expect that the 501st raven we see will also be black? If so, why? Why would seeing 500 black ravens give you any grounds whatsoever to draw such a conclusion?

[7] For a recent discussion of quantum computing skepticism on my blog, see http://www.scottaaronson.com/blog/?p=1211

16 Learning

The puzzle from last chapter is known as Hume's Problem of Induction.

Puzzle: If you observe 500 black ravens, what basis do you have for supposing that the next one you observe will also be black?

Many people's answer would be to apply Bayes's Theorem. For this to work, though, we need to make some assumption such as that all the ravens are drawn from the same distribution. If we don't assume that the future resembles the past at all, then it's very difficult to get anything done. This kind of problem has led to lots of philosophical arguments like the following.

Suppose you see a bunch of emeralds, all of which are green. This would seem to lend support to the hypothesis that all emeralds are green. But then, define the word *grue* to mean "green before 2050 and blue afterwards." Then, the evidence equally well supports the hypothesis that all emeralds are grue, not green. This is known as the grue paradox.

If you want to delve even "deeper," then consider the "gavagai" paradox. Suppose that you're trying to learn a language, and you're an anthropologist visiting an Amazon tribe speaking the language. (Alternatively, maybe you're a baby in the tribe. Either way, suppose you're trying to learn the language from the tribe.) Then, suppose that some antelope runs by and some tribesman points to it and shouts "gavagai!" It seems reasonable to conclude from this that the word "gavagai" means "antelope" in their language, but how do you know that it doesn't refer to just the antelope's horn? Or it could be the name of the specific antelope that ran by. Worse still, it could mean that a specific antelope ran by on some given day of the week! There's any number of situations that the tribesman could be using the word to refer to, and so we conclude that there is no way to learn the language, even if we spend an infinite amount of time with the tribe.

There's a joke about a planet full of people who believe in *anti-induction*: if the sun has risen every day in the past, then today, we should expect that it won't. As a result, these people are all starving and living in poverty. Someone visits the planet and tells them, "Hey, why are you still using this anti-induction philosophy? You're living in horrible poverty!"

"Well, it never worked before . . . "

What we want to talk about here is the efficiency of learning. We've seen all these philosophical problems that seem to suggest that learning is impossible, but we also know that learning does happen, and so we want to give some explanation of *how* it happens. This is sort of a problem in philosophy, but in my opinion, the whole landscape around the problem has been transformed in recent years by what's called "computational learning theory." This is not as widely known as it should be. Even if you're (say) a physicist, it's nice to know something about this theory, since it gives you a framework – different from the better-known Bayesian framework, but related to it, and possibly more useful in some contexts – for deciding when you can expect a hypothesis to predict future data.

I think a key insight that any approach has to take on board – whether it's Bayesianism, computational learning theory, or something else – is that we're never considering all logically conceivable hypotheses on an equal footing. If you have 500 ravens, each either white or black, then in principle there are 2^{500} hypotheses that you have to consider. If the ravens could also be green, that would produce still more hypotheses. In reality, though, you're never considering all of these as equally possible. You're always restricting your attention to some minuscule subset of hypotheses – broadly speaking, those that are "sufficiently simple" – unless the evidence forces you to a more complex hypothesis. In other words, you're always implicitly using what we call Occam's Razor (all though it isn't at all clear if it's what Occam meant).

Why does this work? Fundamentally, because the universe itself is not maximally complicated. We could well ask why it isn't, and maybe there's an anthropic explanation, but whatever the answer, we accept as an article of faith that the universe is reasonably simple, and we do science.

This is all talk and blather, though. Can we actually see what the trade-offs are between the number of hypotheses we consider and how much confidence we can have in predicting the future? One way we do this was formalized by Leslie Valiant in 1984.[1] His framework is called *PAC learning*, where PAC stands for "probably approximately correct." We aren't going to predict everything that happens in the future, nor will we even predict *most* of it with certainty, but with high probability, we'll try to get most of it right.

This might sound like pure philosophy, but you *can* actually connect some of this stuff to experiments. For example, this theory has been used in experiments on things like neural networks and machine learning. When I was writing a paper on PAC learning once, I wanted to find out how the theory was actually used, so I looked on Google Scholar. At the time of this book's publication, the paper by Valiant has been cited about 4000 times. Based on this, we can infer that further papers are likely.

So how does PAC learning work? We'll have a set S which could be finite or infinite, called our *sample space*. For example, we're an infant trying to learn a language, and are given some examples of sentences which are grammatical or ungrammatical. From this, we need to come up with a rule for deciding whether a new sentence

---

[1] L. Valiant, A Theory of the Learnable, *Communications of the ACM* **27**:11 (1984), 1134–1142. http://www.mpi-inf.mpg.de/~mehlhorn/SeminarEvolvability/ValiantLearnable.pdf. For a good introduction see *An Introduction to Computational Learning Theory* by Michael Kearns and Umesh Vazirani, MIT Press, 1994.

is grammatical or not. Here, our sample space is the set of possible sentences.

A *concept* is a Boolean function $f: S \to \{0, 1\}$ that maps each element of the sample space to either 0 or 1. We can later remove the assumption that concepts are Boolean, but for simplicity, we'll stick with it for now. In our example, the concept is the language that we're trying to learn; given a sentence, the concept tells us whether it is or isn't grammatical. Then, we can have a *concept class*, denoted C. Here, C can be thought of as the set of languages which our baby comes in to the world thinking a priori to be possible, before gathering any data as to the actual language spoken.

For now, we're going to say that we have some probability distribution D over the samples. In the infant example, this is like the distribution from which the child's parents or peers draw what sentences to speak. The baby does *not* have to know what this distribution is. We just have to assume that it exists.

So what's the goal? We're given m examples x_i drawn independently from the distribution D, and for each x_i, we're given $f(x_i)$; that is, we're told whether each of our examples is or isn't grammatical. Using this, we want to output a hypothesis language h such that

$$\Pr_{x \sim D} [h(x) = f(x)] \geq 1 - \varepsilon$$

where $\sim$ means that x is drawn from distribution D. That is, we want our hypothesis h to disagree with the concept f no more than ε of the time given examples x drawn from our distribution D. Can we hope to do this with certainty? No? Well, why not?

You might get unlucky with the samples you're given, like getting the sample sampler over and over again. If the only sentence you're ever exposed to as a baby is "what a cute baby!" then you're not going to have any basis for deciding whether "we hold these truths to be self-evident" is also a sentence. In fact, we should assume that there are exponentially many possible sentences, of which the baby only hears a polynomial number.

So, we say that we only need to output an ε-good hypothesis with probability $1 - \delta$ over the choice of samples. Now, we can give the basic theorem from Valiant's paper.

Theorem: In order to satisfy the requirement that the output hypothesis h agrees with $1 - \varepsilon$ of the future data from drawn from D, with probability $1 - \delta$ over the choice of samples, it suffices to find any hypothesis h that agrees with

$$m \geq \frac{1}{\varepsilon} \log\left(\frac{|C|}{\delta}\right)$$

samples chosen independently from D.

The key point about this bound is that it's logarithmic in the number of possible hypotheses $|C|$. Even if there are exponentially many hypotheses, this bound is still polynomial. Now, why do we ask that the distribution D on which the learning algorithm will be tested is the same as the distribution from which the training samples are drawn?

Because if your example space is a limited subset of sample space, then you're hosed.

This is like saying that nothing should be on the quiz that wasn't covered in class. If the sentences that you hear people speaking have support only in English, and you want a hypothesis that agrees with French sentences, this is not going to be very possible. There's going to have to be some assumption about the future resembling the past.

Once you make this assumption, then Valiant's theorem says that for a finite number of hypotheses, with a reasonable number of samples, you can learn. There's really no other assumption involved.

This goes against a belief in the Bayesian religion, that if your priors are different then you'll come to an entirely different conclusion. The Bayesians start out with a probability distribution over the possible hypotheses. As you get more and more data, you update this distribution using Bayes's Rule.

That's one way to do it, but computational learning theory tells us that it's not the only way. You don't need to start out with *any* assumption about a probability distribution over the hypotheses. You can make a worst-case assumption about the hypothesis (which we computer scientists love to do, being pessimists!), and then just say that you'd like to learn *any* hypothesis in the concept class, for any sample distribution, with high probability over the choice of samples. In other words, you can trade the Bayesians' probability distribution over hypotheses for a probability distribution over *sample* data.

In a lot of cases, this is actually preferable: you have no idea what the true hypothesis is, which is the whole problem, so why should you assume some particular prior distribution? We don't have to know what the prior distribution over hypotheses is in order to apply computational learning theory. We just have to assume that there *is* a distribution.

The proof of Valiant's theorem is really simple. Given a hypothesis h, call it *bad* if it disagrees with f for more than an ε fraction of the data. Then, for any specific bad hypothesis h, since $x_1, \ldots, x_m$ are independent we have that

$$\Pr[h(x_1) = f(x_1), \ldots, h(x_m) = f(x_m)] < (1 - \varepsilon)^m.$$

This bounds the probability that this bad hypothesis gave the right prediction on the samples. Now, what is the probability that there *exists* a bad hypothesis $h \in C$ that agrees with all the sample data? We can use the union bound:

$\Pr[$there exists a bad h that agrees with f for all samples$]$ $< |C| (1 - \varepsilon)^m.$

We can set this equal to δ and solve for m. Doing so gives that

$$m = \frac{1}{\varepsilon} \log \left(\frac{|C|}{\delta} \right)$$

QED.

This gives us a bound on the number samples needed for a finite set of hypotheses, but what about *infinite* concept classes? For example, what if we're trying to learn a rectangle in the plane? Then our sample space is the set of points in the plane, and our concept class is the set of all filled-in rectangles. Suppose we're given m points, and for each one are told whether or not it belongs to a "secret rectangle."

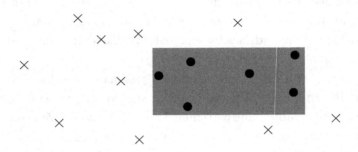

Well, how many possible rectangles are there? There are $2^{\aleph_0}$ possibilities, so we can't apply the previous theorem! Nevertheless, given 20 or 30 random points in the rectangle, and 20 or 30 random points *not* in the rectangle but near it, intuitively it seems like we have a reasonable idea of where the rectangle is. Can we come up with a more general learning theorem to apply when the concept class is infinite? Yes, but first we need a concept called *shattering*.

For some concept class C, we say that a subset of the sample space $\{s_1, s_2, \ldots, s_k\}$ is shattered by C if, for all 2^k possible classifications of $s_1, s_2, \ldots, s_k$, there is some function $f \in C$ that agrees with that classification. Then, define the *VC dimension* of the class C, denoted VCdim(C), as the size of the largest subset shattered by C.

What is the VC dimension of the concept class of rectangles? We need the largest set of points such that, for each possible setting of whether each point is or is not in the rectangle, there is some rectangle that contains only the points we want and not the other ones. The diagram below illustrates how to do this with four points. On the other hand, there's no way to do it with five points (proof: exercise for you!).

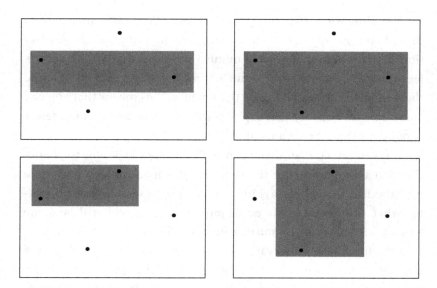

One corollary of the next theorem is that one can perform PAC learning, with a finite number of samples, if and only if the VC dimension of the concept class is finite.

Theorem (Blumer, Ehrenfeucht, Haussler, and Warmuth, 1989):[2] In order to produce a hypothesis h that will explain a $1 - \varepsilon$ fraction of future data drawn from distribution D, with probability $1 - \delta$, it suffices to output any h in C that agrees with

$$m \geq \frac{K}{\varepsilon}\left(\mathrm{VC\,dim}(C)\log\left(\frac{1}{\varepsilon}\right) + \log\left(\frac{1}{\delta}\right)\right)$$

sample points drawn independently from D. Furthermore, this is tight (up to the dependence on ε).

This theorem is harder to prove than the last one, and would take a whole chapter in itself, so we'll skip it here. The intuition behind the proof, however, is simply Occam's Razor. If the VC dimension is finite, then after seeing a number of samples that's larger than

---

[2] A. Blumer, A. Ehrenfeucht, D. Haussler, and M. K. Warmuth, Learnability and the Vapnik-Chernonenkis dimension, *Journal of the ACM* **36**:4 (1989), 929–965.

the VC dimension, the entropy of the data that you've already seen should only go roughly as the VC dimension. You make m observations, after which the possible number of things that you've seen is less than 2^m, otherwise $\text{VCdim}(C) \geq m$. It follows that to describe these m observations takes less than m bits. This means that you can come up with a theory that explains the past data, and that has fewer parameters than the data itself.

If you can do that, then intuitively you should be able to predict the next observation. On the other hand, supposing you had some hypothetical theory in (say) high-energy physics such that, no matter what the next particle accelerator found, there'd still be some way of – I don't know – curling up extra dimensions or something to reproduce those observations – well, in that case you'd have a concept class whose VC dimension was at least as great as the number of observations you were trying to explain. In such a situation, computational learning theory gives you no reason to expect that whatever hypothesis you output will be able to predict the next observation.

The upshot is that this intuitive trade-off between the compressibility of past data and the predictability of future data can actually be formalized and proved; given reasonable assumptions, Occam's Razor is a theorem.

What if the thing that we're trying to learn is a quantum state, say some mixed state ρ? We could have a measurement E with two outcomes. In quantum mechanics, the most general type of measurement is called a *positive operator-valued measurement*, or POVM. A POVM is just an ordinary "projective" measurement – the type of measurement we discussed earlier – except that, before measuring, you get to perform any unitary transformation you want on the state ρ being measured, *together with* some additional "ancilla state" independent of ρ. For present purposes, all you need to know is the following: if you have a two-outcome POVM M, acting on an n-dimensional state mixed state ρ, then you can completely characterize M by an $n \times n$

hermitian matrix E, all of whose eigenvalues belong to $[0, 1]$. Then the probability that M "accepts" ρ is simply $\text{tr}(E\rho)$ (where tr, the trace, is the sum of diagonal entries), and the probability that M "rejects" ρ is $1 - \text{tr}(E\rho)$.

Now, if we're given some state ρ, what we'd like to be able to do is to predict the outcome of any measurement made on the state: that is, to estimate the acceptance probability $\text{tr}(E\rho)$ for any two-outcome POVM measurement E. This is easily seen to be equivalent to *quantum state tomography*, which is recovering the density matrix ρ itself.

But, what is ρ? It's some n-qubit state represented as a $2^n \times 2^n$ matrix with 4^n independent parameters. The number of measurements needed to do tomography on an n-qubit state is well known to grow exponentially with n. Indeed, this is already a serious practical problem for the experimenters. To learn an eight-qubit state, you might have to set your detector in $65\,536$ different ways, and to measure in each way hundreds of times to get a reasonable accuracy.

So again, this is a practical problem for experimenters. But is it a conceptual problem as well? Some quantum computing skeptics seem to think so; we saw in the last chapter that one of the fundamental criticisms of quantum computing is that it involves manipulating these exponentially long vectors. To some skeptics, this is an inherently absurd way of describing the physical world, and either quantum mechanics is going to break down when we try to do this, or there's *something* else that we must not have taken into account, because you clearly can't have 2^n "independent parameters" in your description of n particles.

Now, if you need to make an exponential number of measurements on a quantum state before you know enough to predict the outcome of further measurements on it, then this would seem to be a way of formalizing the above argument and making it more persuasive. After all, our goal in science is always to come up with hypotheses that succinctly explain past observations, and thereby let us predict future observations. We might have other goals, but *at the least* we want to do that. So if, to characterize a general state of 500 qubits, you had to make more measurements than you could in the age of the universe, that would seem to be a problem with quantum mechanics itself, considered as a scientific theory. I'm actually inclined to agree with the skeptics about that.

In 2006, I had a paper[3] where I tried to use computational learning theory to answer this argument. Here's how Umesh Vazirani explained my result. He said, suppose you're a baby trying to learn a rule for predicting whether or not a given object is a chair. You see a bunch of objects labeled "chair" or "not-chair," and based on that you come up with general rules ("a chair has four legs," "you can sit on one," etc.) that work pretty well in most cases. Admittedly, these rules might break down if (say) you're in a modern art gallery, but we don't worry about that. In computational learning theory, we only want to predict *most* of the future observations that you'll actually make. If you're a Philistine, and don't go to MOMA, then you don't worry about any chairlike objects that might be there. We need to take into account the future intentions of the learner, and for this reason, we relax the goal of quantum state tomography to the goal of predicting the outcomes of *most* measurements drawn from some probability distribution D.

More formally, given a mixed state ρ on n qubits, as well as given measurements $E_1, E_2, \ldots, E_m \sim D$ and estimated probabilities $p_j \approx \text{Tr}(E_j \rho)$ for each $j \in \{1, 2, \ldots, m\}$, the goal is to produce a hypothesis state σ that has, with probability at least $1 - \delta$, the property that

$$\Pr_{E \in D}\left[|\text{tr}(E\rho) - \text{tr}(E\sigma)| < \gamma\right] \geq 1 - \varepsilon.$$

For this goal, here's a theorem that bounds the number of sample measurements needed.

[3] S. Aaronson, The learnability of quantum states. *Proceedings of the Royal Society*, A463 (2088), 2007. http://arxiv.org/abs/quant-ph/0608142

Theorem: Fix error parameters ε, δ, and γ and fix $\eta > 0$ such that $\gamma\varepsilon \geq 7\eta$. Call $E = (E_1, \ldots, E_m)$ a "good" training set of measurements if any hypothesis state σ that satisfies $|\mathrm{Tr}(E_i\sigma) - \mathrm{Tr}(E_i\rho)| \leq \eta$ also satisfies

$$\Pr_{E \in D}[|\mathrm{tr}(E\sigma) - \mathrm{tr}(E\rho)| > \gamma] \leq \varepsilon.$$

Then, there exists a constant $K > 0$ such that E is a good training set with probability at least $1 - \delta$ over $E_1, \ldots, E_m$ drawn from D, provided that m satisfies

$$m \geq \frac{K}{\gamma^2\varepsilon^2}\left(\frac{n}{\gamma^2\varepsilon^2}\log^2\frac{1}{\gamma\varepsilon} + \log\frac{1}{\delta}\right).$$

It's important to note that this bound is only linear in the number of qubits n, and so this tells us that the dimensionality is *not* actually exponential in the number of qubits, if we only want to predict most of the measurement outcomes.

Why is this theorem true? Remember the result of Blumer *et al.*, which said that you can learn with a number of samples that grows linearly with the VC dimension of your concept class. In the case of quantum states, we're no longer dealing with Boolean functions. You can think of a quantum state as a real-valued function that takes as input a two-outcome measurement E, and produces as output a real number in $[0, 1]$ (namely, the probability that the measurement accepts). That is, ρ takes a measurement E and returns $\mathrm{Tr}(E\rho)$.

So, can one generalize the Blumer *et al.* result to real-valued functions? Fortunately, this was already done for me by Alon, Ben-David, Cesa-Bianchi, and Haussler, and by Bartlett and Long among others.

Next, recall from Chapter 14 the lower bound on random access codes of Ambainis, Nayak *et al.*, which tells us how many classical bits can be reliably encoded into a state of n qubits. Given an m-bit classical string x, suppose we want to encode x into a quantum state of n qubits, in such a way that any bit x_i of our choice can later be retrieved with probability at least $1 - \varepsilon$. Ambainis *et al.* proved that we really can't get any savings by packing classical bits into a quantum state in this way. That is to say, n still must be linear in m. Since this is a lower bound, we can view it as a limitation of quantum encoding schemes. But we can also turn it on its head, and say: this is actually *good*, as it implies some upper bound on the VC dimension of quantum states considered as a concept class. Roughly speaking, the theorem tells us that the VC dimension of n-qubit states considered as a concept class is at most $m = O(n)$. To make things more formal, we need a real-valued analog of VC dimension (called the "fat-shattering" dimension; don't ask), as well as a theorem saying that we can learn any real-valued concept class using a number of samples that grows linearly with its fat-shattering dimension.

What about actually *finding* the state? Even in the classical case, I've completely ignored the computational complexity of finding a hypothesis. I've said that if you somehow found a hypothesis consistent with the data, then you're set, and can explain future data, but how do you actually find the hypothesis? For that matter, how do you even write down the answer in the quantum case? Writing out the state explicitly would take exponentially many bits! On the other hand, maybe that's not quite so bad, since even in the classical case, it can take exponential time to find your hypothesis.

What this tells us is that, in both cases, if you care about computational and representational efficiency, then you're going to have to restrict the problem to some special case. The results from this chapter, which tell us about sample complexity, are just the beginning of learning theory. They answer the first question, the information-theoretic question, telling us that it suffices to take a linear number of samples. The question of how to find and represent the hypothesis comprises much of the rest of the theory. As yet, very little is known about this part of learning theory in the quantum world.

I can tell you, however, some of what's known in the classical case. Maybe disappointingly, a lot of what's known takes the form of hardness results. For example, with a concept class of Boolean circuits of polynomial size, we believe it's a computationally hard problem to find a circuit (or equivalently, a short efficient computer program) that outputs the data that you've already seen, even supposing such a circuit exists. Of course, we can't actually *prove* that this problem has no polynomial-time algorithm (for that would prove $P \neq NP$), nor, as it turns out, can we even prove in our current state of knowledge that it's NP-complete. What we *do* know is that the problem is at least as hard as inverting one-way functions, and hence breaking almost all modern cryptography. Remember, when we were talking about cryptography in Chapter 8, we talked about one-way functions, which are easy to compute but hard to invert? As we discussed then, Håstad, Impagliazzo, Levin, and Luby[4] proved in 1997 that from any one-way function one can construct a pseudorandom generator, which maps n "truly" random bits to (say) n^2 bits that are indistinguishable from random by any polynomial-time algorithm. And Goldreich, Goldwasser, and Micali had shown earlier[5] that from any pseudorandom generator one can construct a *pseudorandom function family*: a family of Boolean functions $f: \{0, 1\}^n \rightarrow \{0, 1\}$ that are computed by small circuits, but are nevertheless indistinguishable from random functions by any polynomial-time algorithm. And such a family of functions immediately leads to a computationally intractable learning problem.

Thus, we can show based on cryptographic assumptions that these problems of finding a hypothesis to explain data that you've seen are probably hard in general. By tweaking this result a bit, we can say that, if finding a quantum state consistent with measurements that you've made can always be done efficiently, then there's no one-way function secure against quantum attack. What this is saying is that we kind of have to give up hope of solving these learning

---

[4] J. Håstad, R. Impagliazzo, L. A. Levin, and M. Luby, A Pseudorandom Generator from any One-way Function. *SIAM Journal on Computing* **28**:4 (1999), 1364–1396.

[5] O. Goldreich, S. Goldwasser and S. Micali, How to construct random functions. *Journal of the ACM*, 33:4 (1986), 792–807.

problems *in general*, and that we have to just look at special cases. In the classical case, there are special concept classes that we can learn efficiently, such as constant-depth circuits or parity functions. I expect that something similar is true in the quantum world.

PUZZLE

In addition to the aforementioned rectangle learning puzzle, here's another raven puzzle, due to Carl Hempel. Let's say that you want to test our favorite hypothesis that all ravens are black. How do we do this? We go out into the field, find some ravens, and see if they're black. On the other hand, let's take the contrapositive of our hypothesis, which is logically equivalent: "all not-black things are nonravens." This suggests that I can do ornithology research without leaving my office! I just have to look at random objects, note that they are not black, and check if they are ravens. As I go along, I gather data that increasingly confirms that all not-black things are nonravens, confirming my hypothesis. The puzzle is whether this approach works. You're allowed to assume for this problem that I do not go out bird-watching in fields, forests, or anywhere else.

17 Interactive proofs, circuit lower bounds, and more

I ended last chapter by giving you a puzzle problem: can I do ornithology without leaving my office?

I want to know if all ravens are black. The old-fashioned approach would involve *going outside*, looking for a bunch of ravens and seeing if they're black or not. The more modern approach: look around the room at all of the things that are not black and note that they also are not ravens. In this way, I become increasingly convinced that *all* not-black things are not ravens, or equivalently that all ravens are black. Can I be a leader in the field of ornithology this way?

If your answer is "You wouldn't be getting a random sample of nonblack things by just sitting in your office," let me point out that going outside wouldn't get me a random sample of all ravens either.

Something completely tangential that I'm reminded of: there's this game where you're given four cards, each of which you're promised has a letter on one side and a number on the other. If what you can see of the cards is shown in the figure below, which cards do you need to flip over to test the rule that *all cards with a K on one side have a 3 on the other*?

Hand:

Apparently, if you give this puzzle to people, the vast majority get it wrong. In order to test that K $\Rightarrow$ 3, you need to flip the K and the 1. On the other hand, you can give people a completely equivalent problem, where they're a bouncer at a bar and need to know if anyone under 21 (or 19 in Canada) is drinking, and where they're told that there's someone who is drinking, someone who isn't drinking, someone who's over 21 and someone who's under 21. In this scenario, funny enough, most people get it right. You ask the person who's drinking, and the underage customer. This is a completely equivalent problem to the cards, but if you give it to the people in the abstract way, many say (for example) that you have to turn over the 3 and the Q, which is wrong. So, people seem to have this built-in ability to reason logically about social situations, but they have to be painstakingly taught to apply that same ability to abstract mathematical problems.[1]

Anyway, the point is that there are many, many more not-black things then there are ravens, so if there were a pair (raven, not-black), then we would be *much* more likely to find it by randomly sampling a raven then sampling a not-black thing. Therefore, if we sample ravens and fail to find a not-black raven, then we're much more confident in saying that "all ravens are black," because our hypothesis had a much higher chance of being falsified by sampling ravens.

INTERACTIVE PROOFS

"Interactive proofs" have been central objects of study in theoretical computer science and cryptography since the 1980s. Since this book is centered around quantum computing, I'd like to begin our discussion of interactive proofs in an unconventional way: by asking the question *can quantum computers be simulated efficiently by classical computers?*

I was talking to Ed Fredkin a while ago, and he said that he believes that the whole universe is a classical computer and thus everything can be simulated classically. But instead of saying that quantum computing is impossible, he takes things in a very interesting direction, and says that **BQP** must be equal to **P**. Even though we have factoring algorithms for quantum computers that are faster than known classical algorithms, that doesn't mean that there isn't a fast classical factoring algorithm that we don't know about. On the other

---

[1] For more about this, see *How the Mind Works* by Steven Pinker (W. W. Norton & Company, reissue edition, 2009).

side you have David Deutsch, who makes an argument that we've talked about several times before: if Shor's algorithm doesn't involve these "parallel universes," then *how* is it factoring the number?[2] Where was the number factored, if not using these exponentially many universes? I guess one way that you could criticize Deutsch's argument (certainly not the only way), is to say he's assuming that there *isn't* an efficient classical simulation. We believe that there's no way for Nature to perform the same computation using polynomial classical resources, but we don't know that. We can't prove that.

Why can't we prove it? The crucial point is that, if you could prove that $P \neq BQP$, then you would have also proved that $P \neq PSPACE$. Physicists might think it's obvious these classes are unequal and it doesn't even require proof, but that's another matter... As for going in the other direction and proving $P = BQP$, I guess people *have* tried that. I don't know if I should say this on record, but I've even spent a day or two on it. It would at least be nice to put BQP in AM, or the polynomial hierarchy – some preliminary fact like that. Unfortunately, I think we simply don't yet understand efficient computation well enough to answer such questions, completely leaving aside the quantum aspect.

The question is, if $P \neq BQP$, $P \neq NP$, etc., why can't anyone prove these things? There are several arguments that have been given for that. One of them is *relativization*. We can talk about giving a P computer and a BQP computer access to the same oracle. That is, give them the same function that they can compute in a single computation step. There will exist an oracle that makes them equal and there will exist another oracle that makes them unequal. The oracle that makes them equal, for example, could just be a $PSPACE$ oracle which kind of sandwiches everything and just makes everything equal to $PSPACE$. The oracle that makes them unequal could be an oracle for

---

[2] D. Deutsch, *The Fabric of Reality: The Science of Parallel Universes – and Its Implications* (London: Penguin, 1998).

Simon's Problem, or some period-finding problem that the quantum computer can solve but the classical one can't.

If two classes are the same, then intuitively, how can giving them more power make them different? The key is to realize that, when we feed an oracle to a class, we aren't acting on the class itself. We're acting on the *definition* of the class. As an example, even though we believe $P = BPP$ in the real world, it's very easy to construct an oracle O where $P^O \neq BPP^O$. Clearly, if what we were doing was operating on the classes, then operating on two equal classes would give two results that were still equal. But that's *not* what we're doing, and maybe the notation is confusing that way. A rough analogy: it's true that Obama is the president, and it's also true that, if Romney had won the election, then *he* would've been the president. But we can't just blindly substitute the first of these equations into the second in order to conclude that, if Romney had won the election, then he would've been Obama.

So, the message of relativization is that any technique for solving P versus NP, or most of the other great problems of complexity theory, is going to have to be sensitive to the presence of these oracles. This doesn't sound like such a big deal until you realize that almost every proof technique we have is *not* sensitive to the presence of oracles. It's very hard to come up with a technique that *is* sensitive, and that – to me – is why interactive proofs are interesting. This is the one clear and unambiguous example I can show you of a technique we have that doesn't relativize. In other words, we can prove that something is true, which wouldn't be true if you just gave everything an oracle. You can see this as the foot in the door or the one distant point of light in this cave that we're stuck in. Through the interactive proof results, we can get a tiny glimmer of what the separation proofs would eventually have to look like if we ever came up with them. The interactive proof techniques seem much too weak to prove anything like $P \neq NP$, or else you would have heard about it. Already, though, we can use these techniques to get some nonrelativizing separation results. I'll show you some examples of that also.

What about **P** versus **BPP**? The consensus there is that **P** and **BPP** actually *are* equal. We know from Impagliazzo and Wigderson[3] that, if we could prove that there exists a problem solvable in 2^n time that requires circuits of size 2^{cn}, for some $c > 0$, then we could construct a very good pseudorandom generator; that is, one which cannot be distinguished from random by any circuit of fixed polynomial size. Once you have such a generator, you can use it to derandomize any probabilistic polynomial-time algorithm. You can feed your algorithm the output of the pseudorandom generator, and your algorithm won't be able to tell the difference between it and a truly random string. Therefore, the probabilistic algorithm could be simulated deterministically. So we really seem to be seeing a difference between classical randomness and *quantum* randomness. It seems like classical randomness really can be efficiently simulated by a deterministic algorithm, whereas quantum "randomness" can't. One intuition for this is that, with a classical randomized algorithm, you can always just "pull the randomness out," i.e., treat the algorithm as deterministic and the random bits as part of its input. On the other hand, if we want to simulate a quantum algorithm, what does it mean to "pull the quantumness out"?

So let's see this one example of a nonrelativizing technique. So we've got a Boolean formula (like the ones used in SAT) in n variables which is *not* satisfiable. What we'd like is a *proof* that it's not satisfiable. That is, we'd like to be convinced that there is no setting of the n variables that causes our formula to evaluate to TRUE. This is what we saw before as an example of a **coNP**-complete problem. The trouble is that we don't have enough time to loop through every possible assignment and check that none of them work. Now the question that was asked in the 1980s was "what if we have some super-intelligent alien that

[3] R. Impagliazzo and A. Wigderson, P = BPP if E requires exponential circuits: Derandomizing the XOR lemma. In *Proceedings of ACM Symposium on Theory of Computing* (1997), pp. 220–229.

comes to Earth and can interact with us?" We don't trust the alien and its technology, but we'd like it to *prove* to us that the formula is unsatisfiable in such a way that we don't *have* to trust it. Is this possible?

In computational complexity, when we have no idea how to answer a question, we often settle for finding an "oracle" that makes the answer either yes or no. For example, let's say we want to reassure ourselves that, given a Boolean circuit computing some function f, there's no polynomial-time algorithm that takes as input a description of the circuit, and that reliably discovers some particular kind of pattern or regularity in f. (Note that much of modern cryptography is based on beliefs of that kind!) The trouble is that usually, there's no hope of proving such a conjecture, without proving $P \neq NP$ as a *first step*! On the other hand, very often we *can* prove a weaker claim: that no polynomial-time algorithm can discover the pattern or regularity in question, *so long as it accesses f as a black box only.* That is, we can prove that, if the algorithm only learns about f by picking various x and then asking a magic subroutine for the value of $f(x)$, then it will need to access the subroutine exponentially many times.

It's probably like what physicists do when they do perturbative calculations. You do it because it you *can*, and because it at least provides a consistency check on what you really care about. (If *even the black-box version* turned out to be false, then your "real" conjecture would be in deep doodoo!)

So, this is what Fortnow and Sipser did in the late 1980s.[4] They said, alright, suppose you have an exponentially long string, and an alien wants to convince you that this exponentially long string is the all-zero string. That is, that there are no 1s anywhere. So can this prover do it? Let's think of what could happen. The prover could say, "the string is all zeros."

---

[4] L. Fortnow and M. Sipser, Are there interactive protocols for CO-NP languages? *Information Processing Letters*, **28**:5 (1988), 249–51.

"Well, I don't believe you. Convince me."

"Here, this location's a zero. This one's also a zero. So is this one..."

OK, now there's only $2^{10\,000}$ bits left to check, and so the alien says "trust me, they're all zeros." There's not a whole lot the prover can do. Fortnow and Sipser basically formally proved this obvious intuition. Take any protocol of messages between you and the prover that terminates with you saying "yes" if you're convinced and "no" if you aren't. What we could then do is to then pick one of the bits of the string at random, surreptitiously change it to a 1, and almost certainly, the entire protocol goes through as before. You'll still say that the string is all zeros.

As always, we can define a complexity class: **IP**. This is the set of problems where you can be convinced of a "yes" answer by interacting with the prover. So we talked before about these classes like **MA** and **AM** – those are where you have a *constant* number of interactions. **MA** is where the prover sends a message to you and you perform a probabilistic computation to check it. In **AM**, you send a message to the prover, and then the prover sends a message back to you and you run a probabilistic computation. It turns out that with any *constant* number of interactions you get the same class **AM**, so let's be generous and allow polynomially many interactions. The resulting class is **IP**. So what Fortnow and Sipser did is they gave a way of constructing an oracle relative to which **coNP** $\not\subset$ **IP**. They showed that, relative to this oracle, you *cannot* verify the unsatisfiability of a formula via a polynomial number of interactions with a prover. Following the standard paradigm of the field, of course we can't prove unconditionally that **coNP** $\not\subset$ **IP**, but this gives us some evidence; that is, it tells us what we might *expect* to be true.

Now for the bombshell (which was discovered by Lund, Fortnow, Karloff, and Nisan):[5] in the "real," unrelativized world, how do

---

[5] C. Lund, L. Fortnow, H. J. Karloff, and N. Nisan, Algebraic methods for interactive proof systems. *Journal of the ACM*, **39**:4 (1992), 859–68.

we show that a formula is unsatisfiable? We're going to somehow have to use the *structure* of the formula. We'll have to use that it's a Boolean formula that was explicitly given to us, and not just some abstract Boolean function. What will we do? Let's assume this is a 3SAT problem. Since 3SAT is **NP**-complete, that assumption is without loss of generality. There's a bunch of clauses (n of them) involving three variables each, and we want to verify that there's no way to satisfy all the clauses.

Now what we'll do is map this formula to a polynomial over a finite field. This trick is called *arithmetization*. Basically, we're going to convert this logic problem into an algebra problem, and that'll give us more leverage to work with. This is how it works: we rewrite our 3SAT instance as a product of degree-3 polynomials. Each clause – that is, each OR of three literals – just becomes 1 minus the product of 1 minus each of the literals: e.g., (x OR y OR z) becomes

$$1 - (1 - x)(1 - y)(1 - z).$$

Notice that, so long as x, y, and z can only take the values 0 and 1, corresponding to FALSE and TRUE, this polynomial is exactly equivalent to the logic expression that we started with. But now, what we can do is reinterpret the polynomial as being over some much *larger* field. Pick some reasonably large prime number N, and we'll interpret the polynomial as being over GF_N (the field of N elements). I'll call the polynomial $P(x_1, \ldots, x_n)$.

If the formula is unsatisfiable, then no matter what setting $x_1, \ldots, x_n$ you pick for the variables, there's going to be some clause in the formula that isn't satisfied. Hence, one of the degree-3 polynomials that we're multiplying together will be zero, and hence the product will itself be zero. So, there being no satisfying assignments is equivalent to getting zero when you take the sum of the $P(x_1, \ldots, x_n)$ over all 2^n possible Boolean settings of $x_1, \ldots, x_n$.

The problem, of course, is that this doesn't seem any easier than what we started with! We've got this sum over exponentially many terms, and we have to check every one of them and make sure that they're all zero. But now, we can have the prover help us. If we just have this string of all zeros, and he just tells us that it's all zeros, we don't believe him. But now, we've lifted everything to a larger field, and we have some more structure to work with.

So now what can we do? What we ask the prover to do is to sum for us over all 2^{n-1} possible settings of the variables $x_2, \ldots, x_n$, leaving x_1 unfixed. Thus, the prover sends us a univariate polynomial Q_1 in the first variable. Since the polynomial we started with had poly(n) degree, the prover can do this by sending us a polynomial number of coefficients. He can send us this univariate polynomial. Then, what we have to verify is that $Q_1(0) + Q_1(1) = 0$ (everything being mod N). How can we do that? The prover has given us the claimed value of the entire polynomial. So just pick an r_1 at random from our field. Now, what we would like to do is verify that $Q_1(r_1)$ equals what it's supposed to. Forget about 0 and 1, we're just going to go somewhere else in the field. Thus, we send r_1 to the prover. Now, the prover sends a new polynomial Q_2, where the first variable is fixed to be r_1, but where x_2 is left unfixed and $x_3, \ldots, x_n$ are summed over all possible Boolean values (like before). We still don't know that the prover hasn't been lying to us and sending bullshit polynomials. So what can we do?

Check that $Q_2(0) + Q_2(1) = Q_1(r_1)$, then pick another element r_2 at random and send it to the prover. In response, he'll send us a polynomial $Q_3(X)$. This will be a sum of $P(x_1, \ldots, x_n)$ over all possible Boolean settings of x_4 up to x_n, with x_1 set to r_1 and x_2 set to r_2, and x_3 left unfixed. Again, we'll check and make sure that $Q_3(0) + Q_3(1) = Q_2(r_2)$. We'll continue by picking a random r_3 and sending it along to the prover. This keeps going for n iterations, when we reach the last variable. What do we do when we reach the last iteration? At that point, we can just evaluate $P(r_1, \ldots, r_n)$ ourselves without the prover's help, and check directly if it equals $Q_n(r_n)$.

We have a bunch of tests that we're doing along the way. My first claim is that, if there is no satisfying assignment, and if the prover was not lying to us, then each of the n tests accepts with certainty.

The second claim is that, if there was a satisfying assignment, then with high probability, at least one of these tests would fail. Why is that the case? The way I think of it is that the prover is basically like the girl in Rumpelstiltskin. The prover is just going to get trapped in bigger and bigger lies as time goes on until finally the lies that will be so preposterous that we'll be able to catch them. This is what's going on. Why? Let's say that, for the first iteration, the real polynomial that the prover should give us is Q_1, but that the prover gives us Q_1' instead. Here's the thing: these are polynomials of not too large a degree. The final polynomial, P, has degree at most three times the number of clauses. We can easily fix the field size to be larger. So let the degree d of the polynomial be much smaller than the field size N.

A quick question: suppose we have two polynomials P_1 and P_2 of degree d. How many points can they be equal at (assuming they aren't identical)? Consider the difference $P_1 - P_2$. Since this is also a polynomial of degree at most d, by the Fundamental Theorem of Algebra, it can have at most d distinct roots (again, assuming it's not identically zero). Thus, two polynomials that are not equal can agree in at most d places, where d is the degree. This means that, if these are polynomials over a field of size N, and we pick a random element in the field, we can bound the probability that the two will agree at that point: it's at most d/N.

Going back to the protocol, we assumed that d is much less than N, and so the probability that Q_1 and Q_1' agree at some random element of the field is much less than unity. So when we pick r_1 at random, the probability that $Q_1(r_1) = Q_1'(r_1)$ is at most d/N. Only if we've gotten very unlucky will we pick r_1 such that these are equal, so we can go on and assume that $Q_1(r_1) \neq Q_1'(r_1)$. Now, you can picture the prover sweating a little. He's trying to convince us of a lie, but maybe he can still recover. But next, we're going to pick an r_2 at random. Again, the probability that he'll be able to talk himself out of the next lie is going to be at most d/N. This is the same in each of the iterations, so the probability that he can talk himself out of *any* of the lies is at most nd/N. We can just choose N to be big enough that this will be much smaller than unity.

Why not just run this protocol over the positive integers? Because we don't have a way of generating a *random* positive integer, and we need to be able to do that. So we just pick a very large finite field.

So this protocol gives us that **coNP** $\subseteq$ **IP**. Actually, it gives us something stronger.

A standard kind of argument shows us that the biggest **IP** could possibly be in our wildest dreams would be **PSPACE**. You can prove that anything you can do with an interactive protocol, you can *simulate* in **PSPACE**. Can we bring **IP** up? Make it bigger? What we were trying to verify was that all of these values of $P(x_1, \ldots, x_n)$ summed to zero, but the same proof would go through as before if we were trying to verify that they summed to some other constant (whatever we want).

In other words, with Merlin's help, Arthur can actually *count* the number of Boolean strings $x_1, \ldots, x_n$ such that $P(x_1, \ldots, x_n) = 1$, not just decide whether the number is zero. More formally, Arthur can solve any problem in the complexity class **#P** (pronounced "sharp-P"), which was defined by Valiant in 1979.[6]

OK, time for a digression. Unlike the other complexity classes we've seen so far, **#P** consists not of yes-or-no decision problems but of functions. A function f, mapping binary strings to nonnegative integers, is said to be in **#P** if there exists a polynomial-time algorithm V and polynomial p, such that $f(x)$ is equal to the number of $p(n)$-bit strings w that cause $V(x, w)$ to accept. In simpler terms, **#P** is the class of all problems that can be phrased as *counting* the number of solutions to an **NP** problem. Now, if we ask where **#P** fits in with the

---

[6] L. G. Valiant, The complexity of enumeration and reliability problems, *SIAM Journal on Computing*, 8:3 (1979), 410–421.

complexity classes we've already seen, we encounter an apples-and-oranges issue: how do we compare a class of functions with classes of languages? But a simple solution, often used in practice, is to consider the class $P^{\#P}$, which consists of all languages decidable by a P machine with access to a $\#P$ *oracle*.

Exercise for the non-lazy reader: Show that $P^{\#P} = P^{PP}$, where PP is the "majority-vote" class defined in Chapter 7. (That is, in some sense, PP "already contains the power of $\#P$ latent within it.")

Now, an extremely important result proved in 1990, called *Toda's Theorem*, says that $P^{\#P}$ contains the entire polynomial hierarchy PH. If it's not intuitively obvious to you why a counting oracle is so powerful – well, it *shouldn't* be obvious! Toda's Theorem came as a great surprise to everyone. Sadly, I don't have time here to discuss the proof of Toda's Theorem,[7] but I'll have several occasions to use the theorem in the remainder of this book.

Anyway, in terms of complexity classes, what we observed earlier means that $P^{\#P} \subseteq IP$: in an interactive protocol, Merlin can convince Arthur of the solution to any $\#P$ problem, and therefore any $P^{\#P}$ problem as well (since Arthur can simply use Merlin in place of the $\#P$ oracle). By Toda's Theorem, this in turn means that IP contains PH.

After this "LFKN Theorem" came out, a number of people carried out a discussion by email, and a month later Shamir figured out that $IP = PSPACE$ – that is, IP actually "hits the roof."[8] I won't go through Shamir's result here, but this means that, if a super-intelligent alien came to Earth, it could prove to us whether white or black has the winning strategy in chess, or if chess is a draw. It could play us and beat us, of course, but then all we'd know is that it's a better chess player. But it could prove to us which player has the winning strategy by reducing chess to this game of summing polynomials over large

---

[7] For nice proofs, see for example Lance Fortnow's "A Simple Proof of Toda's Theorem" (http://theoryofcomputing.org/articles/v005a007/v005a007.pdf), or the book *Gems of Theoretical Computer Science* by Uwe Schöning (Springer, 1998).

[8] A. Shamir, IP = PSPACE. *Journal of the ACM*, **39**:4 (1992), 869–77.

finite fields. (Technical note: this only works for chess with some reasonable limit on the number of moves, like the "50-move rule" used in tournament play.)

This is already something that is – to me – pretty counterintuitive. Like I said, it gives us a very small glimpse of the kinds of technique we'd need to use to prove nonrelativizing results like **P** $\neq$ **NP**. A lot of people seem to think that the key is somehow to transform these problems from Boolean to algebraic ones. The question is how to do that. I can show you, though, how these techniques already let you get some new lower bounds. Heck, even some quantum circuit lower bounds.

First claim: if we imagine that there are polynomial-size circuits for counting the number of satisfying assignments of a Boolean formula, then there's also a way to *prove* to someone what the number of solutions is. Do you see why this would follow from the interactive proof result? Well, notice that, to convince the verifier about the number of satisfying assignments of a Boolean formula, the prover *itself* does not need to have more computational power than is needed to count the number of assignments. After all, the prover just keeps having to compute these exponentially large sums! In other words, the prover for **#P** can be implemented in **#P**. If you had a **#P** oracle, then you too could be the prover. Using this fact, Lund *et al.* pointed out that if **#P** $\subset$ **P/poly** – that is, if there's some circuit of size polynomial in n for counting the number of solutions to a formula of size n – then **P**$^{\#P}$ = **MA**. For in **MA**, Merlin can give Arthur the polynomial-size circuit for solving **#P** problems, and then Arthur just has to verify that it works. To do this, Arthur just runs the interactive protocol from before, but where he plays the part of both the prover and the verifier, and uses the circuit itself to simulate the prover. This is an example of what are called *self-checking programs*. You don't have to trust an alleged circuit for counting the number the solutions to a formula, since you can put it in the role of a prover in an interactive protocol.

Now, we can prove that the class **PP**, consisting of problems solvable in probabilistic polynomial-time with unbounded error, does not have linear-sized circuits. This result is originally due to Vinodchandran.[9] Why? Well, there are two cases. If **PP** doesn't even have polynomial-sized circuits, then we're done. On the other hand, if **PP** *does* have polynomial-sized circuits, then so does $\mathbf{P}^{\#P}$, by the basic fact (which you might enjoy proving) that $\mathbf{P}^{\#P} = \mathbf{P}^{PP}$. Therefore, $\mathbf{P}^{\#P} = \mathbf{MA}$ by the LFKN Theorem, so $\mathbf{P}^{\#P} = \mathbf{MA} = \mathbf{PP}$, since **PP** is sandwiched in between **MA** and $\mathbf{P}^{\#P}$. But one can prove (and we'll do this shortly) that $\mathbf{P}^{\#P}$ doesn't have linear-sized circuits, using a direct diagonalization argument. Therefore, **PP** doesn't have linear-sized circuits either.

In fact, the conclusion is stronger: for any fixed k, we can find a language L in $\mathbf{P}^{\#P}$, or even **PP**, such that L cannot be decided by a circuit of size $O(n^k)$. That's very different from saying that there is a *single* language in **PP** that does not have circuits of polynomial size. The second statement is almost unimaginably harder to show! If you give me your (polynomial) bound, then I find a **PP** problem that defeats circuits constrained by your bound, but the problem might be solvable by circuits with some larger polynomial bound. To defeat that larger polynomial bound, I'd have to construct a different problem, and so on indefinitely.

Let's go back now and fill in the missing step in the argument. We want to show for some fixed k that $\mathbf{P}^{\#P}$ doesn't have circuits of size n^k. How many possible circuits are there of size n^k? Something like n^{2n^k}. Now what we can do is define a Boolean function f by looking at the behavior of all circuits of size n^k. Order the possible inputs of size n as $x_1, \ldots, x_{2^n}$. If at least half of the circuits accept x_1, then set $f(x_1) = 0$, while if more than half of the circuits reject x_1, then set $f(x_1) = 1$. This kills off at least half of the circuits of size n^k (i.e., causes them to fail at computing f on at least one input). Now, of those circuits that got the "right answer" for x_1, do the majority of them accept or reject x_2? If the majority accept, then set $f(x_2) = 0$. If the majority reject,

[9] N. V. Vinodchandran, A note on the circuit complexity of PP. *Theoretical Computer Science*, **347**:1/2 (2005), 415–18.

then set $f(x_2) = 1$. Again, this kills off at least half of those circuits remaining. We continue this Darwinian process, where each time we define a new value of our function we kill off at least half of the remaining circuits of size n^k. After $\log_2(n^{2n^k}) + 1 \approx 2n^k \log(n)$ steps, we will have killed off *all* of the circuits of size n^k. Furthermore, the process of constructing f involves a polynomial number of counting problems, each of which we can solve in $\mathbf{P^{\#P}}$. So the end result is a problem that is in $\mathbf{P^{\#P}}$, but which by construction does not have circuits of size n^k (for any fixed k of our choice). This is an example of a relativizing argument, because we paid no attention to whether these circuits had any oracles or not. To get this argument to go down from $\mathbf{P^{\#P}}$ to the smaller class $\mathbf{PP}$, we had to use a nonrelativizing ingredient: namely, the interactive proof result of LFKN.

But does this actually give us a nonrelativizing circuit lower bound? That is, does there exist an oracle relative to which $\mathbf{PP}$ *has* linear-sized circuits? Years ago, I was able to construct such an oracle.[10] This shows that Vinodchandran's result was nonrelativizing – indeed, it's one of the few examples in all of complexity theory of an indisputably nonrelativizing separation result. In other words, the relativization barrier – which is one of the main barriers to showing $\mathbf{P} \neq \mathbf{NP}$ – can be overcome in some very limited cases.

New developments

Anyway, that's where things stood when I first wrote this chapter in 2006. Since then, there have been some exciting developments. First, in 2007, Rahul Santhanam[11] improved on Vinodchandran's result, to show that **PromiseMA** – the class of all promise problems with Merlin–Arthur proof protocols – does not have circuits of size n^k, for any fixed k.

---

[10] S. Aaronson, Oracles are subtle but not malicious. In *Proceedings of IEEE Conference on Computational Complexity* (2006), pp. 340–54. http://arxiv.org/pdf/cs.CC/0504048.pdf

[11] R. Santhanam, Circuit lower bounds for Merlin–Arthur classes. *SIAM Journal on Computing*, 39:3 (2009), 1038–61.

Shortly afterward, inspired by Santhanam's result, Avi Wigderson and I[12] discovered a *new* barrier to further progress in complexity theory, which we called "algebrization." Basically, algebrization extends the original relativization barrier of Baker, Gill, and Solovay, in that, when we study a question about complexity classes relative to some oracle A, we now let one of the complexity classes access a "low-degree polynomial extension" of A, rather than just A itself. This more powerful kind of oracle access gives us some additional leverage; in particular, it lets us mimic all the standard non-relativizing results based on arithmetization. So, for example, while (as we discussed before) it's not true that $\mathbf{IP}^A = \mathbf{PSPACE}^A$ for every oracle A, it turns out it *is* true that $\mathbf{PSPACE}^A \subseteq \mathbf{IP}^{\sim A}$, where $\sim$A denotes a low-degree polynomial over a large finite field that happens to equal A when restricted to Boolean inputs. Thus, we say that the $\mathbf{IP} = \mathbf{PSPACE}$ theorem does "algebrize," even though it doesn't relativize. On the other hand, Avi and I also showed that, for most of the famous open problems – including not only **P** vs. **NP**, but also **P** vs. **BPP**, **NEXP** vs. **P/poly**, and others – any solution will require "non-algebrizing techniques," which fail relative even to these new algebraic oracles, in the same sense that the $\mathbf{IP} = \mathbf{PSPACE}$ theorem failed relative to ordinary oracles. So, the upshot is that the techniques used for the interactive proof breakthroughs can only get us so far: sure, they evade the relativization barrier, but only to smack headfirst into a "generalized" relativization barrier waiting a few steps beyond.

Are there lower bound techniques that avoid the relativization *and* algebrization barriers? Yes; in fact they've existed for decades. In the early 1980s, Furst, Saxe, and Sipser[13] and (independently) Ajtai[14] discovered a revolutionary technique for lower-bounding the sizes

[12] S. Aaronson and A. Wigderson, Algebrization: a new barrier in complexity theory. *ACM Transactions on Computing Theory*, 1:1 (2009), 2:1–54.

[13] M. L. Furst, J. B. Saxe, and M. Sipser, Parity, circuits, and the polynomial-time hierarchy. *Mathematical Systems Theory*, 17:1 (1984), 13–27.

[14] M. Ajtai. Sigma_1^1-formulae on finite structures. *Annals of Pure and Applied Logic*, 24 (1983), 1–48.

of *constant-depth circuits*: for example, AC^0 circuits, which consist of AND, OR, and NOT gates arranged into $O(1)$ layers (where each AND and OR gate can have an arbitrary number of inputs). Furst *et al.* and Ajtai showed that, for certain functions like the parity of n bits, any AC^0 circuit must have an exponential number of gates. Since their techniques were highly combinatorial – based on looking at the behaviors of actual, individual gates – they evaded the relativization barrier. Since then, other lower bounds have been proved along these lines – most notably those of Razborov[15] and Smolensky[16] for AC^0 circuits augmented with the ability to perform arithmetic mod p (where p is some fixed prime).

Alas, in 1993, Razborov and Rudich pointed out[17] that almost all of these "combinatorial-style" lower bounds ran up against a barrier they called "Natural Proofs" – one that, in some ways, is complementary to the relativization barrier. To summarize in a paragraph: the combinatorial lower bound techniques work by showing that certain functions (e.g., PARITY) are hard for small circuits, *because* these functions "look like random functions" in some efficiently computable respect, whereas any function computed by a small circuit must look *non*-random in that respect. However, any argument of this sort can be turned on its head, and used to distinguish "truly" random functions from pseudorandom functions – thereby, ironically, solving some of the same problems we wanted to prove were hard! The Furst *et al.* and Ajtai arguments were able to work *because* AC^0 circuits are too weak to compute pseudorandom functions – in fact, the impossibility of pseudorandomness in AC^0 can be derived as a *consequence* of their lower-bound proofs. But we can't expect any similar arguments to work for proving lower bounds against more

[15] A. A. Razborov, On the method of approximations. In *Proceedings of ACM Symposium on Theory of Computing* (New York: ACM, 1989), pp. 167–76.
[16] R. Smolensky, Algebraic methods in the theory of lower bounds for Boolean circuit complexity. In *Proceedings of ACM Symposium on Theory of Computing* (New York: ACM, 1987), pp. 77–82.
[17] A. A. Razborov and S. Rudich, Natural proofs. *Journal of Computer and System Sciences*, 55:1 (1997), 24–35.

powerful circuit classes like **P/poly** – assuming, as almost all of us believe, that those classes *do* have pseudorandom functions. (In slogan form, the *fact* of computational hardness is what makes *proving* computational hardness so hard!) Furthermore, Naor and Reingold have shown[18] that, under plausible cryptographic assumptions, even the class **TC⁰** – consisting of constant-depth circuits with MAJORITY gates – is able to compute pseudorandom functions. So the Razborov–Rudich natural proof barrier really seems to kick in just "slightly" above **AC⁰**.

If you want to evade the natural proof barrier, you seem to need techniques that "zero in" on some *special* property of the function f that you're trying to prove is hard – a property that f *doesn't* share with a random function. The obvious example of a technique that *does* zero in on such a special property is "diagonalization" – i.e., the sort of technique we used earlier to prove that $\mathbf{P^{\#P}}$ doesn't have linear-size circuits. (Recall that our proof used the ability of a #P machine to simulate all possible linear-size circuits, and evade simulation by any of them.) Alas, while these sorts of technique evade the natural proof barriers, they're also precisely the ones that *don't* evade the relativization barrier! I mean, yes, they evade relativization if you soup them up with interactive proof techniques – but even then, they're still subject to the algebrization barrier.

So, let's ask the obvious next question: is there a circuit lower bound that evades *all three barriers simultaneously* – relativization, algebrization, and natural proofs? In my opinion, the first convincing example of such a lower bound came very recently, in 2010, with Ryan Williams' breakthrough result[19] that **NEXP** $\not\subset$ **ACC⁰**. Here **NEXP** is nondeterministic exponential time, while **ACC⁰** is a slight extension of **AC⁰**, to allow modular arithmetic in *any* base (recall that we already

[18] M. Naor and O. Reingold, Number-theoretic constructions of efficient pseudo-random functions. *Journal of the ACM*, **51**:2 (2004), 231–62.

[19] R. Williams, Non-uniform ACC circuit lower bounds. In *Proceedings of IEEE Conference on Computational Complexity* (Silver Springs, MD: IEEE Computer Society Press, 2011), pp. 115–25.

knew lower bounds, if $\mathbf{AC^0}$ were extended to allow arithmetic modulo a *specific* prime). You might notice that this result seems pathetically weak, compared with what we believe is true! Yet it's a real milestone because of its evasion of all the known barriers (strictly speaking, we don't know whether the natural proof barrier applies to $\mathbf{ACC^0}$, but *if it does* then Williams's proof evades it!). To pull this off, Williams had to use the "kitchen sink": diagonalization, insights from interactive proofs, *and* various new and old results about nontrivial structure in $\mathbf{ACC^0}$ functions.

Is there a *fourth* barrier, to which even Williams's new result is subject? Well, I don't know! As a general rule, before we can think about the barriers to a given technique, I'd say we need *at least two successful examples* of applications of the technique, for pretty much the same reason why we need at least two points to fit a line.

In any case, one thing the existing lower bounds have made clear is the depth of ideas needed even to prove things ridiculously weaker than $\mathbf{P} \neq \mathbf{NP}$. *This* is the reason why I don't get heart palpitations every time another claimed $\mathbf{P} \neq \mathbf{NP}$ proof arrives in my in-box (and they *do* show up at least once per month)! It's not just because I've seen so many previous attempts crash and burn; it's because I ask myself, how does this generalize, or subsume, or build on the nontrivial solutions we already know to tiny subproblems of $\mathbf{P}$ versus $\mathbf{NP}$?

The (secret?) fear of many of us is that, to make further progress in circuit lower bounds, it will be necessary to ramp up this field's mathematical sophistication by orders of magnitude. At any rate, that's a central contention of Ketan Mulmuley's "Geometric Complexity Theory" (GCT) program,[20] which tries to tackle circuit lower bounds using algebraic geometry, representation theory, and seemingly every other subject about which yellow books have been written.

---

[20] For more information, see K. Mulmuley, The GCT program toward the P vs. NP problem. *Communications of the ACM*, 55:6 (2012), 98–107, http://ramakrishnadas.cs .uchicago.edu/, or Joshua Grochow's beautiful PhD thesis (Symmetry and equivalence relations in classical and geometric complexity theory. Doctoral dissertation, University of Chicago (2012). http://people.cs.uchicago.edu/~joshuag/grochow-thesis.pdf).

GCT is a whole topic to itself, and it would take me too far afield to even *start* to explain it. I'll simply say that, personally, I like to call GCT "the string theory of computer science": on the one hand, it's made such striking mathematical connections that, once you see them, you feel like the program *must* be somehow on the right track. On the other hand, if you judge the program by how many answers it's delivered to the sorts of question it originally sought to address – questions *not* internal to the program itself – then the results haven't yet lived up to the early hopes.

Quantum interactive proofs

While we're waiting for better classical circuit lower bounds, let me now circle back and tell you something about *quantum* interactive proof systems. Well, I guess a first thing to say is that even results about classical interactive proof systems – the results we already saw – can be used to get quantum circuit lower bounds. So, for example, by slightly modifying our proof that **PP** does not have circuits of size n^k, one can prove that **PP** doesn't even have *quantum* circuits of size n^k. OK, but this is peanuts. Let's try to throw in quantum to something and get a *different* answer.

We can define a complexity class **QIP**: Quantum Interactive Proofs. This is the same as **IP**, except that now you're a quantum polynomial-time verifier, and instead of exchanging classical messages with the prover, you can exchange quantum messages. For example, you could send the prover half of an EPR pair and keep the other half, and play whatever other such games you want.

Certainly, this class is at least as powerful as **IP**. You could just restrict yourself to classical messages if that's what you wanted to do. Since **IP** = **PSPACE**, we also know **QIP** has to be *at least* as big as **PSPACE**. Using a semidefinite programming argument, Kitaev and Watrous[21] also proved early on that **QIP** ⊆ **EXP**. In 2006, when I first wrote this chapter, this was actually all we knew about where

---

[21] http://www.cpsc.ucalgary.ca/~jwatrous/papers/qip2.ps

QIP lies. But in 2009, Jain, Ji, Upadhyay, and Watrous had this break-through where they showed[22] that QIP can even be simulated in PSPACE, and hence QIP = IP = PSPACE. So, ultimately, quantum interactive proof systems turned out to have exactly the same power as classical ones. Amusingly, in the classical case, the big surprise was that these systems can simulate PSPACE, while in the quantum case, the big surprise was that PSPACE can simulate *them!*

So, is there any way in which quantum interactive proof systems are *interestingly* different from classical ones? Well, one amazing fact – which was proved by Kitaev and Watrous,[23] and which played a crucial role in the proof of the QIP = PSPACE theorem – is that any quantum interactive protocol can be simulated by one that takes place in *three rounds*. In the classical case, we had to play this whole Rumpelstiltskin game, where we kept asking the prover one question after another until we finally caught him in a lie. We had to ask the prover polynomially many questions. But in the quantum case, it's no longer necessary to do that. The prover sends you a message, you send a message back, then the prover sends you one more message and that's it. That's all you ever need.

We won't prove here why that's true, but I can give you some intuition. Basically, the prover prepares a state that looks like $\sum_r |r\rangle |q(r)\rangle$. This r is the sequence of all the random bits that you would use in the classical interactive protocol. Let's say that we're taking the classical protocol for solving coNP or PSPACE, and we just want to simulate it by a three-round quantum protocol. We sort of glom together all the random bits that the verifier would use in the entire protocol and take a superposition over all possible settings of those random bits. Now what's $q(r)$? It's the sequence of messages that the prover would send back to you if you were to feed it the random bits

[22] R. Jain, Z. Ji, S. Upadhyay, and J. Watrous, QIP = PSPACE. *Journal of the ACM*, 58:6 (2011), 30.

[23] A. Kitaev and J. Watrous, Parallelization, amplification, and exponential time simulation of quantum interactive proof systems. In *Proceedings of Annual ACM Symposium on Theory of Computing* (New York: ACM, 2000), pp. 608–17.

in r. Now, the prover will just take the q register and second r register and will send it to you. Certainly, the verifier can check that then $q(r)$ is a valid sequence of messages given r. What's the problem? Why isn't this a good protocol?

The superposition could be over a subset of the possible random bits! How do we know that the prover didn't just cherry-pick r to be only drawn from those that he could successfully lie about? The *verifier* needs to pick the challenges. You can't have the prover picking them for you. But now, we're in the quantum world, so maybe things are better. If you imagine in the classical world that there was some way to verify that a bit is random, then maybe this would work. In the quantum world, there *is* such a way. For example, if you were given a state like

$$\frac{|0\rangle + |1\rangle}{\sqrt{2}}$$

you could just rotate it and verify that, had you measured in the standard basis, you would have gotten 0 and 1 with roughly equal probability. More precisely: if the outcome in the standard basis would have been random, then you'll accept with unit probability; if the outcome would have been far from random, then you'll reject with noticeable probability.

Still, the trouble is that our $|r\rangle$ is entangled with the $|q(r)\rangle$ qubits. So we can't just apply Hadamard operations to $|r\rangle$ – if we did, we'd just get garbage out. However, it turns out that what the verifier can do is to pick a random round i of the protocol being simulated – say, there are n such rounds – and then ask the prover to *uncompute* everything after round i. Once the prover has done that, he's eliminated the entanglement, and the verifier can then check by measuring in the Hadamard basis that the bits for round i really were random. If the prover cheated in some round and didn't send random bits, this lets the verifier detect that with probability that scales inversely with the number of rounds. Finally, you can repeat the whole protocol in parallel a polynomial number of times to increase your confidence.

(I'm skipping a whole bunch of details – my goal here was just to give some intuition.)

Let's compare the quantum situation to the classical world. There, you've just got **MA** and **AM**: every proof protocol between Arthur and Merlin with a larger constant number of rounds collapses to **AM**. If you allow a polynomial number of rounds, then you go up to **IP** (which equals **PSPACE**). In the quantum world, you've **QMA**, **QAM**, and then **QMAM**, which is the same as **QIP** = **PSPACE**. There's also another class, **QIP[2]**, which is different from **QAM** in that Arthur can send any arbitrary string to Merlin (or even a quantum state) instead of just a random string. In the classical case, **AM** and **IP[2]** are the same, but in the quantum case, we don't know that.

That's our tour of interactive proofs, so I'll end with a puzzle for next chapter. God flips a fair coin. Assuming that the coin lands tails, She creates a room with a red-haired person. If the coin lands heads, She creates two rooms: one has a person with red hair and the other has a person with green hair. Suppose that you know that this is the whole situation, then wake up to find a mirror in the room. Your goal is to find out which way the coin landed. If you see that you've got green hair, then you know right away how the coin landed. Here's the puzzle: if you see that you have red hair, what is the probability that the coin landed heads?

18 Fun with the Anthropic Principle[1]

This is a chapter about the Anthropic Principle, and how you apply Bayesian reasoning where you have reason about the probability of your own existence, which seems like a very strange question. It's a *fun* question, though – which you can almost define as a question where it's much easier to have an opinion than to have a result. But then, we can try to at least clarify the issues, and there are *some* interesting results that we can get.

There's a central proposition that many people interested in rationality believe they should base their lives around – even though they generally don't in practice. This is Bayes's Theorem.

$$P[H|E] = \frac{P[E|H]P[H]}{P[E]}$$
$$= \frac{P[E|H]P[H]}{\sum_{H'} P[E|H']P[H']}.$$

If you talk to philosophers, this is often the one mathematical fact that they know. (Kidding!) As a theorem, Bayes's Theorem is completely unobjectionable. The theorem tells you how to update your probability of a hypothesis H being true, given some evidence E.

The term $P[E|H]$ describes how likely you are to observe the evidence E in the case that the hypothesis H holds. The remaining terms on the right-hand side, $P[H]$ and $P[E]$, are the two tricky ones. The first one describes the probability of the hypothesis being true, independent of any evidence, while the second describes the probability of the evidence being observed, averaged over all possible hypotheses (and weighted by the hypotheses' probabilities). Here, you're making

---

[1] Starting in this chapter, we've included some Q&A dialogues with students who took the class.

a commitment that there *are* such probabilities in the first place – in other words, that it makes sense to talk about what the Bayesians call a *prior*. When you're an infant first born into the world, you estimate there's some chance that you're living on the third planet around the local star, some other chance you're living on the fourth planet and so on. That's what a prior means: your beliefs before you're exposed to any evidence about anything. You can see already that maybe that's a little bit of a fiction, but supposing you have such a prior, Bayes's Theorem tells you how to update it given new knowledge.

The proof of the theorem is trivial. Multiply both sides by $P[E]$, and you get that $P[H|E]P[E] = P[E|H]P[H]$. This is clearly true, since both sides are equal to the probability of the evidence and the hypothesis together.

So if Bayes's Theorem seems unobjectionable, then I want to make you feel queasy about it. That's my goal. The way to do that is to take the theorem very, very seriously as an account of how we should reason about the state of the world.

I'm going to start with a nice thought experiment which is due to the philosopher Nick Bostrom.[2] This is called *God's Coin Toss*. Last chapter, I described the thought experiment as a puzzle.

Imagine that, at the beginning of time, God flips a fair coin (one that lands heads or tails with equal probability). If the coin lands heads, then God creates two rooms: one has a person with a red hair and the other has a person with green hair. If the coin lands tails, then God creates one room with a red-haired person. These rooms are the entire universe, and these are the only people in the universe.

We also imagine that everyone knows the entire situation and that the rooms have mirrors. Now, suppose that you wake up, look in the mirror and learn your hair color. What you really want to know is which way the coin landed. Well, in one case, this is easy. If you have

---

[2] See for example Nick Bostrom, *Anthropic Bias: Observation Selection Effects in Science and Philosophy*, Routledge, 2010.

green hair, then the coin *must* have landed heads. Suppose that you find you have red hair. Conditioned on that, what probability should you assign to the coin having landed heads?

One half is the first answer that someone could suggest. You could just say, "look, we know the coin was equally likely going to land heads or tails, and we know that in either case that there was going to be a red-haired person, so being red-haired doesn't really tell you anything about which way it landed, and therefore it should be a half." Could someone defend a different answer?

Student: It seems more likely for it to have landed tails, since in the heads case, the event of waking up with red hair is diluted with the other possible event of waking up with green hair. The effect would be more dramatic if there were a hundred rooms with green hair.

Scott: Exactly.

Student: It isn't clear at all that the choice of whether you're the red- or the green-haired person in the heads case is at all probabilistic. We aren't guaranteed that.

Scott: Right. This is a question.

Student: It could have been that, before flipping the coin, God wrote down a rule saying that if the coin lands heads, make you red-haired.

Scott: Well, then we have to ask, what do we mean by "you?" Before you've looked in the mirror, you really don't know what your hair color is. It really could be either, unless you believe that it's really a part of the "essence" of you to have red hair. That is, if you believe that there is no possible state of the universe in which you had green hair, but otherwise would have been "you."

Student: Are both people asked this question, then?

Scott: Well, the green-haired person certainly knows how to answer, but you can just imagine that the red-haired people in the heads and tails cases are both asked the question.

To make the argument a little more formal, you can just plop things into Bayes's Theorem. We want to know the probability $P[H|R]$ that the coin landed heads, given that you have red hair. We could do the calculation, using that the probability of us being red haired given that the coin lands heads is $1/2$, all else being equal. There are two people, and you aren't a priori more likely to be either the red-haired or the green-haired person. Now, the probability of heads is also $1/2$ – that's no problem. What's the total probability of your having red hair? That's just given by $P[R|H]P[H] + P[R|T]P[T]$. As we've said before, if the coin lands tails, you certainly have red hair, so $P[R|T] = 1$. Moreover, we've already assumed that $P[R|H] = 1/2$. Thus, what you get out is that $P[H|R] = 1/4 / 3/4 = 1/3$. So, if we do the Bayesian calculation, it tells us that the probability should be $1/3$ and not $1/2$.

Do you see an assumption that you could make that would get the probability back to $1/2$?

Student: You could make the assumption that whenever you exist, you have red hair.

Scott: Yeah, that's one way, but is there a way to do it that doesn't require a prior commitment about your hair color?

Well, there is a way to do it, but it's going to be a little bit strange. One way is to point out that in the heads world, there are twice as many people in the first place. So you could say that you're a priori twice as likely to exist in a world with twice as many people. In other words, you could say that your own existence is a piece of evidence you should condition upon. I guess if you want to make this concrete, the metaphysics that this would correspond to is that there's some sort of a warehouse full of souls, and depending on how many people there are in the world, a certain number of souls get picked out and placed into bodies. You should say that in a world with more people, it would be more likely that you'd be picked at all.

If you do make that assumption, then you can run through the same Bayesian ringer. You find that the assumption precisely negates the effect of reasoning that if the coin landed heads, then you could have had green hair. So you get back to $1/2$.

Thus, we see that depending on how you want to do it, you can get an answer of either a third or a half. It's possible that there are other answers that could be defended, but these seem like the most plausible two.

That was a fairly serene thought experiment. Can we make it more dramatic? Part of the reason that this feels like philosophy is that there aren't any real stakes here. Let's get closer to something with real stakes in it.

The next thought experiment is, I think, due to the philosopher John Leslie.[3] Let's call it the *Dice Room*. Imagine that there's a very, very large population of people in the world, and that there's a madman. What this madman does is, he kidnaps ten people and puts them in a room. He then throws a pair of dice. If the dice land snake-eyes (two ones), then he simply murders everyone in the room. If the dice do not land snake-eyes, then he releases everyone, then kidnaps 100 people. He now does the same thing: he rolls two dice; if they land snake-eyes, then he kills everyone, and if they don't land snake-eyes, then he releases them and kidnaps 1000 people. He keeps doing this until he gets snake-eyes, at which point he's done. So now, imagine that you've been kidnapped. You have been watching the news and you know the entire situation. You can assume either that you do or do not know how many other people are in the room.

So you're in the room. Conditioned on that fact, how worried should you be? How likely is it that you're going to die?

One answer is that the dice have a 1/36 chance of landing snake-eyes, so you should be only a "little bit" worried (considering). A second reflection you could make is to consider, of people who enter the room, what the fraction is of people who ever get out. Let's say that it ends at 1000. Then, 110 people get out and 1000 die. If it ends at 10000, then 1110 people get out and 10000 die. In either case, about 8/9 of the people who ever go into the room will die.

---

[3] See for example John Leslie, *The End of the World: The Science and Ethics of Human Extinction*, Routledge, 1998.

Student: But that's not conditioning on the full set of information. That's just conditioning on the fact that I'm in the room at *some* point.

Scott: But you'll basically get the same answer, no matter what time you go into the room. No matter when you assume the process terminates, about 8/9 of the people who ever enter the room will be killed. For each termination point, you can imagine being a random person in the set of rooms leading up to that point. In that case, you're much more likely to die.

Student: But aren't you conditioning on future events?

Scott: Yes, but the point is that we can *remove* that conditioning. We can say that we're conditioning on a specific termination point, but that no matter what that point is, we get the same answer. It could be 10 steps or 50 steps, but no matter what the termination point is, almost all the people who go into the room are going to die, because the number of people is increasing exponentially.

If you're a Bayesian, then this kind of seems like a problem. You could see this as a bizarre madman-related thought experiment, or if you preferred to, you could defend the view that this is the actual situation the human species is in. We'd like to know what the probability is that we're going to suffer some cataclysm or go extinct for some reason. It could be an asteroid hitting the earth, nuclear war, global warming, or whatever else. So there are kind of two ways of looking at it. The first way is to say that all of these risks seem pretty small – they haven't killed us *yet*! There have been many generations, and in each one, there have been people predicting imminent doom and it never materialized. So we should condition on that and assign a relatively small probability to *our* generation being the one to go extinct. That's the argument that conservatives like to make; I'll call it the Chicken Little Argument.

Against that, there's the argument that the population has been increasing exponentially, and that if you imagine that the population increases exponentially until it exhausts the resources of the planet and collapses, then the vast majority of the people who ever lived will live close to the end, much like in the dice room. Even supposing that with each generation there's only a small chance of doom, the vast majority of the people ever born will be there close to when that chance materializes.

Student: But it still seems to me like that's conditioning on a future event. Even if the answer is the same no matter which future event you choose, you're still conditioning on one of them.
Scott: Well, if you believe in the axioms of probability theory, then if $p = P[A|B] = P[A|\neg B]$, then $P[A] = p$.
Student: Yes, but we're not talking about B and $\neg B$, we're talking about an infinite list of choices.
Scott: So you're saying the infiniteness makes a difference here?
Student: Basically. It's not clear to me that you can just take that limit, and not worry about it. If your population is infinite, maybe the madman gets really unlucky and is just failing to roll snake-eyes for all eternity.
Scott: OK, we can admit that the lack of an upper bound on the number of rolls could maybe complicate matters. However, one can certainly give variants of this thought experiment that don't involve infinity.

The argument that I've been talking about goes by the name of *the Doomsday Argument*.[4] What it's basically saying is that you should assign to the event of a cataclysm in the near future a much higher probability that you might naïvely think, because of this sort of reasoning. One can give a completely finitary version of the Doomsday Argument. Just imagine for simplicity that there are only two possibilities: Doom Soon and Doom Late. In one, the human race goes extinct very soon, whereas in the other it colonizes the galaxy. In each case, we can write down the number of people who will ever have existed. For the sake of discussion, suppose 80 billion people will have existed in the Doom Soon case, as opposed to 80 quadrillion in

---

[4] There is a huge literature on the Doomsday Argument, but the books by Bostrom and Leslie referenced earlier are good starting points, as is http://en.wikipedia.org/wiki/Doomsday_argument.

the Doom Late case. So now, suppose that we're at the point in history where almost 80 billion people have lived. Now, you basically apply the same sort of argument as in God's Coin Toss. You can make it stark and intuitive. If we're in the Doom Late situation, then the vast, vast majority of the people who will ever have lived will be born after us. We're in the very special position of being in the first 80 billion humans – we might as well be Adam and Eve! If we condition on that, we get a much lower probability of being in the Doom Late case than of being in the Doom Soon case. If you do the Bayesian calculation, you'll find that if you naïvely view the two cases as equally likely, then after applying the Doomsday reasoning we're almost certainly in the Doom Soon case. For, conditioned on being in the Doom Late case, we almost certainly would not be in the special position of being among the first 80 billion people.

Maybe I should give a little history. The Doomsday Argument was introduced by an astrophysicist named Brandon Carter in 1974. The argument was then discussed intermittently throughout the 1980s. Richard Gott,[5] who was also an astrophysicist, proposed the "mediocrity principle": if you view the entire history of the human race from a timeless perspective, then all else being equal we should be somewhere in the middle of that history. That is, the number of people who live after us should not be too much different from the number of people who lived before us. If the population is increasing exponentially, then that's very bad news, because it means that humans are not much longer for the world. This argument seems intuitively appealing, but has been largely rejected because it doesn't really fit into the Bayesian formalism. Not only is it not clear what the prior distribution is, but you may have special information that indicates that you *aren't* likely to be in the middle.

So the modern form of the Doomsday Argument, which was formalized by Bostrom, is the Bayesian form where you just assume

---

[5] J. R. Gott III, Implications of the Copernican principle for our future prospects. *Nature*, **363**:6427 (1993), 315–319.

that you have *some* prior over the possible cases. Then, all the argument says is that you have to take your own existence into account and adjust the prior. Bostrom, in his book about this, concludes that the resolution of the Doomsday Argument really depends on how you'd resolve the God's Coin Toss puzzle. If you give $1/3$ as your answer to the puzzle, that corresponds to the Self-Sampling Assumption (SSA) that you can sample a world according to your prior distribution and then sample a person within that world at random. If you make that assumption about how to apply Bayes's Theorem, then it seems very hard to escape the doomsday conclusion.

If you want to negate that conclusion, then you need an assumption he calls the *Self-Indication Assumption* (SIA). That assumption says that you are more likely to exist in a world with more beings than one with fewer beings. You would say in the Doomsday Argument that, if the "real" case is the Doom Late case, then while it's true that you are much less likely to be one of the first 80 billion people, it's also true that because there are so many more people you're much more likely to exist in the first place. If you make both assumptions, then they cancel each other out, taking you back to your original prior distribution over Doom Soon and Doom Late, in exactly the same way that making the SIA led us to get back to $1/2$ in the coin toss puzzle.

In this view, it all boils down to which of the SSA and SIA you believe. There are some arguments against Doomsday that don't accept these presuppositions at all, but those arguments are open to different objections. One of the most common counterarguments against Doomsday that you hear is that cavemen could have made the same argument, but that they would have been completely wrong. The problem with that counterargument is that the Doomsday Argument doesn't at all ignore that effect. Sure, some people who make the argument will be wrong; the point is that the vast majority will be right.

Student: It seems as though there's a tension between wanting to be right yourself and wanting to design policies that try to maximize the number of people who are right.

Scott: That's interesting.

Student: Here's a variation I want to make on the red room business: suppose that God has a biased coin such that with 0.9 probability, there's one red and many, many greens. With 0.1 probability, there's just a red. In either case, in the red-haired person's room, there's a button. You have the option to push the button or not. If you're in the no-greens case, you get a cookie if you press the button, whereas if you're in the many-greens case, you get punched in the face if you press the button. You have to decide whether to press the button. So now, if I use the SSA and find that I'm in a red room, then probably we're in the no-greens case and I should press the button.

Scott: Absolutely. It's clear that what probabilities you assign to different states of the world can influence what decisions you consider to be rational. That, in some sense, is why we care about any of this.

There's also an objection to the Doomsday Argument that denies that it's valid at all to talk about you being drawn from some class of observers. "I'm not a random person, I'm just me." The response to that is that there are clearly cases where you think of yourself as a random observer. For example, suppose there's a drug that kills 99% of the people who take it, but such that 1% are fine. Are you going to say that since you aren't a random person, the fact that it kills 99% of the people is completely irrelevant? Are you going to just take it anyway? So for many purposes, you do think of yourself as being drawn over some distribution of people. The question is when is such an assumption valid and when isn't it?

Student: I guess to me, there's a difference between being drawn from a uniform distribution of people and a uniform distribution of time. Do you weight the probability of being alive in a given time by the population at that time?

Scott: I agree; the temporal aspect does introduce something unsettling into all of this. Later, we'll get to puzzles that don't have a temporal aspect. We'll see what you think of those.

Student: I've also sometimes wondered "why am I a human?" Maybe I'm not a random human, but a random fragment of consciousness. In that case, since humans have more brain matter than other animals, I'm much more likely to be a human.

Scott: Another question is if you're more likely to be someone who lives for a long time. We can go on like this. Suppose that there's lots of extraterrestrials. Does that change the Doomsday reasoning? Almost certainly, you wouldn't have been a human at all.

Student: Maybe this is where you're going, but it seems like a lot of this comes down to what you even mean by a probability. Do you mean that you're somehow encoding a lack of knowledge, or that something is truly random? With the Doomsday Argument, has the choice of Doom Soon or Doom Late already been fixed? With that drug argument, you could say, "no, I'm not randomly chosen – I am me – but I don't know this certain property of me."

Scott: That is one of the issues here. I think that as long as you're using Bayes's Theorem in the first place, you may have made some commitment about that. You certainly made a commitment that it makes sense to assign probabilities to the events in question. Even if we imagine the world was completely deterministic, and we're just using all this to encode our own uncertainty, then a Bayesian would tell you that's what you should do anytime you're uncertain about *anything*, no matter what the reason. You must have some prior over the possible answers, and you just assign probabilities and start updating them. Certainly, if you take that view and try to be consistent about it, then you're led to strange situations like these.

As the physicist John Baez has pointed out, anthropic reasoning is kind of like science on the cheap.[6] When you do more experiments,

---

[6] http://math.ucr.edu/home/baez/week246.html

you can get more knowledge, right? Checking to see if you exist is always an experiment that you can do easily. The question is, what can you learn from having done it? It seems like there are some situations where it's completely unobjectionable and uncontroversial to use anthropic reasoning. One example is asking why the Earth is 93 million miles away from the Sun and not some other distance. Can we derive 93 million miles as some sort of a physical constant, or get the number from first principles? It seems clear that we can't, and it seems clear that to the extent that there's an explanation, it has to be that if Earth were much closer, it would be too hot and life wouldn't have evolved, whereas if it were much further, it'd be too cold. This is called the "Goldilocks Principle": of course, life is going to only arise on those planets that are at the right temperature for life to evolve. It seems like even if there's a tiny chance of life evolving on a Venus or a Mars, it'd still be vastly *more* likely that life would evolve on a planet roughly our distance from the Sun, and so the reasoning still holds.

Then there are much more ambiguous situations. This is actually a current issue in physics, which the physicists argue over. Why is the fine structure constant roughly 1/137 and not some other value? You can give arguments to the effect that if it were much different, we wouldn't be here.

Student: Is that the case like with the inverse-square law of gravity? If it weren't r^2, but just a little bit different, would the universe be kind of clumpy?
Scott: Yes. That's absolutely right. In the case of gravity, though, we can say that general relativity explains why it's an inverse square and not anything else, as a direct consequence of space having three dimensions.
Student: But we wouldn't need that kind of advanced explanation if we just did science on the cheap and said "it's gotta be this way by the Anthropic Principle."
Scott: This is exactly what people who object to the Anthropic Principle are worried about – that people will just get lazy and decide that there's

no need to do any experiment about anything, because the world just is how it is. If it were any other way, we wouldn't be us; we'd be observers in a different world.

Student: But the Anthropic Principle doesn't *predict*, does it?

Scott: Right, in many cases, that's exactly the problem. The principle doesn't seem to constrain things before we've seen them. The *reductio ad absurdum* that I like is where a kid asks her parents why the moon is round. "Clearly, if the moon were square, you wouldn't be you, but you would be the counterpart of you in a square-moon universe. Given that you are you, clearly the moon has to be round." The problem being that if you hadn't seen the moon yet, you couldn't make a prediction. On the other hand, if you knew that life was much more likely to evolve on a planet that's 93 million miles away from the Sun than 300 million miles away, then even before you'd measured the distance, you could make such a prediction. In some cases, it seems like the principle really does give you predictions.

Student: So apply the principle exactly when it gives you a concrete prediction?

Scott: That would be one point of view, but I guess one question would be what if the prediction comes out wrong?

As was mentioned before, this *does* feel like it's "just philosophy." You can set things up, though, so that real decisions depend on it. Maybe you've heard of the surefire way of winning the lottery: buy a lottery ticket and if it doesn't win, then you kill yourself. Then, you clearly have to condition on being alive to ask the question of whether you are alive or not, and so because you're asking the question, you must be alive, and thus must have won the lottery. What can you say about this? You can say that in actual practice, most of us don't accept as a decision-theoretic axiom that you're allowed to condition on remaining alive. You could jump off a building and condition on there happening to be a trampoline or something that will rescue you. You have to take into account the possibility that your choices are going to kill you. On the other hand, tragically, some people do kill themselves. Was this in fact what they were doing? Were they eliminating the worlds where things didn't turn out how they wanted?

Of course, everything must come back to complexity theory at some point. And indeed, certain versions of the Anthropic Principle should have consequences for computation. We've already seen how it could be the case with the lottery example. Instead of winning the lottery, you want to do something even better: solve an **NP**-complete problem. You could use the same approach. Pick a random solution, check if it's satisfying, and if it isn't, kill yourself. Incidentally, there is a technical problem with that algorithm. Do you see what it is?

Right. If there's no solution then you'd seem to be in trouble. On the other hand, there's actually a very simple fix to this problem: Add some dummy string like $*^n$ that acts as a "get out of jail free" card.

So we say that there are these 2^n possible solutions, and that there's also this dummy solution you pick with some tiny probability like 2^{-2n}. If you pick the dummy solution, then you do nothing. Otherwise, you kill yourself if and only if the solution you picked is unsatisfying. Conditioned upon there being no solution and your being alive, then you'll have picked the dummy solution. Otherwise, if there is a solution, you'll almost certainly have picked a satisfying solution, again conditioned on your being alive.

As you might expect, you can define a complexity class based on this principle: **BPP**$_{path}$. Recall the definition of **BPP**: the class of problems solvable by a probabilistic polynomial time algorithm with bounded error. That is, if the answer to a problem is "yes," at least $2/3$ of the **BPP** machine's paths must accept, whereas if the answer is "no," then at most $1/3$ of the paths must accept. So **BPP**$_{path}$ is just the same, except that the computation paths can all have different lengths.[7] They have to be polynomial, but they can be different.

[7] **BPP**$_{path}$ was defined in Y. Han, L. A. Hemaspaandra, and T. Thierauf, Threshold computation and cryptographic security. *SIAM Journal on Computing*, 26:1 (1997), 59–78.

Here's the point: in $\mathbf{BPP_{path}}$, if a choice leads to more different paths, then it can get counted more. Let's say, for example, that in $2^n - 1$ branches we just accept or reject – that is, we just halt, but in one branch we're going to flip more coins and do something more. In $\mathbf{BPP_{path}}$, we can make one branch completely dominate all the other branches. I've shown an example of this in the figure below – suppose we want the branch colored in gray to dominate everything else. Then we can hang a whole tree from that path, and it will dominate the paths we don't want (colored in black).

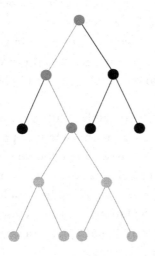

A simple argument shows that $\mathbf{BPP_{path}}$ is equivalent to a class that I'll call **PostBPP** (**BPP** with postselection). **PostBPP** is again the set of problems solvable by a polynomial time probabilistic algorithm where again you have the $\frac{2}{3}$ versus $\frac{1}{3}$ acceptance condition, but now, if you don't like your choice of random bits, then you can just kill yourself. You can condition on having chosen random bits such that you remain alive. Physicists call this postselection. You can postselect on having random bits with some very special property. Conditioned on having that property, a "yes" answer should cause $\frac{2}{3}$ of the paths to accept, and a "no" answer should cause no more than $\frac{1}{3}$ to accept.

If you want a formal definition, **PostBPP** is the class of all languages L for which there exist polynomial time Turing machines A and B the thing that decides to postselect) such that:

1. For every $x \in L$, $\Pr_r[A(x, r)|B(x, r)] \geq 2/3$.
2. For every $x \notin L$, $\Pr_r[A(x, r)|B(x, r)] \leq 1/3$.

As a technical issue, we also require that $\Pr[B(x, r)] > 0$.

Can you see why this is equivalent to **BPP**$_{\text{path}}$?

First, here's a proof that **PostBPP** $\subseteq$ **BPP**$_{\text{path}}$. Given an algorithm with postselection, you make a bunch of random choices, and if you like them you make a bunch more random choices, and those paths overwhelm the paths where you didn't like your random bits. What about the other direction? **BPP**$_{\text{path}}$ $\subseteq$ **PostBPP**?

The point is, in **BPP**$_{\text{path}}$ we've got this tree of paths that have different lengths. What we could do is complete it to make a balanced binary tree. Then, we could use postselection to give all these ghost paths suitably lower probabilities than the true paths, and thereby simulate **BPP**$_{\text{path}}$ in **PostBPP**.

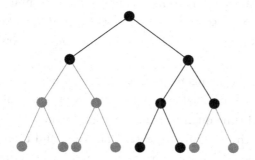

Now that we know that **PostBPP** = **BPP**$_{\text{path}}$, we can ask how big **BPP**$_{\text{path}}$ is. By the argument we gave before, **NP** $\subseteq$ **BPP**$_{\text{path}}$.

On the other hand, is **NP** = **BPP**$_{\text{path}}$? Certainly, that's going to be hard to show, even if it's true. One reason is that **BPP**$_{\text{path}}$ is closed under complement. Another reason is that it contains **BPP**. In fact, you can show that **BPP**$_{\text{path}}$ contains **MA** and **P**$^{\|\text{NP}}$ (**P** with *parallel* queries to an **NP** oracle, meaning queries that can't depend on the answers to previous queries). I'll leave that as an exercise. In the other direction, it's possible to show that $\mathbf{BPP_{path}}$ is contained in $\mathbf{BPP^{\|NP}}$, and thus in the polynomial hierarchy. Under a derandomization hypothesis, then, we find that the Anthropic Principle gives you the same computational power as $\mathbf{P^{\|NP}}$.

How about an upper bound? Let's show that $\mathbf{BPP_{path}} \subseteq \mathbf{PP}$? Deciding whether to accept or reject is kind of this exponential summation problem. You can say that each of the paths, which is a dummy path, contributes both an accept and a reject, while each of the accepting paths contributes two accepts and each of the rejecting paths contributes two rejects. Then, just ask if there are more accepts than rejects. That will simulate it in $\mathbf{PP}$.

Of course, none of this would be complete if we didn't consider *quantum* postselection. That's what I wanted to end with. In direct analogy to **PostBPP**, we can define **PostBQP** as the class of decision problems solvable in polynomial time by a quantum computer with the power of postselection. What I mean is that this is the class of problems where you get to perform a polynomial-time quantum computation and then you get to make some measurement. If you don't like the measurement outcome, you get to kill yourself and condition on still being alive.

In **PostBQP**, we're going to have to define things a bit differently, because there's no analog of r. Instead, we'll say that you perform a polynomial-time quantum computation, make a measurement that accepts with probability greater than zero, and then condition on the outcome of that measurement. Finally, you perform a further measurement on the reduced quantum state that tells you whether to accept or reject. If the answer to the problem is "yes," then the second measurement should accept with probability at least $\frac{2}{3}$, conditioned on the first measurement accepting. Likewise, if the answer to the problem is "no," the second measurement should accept with probability at most $\frac{1}{3}$, conditioned on the first measurement accepting.

Then, we can ask how powerful **PostBQP** is. One of the first things you can say is that, certainly, **PostBPP** ⊆ **PostBQP**. That is, we can simulate a classical computer with postselection. In the other direction, we have **PostBQP** ⊆ **PP**. There's this proof that **BQP** ⊆ **PP** due to Adleman, DeMarrais and Huang.[8] In that proof, they basically do what physicists would call a Feynman path integral, where you sum over all possible contributions to each of the final amplitudes. It's just a big **PP** computation. From my point of view, Feynman won the Nobel Prize in Physics for showing that **BQP** ⊆ **PP**, though he didn't state it that way. Anyway, the proof easily generalizes to show that **PostBQP** ⊆ **PP**, because you just have to restrict your summation to those paths where you end up in one of those states you postselect on. You can make all the other paths not matter by making them contribute an equal number of pluses and minuses.

Can you simulate multiple postselections with one postselection? That's another great question. The answer is yes. We get to that by using the so-called Principle of Deferred Measurement, which tells us that, in any quantum computation, we can assume without loss of generality that there's only one measurement at the end. You can simulate all the other measurements using controlled-NOT gates, and then just not look at the qubits containing the measurement outcomes. The same thing holds for postselection. You can save up all the postselections until the end.

What I showed some years ago is that the other direction holds as well: **PP** ⊆ **PostBQP**.[9] In particular, this means that quantum postselection is much more powerful than classical postselection, which seems kind of surprising. Classical postselection leaves you in the polynomial hierarchy while quantum postselection takes you up to the counting classes, which we think of as much larger.

[8] L. M. Adleman, J. DeMarrais, and M.-D. A. Huang, Quantum computability. *SIAM Journal on Computing*, 26:5 (1997), 1524–40.

[9] S. Aaronson, Quantum computing, postselection, and probabilistic polynomial-time. *Proceedings of the Royal Society* A, 461:2063 (2005), 3473–82. http://arxiv.org/abs/quant-ph/0412187

Let's run through the proof. So we've got some Boolean function $f:\{0, 1\}^n \to \{0, 1\}$ where f is efficiently computable. Let s be the number of inputs x for which $f(x) = 1$. Our goal is to decide whether $s \geq 2^{n-1}$. This is clearly a **PP**-complete problem. For simplicity, we will assume without loss of generality that $s > 0$. Now using standard quantum computing tricks (which I'll skip), it's relatively easy to prepare a single-qubit state like

$$|\psi\rangle = \frac{(2^n - s)|0\rangle + s|1\rangle}{\sqrt{(2^n - s)^2 + s^2}}.$$

This also means that we can prepare the state

$$\frac{\alpha|0\rangle|\psi\rangle + \beta|1\rangle H|\psi\rangle}{\sqrt{\alpha^2 + \beta^2}}.$$

That is, essentially a conditional Hadamard applied to $|\psi\rangle$, for some real α and β to be specified later. Let's write out explicitly what $H|\psi\rangle$ is:

$$H|\psi\rangle = \frac{\sqrt{\frac{1}{2}2^n}|0\rangle + \sqrt{\frac{1}{2}(2^n - 2s)}|1\rangle}{\sqrt{(2^n - s)^2 + s^2}}.$$

So now, I want to suppose that we take the two-qubit state above and postselect on the second qubit being 1, then look at what that leaves in the first qubit. You can just do the calculation and you'll get the following state, which depends on what values of α and β we chose before:

$$|\psi_{\alpha,\beta}\rangle = \alpha s|0\rangle + \beta \frac{2^n - 2s}{\sqrt{2}}|1\rangle.$$

Using postselection, we can prepare a state of that form for any fixed α and β that we want. So now, given that, how do we simulate **PP**? What we're going to do is keep preparing different versions of that state, varying the ratio β/α through $\{2^{-n}, 2^{-n+1}, \ldots, \frac{1}{2}, 1, 2, \ldots, 2^n\}$. Now, there are two cases: either $s < 2^{n-1}$ or $s \geq 2^{n-1}$. Suppose that the first case holds. Then, s and $2^n - 2s$ have the same sign. Since α and β are real, the states $|\psi_{\alpha,\beta}\rangle$ lie along the unit circle:

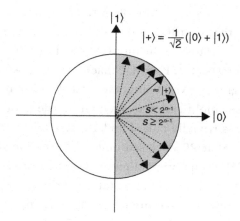

If $s < 2^{n-1}$, then as we vary β/α the state $|\psi_{\alpha,\beta}\rangle$ will always have a positive amplitude for both $|0\rangle$ and $|1\rangle$ (it will lie in the upper-right quadrant). It's not hard to see that, at some point, the state is going to become reasonably balanced. That is, the amplitudes of $|0\rangle$ and $|1\rangle$ will come within a constant of each other, as is shown by the solid vector in the figure above. If we keep measuring these states in the $\{|+\rangle, |-\rangle\}$ basis, then one of the states will yield the outcome $|+\rangle$ with high probability.

In the second case, where $s \geq 2^{n-1}$, the amplitude of $|1\rangle$ is never positive, no matter what α and β are set to, while the amplitude of $|0\rangle$ is always positive. Hence, the state always stays in the lower-right quadrant. Now, as we vary β/α through a polynomial number of values, $|\psi_{\alpha,\beta}\rangle$ never gets close to $|+\rangle$. This is a detectable difference.

So I wrote this up, thinking it was a kind of cute proof. A year later, I realized that there's this Beigel–Reingold–Spielman Theorem,[10] which showed that **PP** is closed under intersection. That is, if two languages are both in **PP**, then the AND of those two

---

[10] R. Beigel, N. Reingold, and D. A. Spielman, PP is closed under intersection. *Journal of Computer and System Sciences*, 50:2 (1995), 191–202.

languages is also in **PP**. This solved a problem that was open for 20 years. What I noticed is that **PostBQP** is trivially closed under intersection, because, if you want to find the intersection of two **PostBQP** languages, just run their respective **PostBQP** machines and postselect on both computations giving a valid outcome, and then see if they both accept. You can use amplification to stay within the right error bounds. Since **PostBQP** is trivially closed under intersection, it provides an alternate proof that **PP** is closed under intersection that I think is much simpler than the original proof. The way to get this simpler proof really is by thinking about quantum anthropic postselection. It's like a higher-level programming language for constructing the "threshold polynomials" that Beigel–Reingold–Spielman needed for their theorem to work. It's just that quantum mechanics and postselection give you a much more intuitive way of constructing these polynomials.

Let me bring out another interesting consequence of the **PostBQP** = **PP** theorem, this one for quantum computing. We saw before that **PostBPP** = **BPP**$_{\text{path}}$ is contained in the polynomial hierarchy. On the other hand, suppose **PostBQP** = **PP** were contained in the polynomial hierarchy. Then $\mathbf{P^{PP}} = \mathbf{P^{\#P}}$ would also be contained in **PH** – but, by Toda's Theorem (that $\mathbf{PH} \subseteq \mathbf{P^{\#P}}$), that would mean **PH** would collapse to a finite level! So the conclusion is that, unless **PH** collapses, **PostBQP** really is *strictly* larger than **PostBPP**. Yes, quantum and classical postselection are both absurdly powerful, but we can be quite confident that the quantum kind is *more* powerful! Indeed, I'd say we can be much *more* confident of that inequality than of the more familiar conjecture **BPP** $\neq$ **BQP**, which is "merely" based on stuff like the presumed classical hardness of factoring, nothing as "robust" as the infinitude of the polynomial hierarchy.

But does any of this imply anything about the power of quantum computers in the "real" world, as opposed to hypothetical postselected worlds? Since I first wrote this chapter in 2006, there have been some new developments that strongly suggest the answer is yes.

Namely, Bremner, Jozsa, and Shepherd (2011)[11] pointed out that, if every distribution that can be sampled in quantum polynomial time could also be sampled in *classical* polynomial time, then **PostBPP** would equal **PostBQP**, which (by the above reasoning) would cause the polynomial hierarchy to collapse. Furthermore, this conclusion holds even if we tie quantum computing's hands behind its back, and consider only the distributions that can be sampled by extremely rudimentary, almost certainly nonuniversal, *kinds* of quantum computers. The example of Bremner *et al.* was what they called the "instantaneous quantum computer," whose only ability is to apply a Hamiltonian that's a sum of tensor products of Pauli operators on various subsets of qubits. In independent work, Alex Arkhipov and I[12] obtained the same conclusion for *linear-optical quantum computers*, in which the only thing you're allowed to do is generate a bunch of identical photons, send them through a complicated network of "passive optical elements" (i.e., beamsplitters and phaseshifters), then measure how many photons ended up at every possible location. In both cases, you end up with a quantum computing model that probably *can't* implement Shor's algorithm, Grover's algorithm, or any other "standard" quantum algorithm – and, for that matter, that probably can't even do universal *classical* computation! Yet, in these models, you can easily generate samples from a probability distribution that can't be efficiently sampled by a classical computer, unless **PostBPP = PostBQP** and the polynomial hierarchy collapses. Furthermore, from a technological standpoint, these models might be easier to realize than universal quantum computing.[13]

[11] M. Bremner, R. Jozsa, and D. Shepherd, Classical simulation of commuting quantum computations implies collapse of the polynomial hierarchy. *Proceedings of the Royal Society* A, **467**:2126 (2010), 459–72. http://arxiv.org/abs/1005.1407

[12] S. Aaronson and A. Arkhipov, The computational complexity of linear optics. In *Proceedings of Annual ACM Symposium on Theory of Computing* (2011), pp. 333–42. http://arxiv.org/abs/1011.3245

[13] Indeed, as this book was undergoing final revisions, four quantum optics groups announced the first experimental demonstrations of my and Arkhipov's "Boson-Sampling" proposal, albeit so far only with 3 identical photons. See http://www.scottaaronson.com/blog/?p=1177 for more information.

Right now, the biggest theoretical challenge in this area is to show that, even if a classical computer could generate samples from *approximately* the same probability distribution as a quantum computer, this would *still* cause the polynomial hierarchy to collapse. The main thing Arkhipov and I did in our paper was to give evidence that even this stronger statement is true. But making it rigorous seems to require a significant advance in classical complexity theory: appealing to the **PostBPP = PostBQP** theorem is no longer enough. In case you care, Arkhipov and I showed that it would suffice to prove that estimating the *permanent* of an $n \times n$ matrix of independent, complex Gaussian entries, with high probability over the matrix, is a **#P**-complete problem. It's already known that approximating the permanent of an *arbitrary* complex matrix is **#P**-complete, and also that *exactly* computing the permanent of a Gaussian random matrix is **#P**-complete. So "all that's left" is to show that the problem is *still* **#P**-complete, even after we combine approximation and average-case in the same problem!

I just had a couple puzzles to leave you with. We discussed the temporal aspect and how it introduced additional confusion into the Doomsday Argument. There's one puzzle that doesn't involve any of that, but which is still quite unsettling. This puzzle – also due to Bostrom – is called the Presumptuous Philosophers. Imagine that physicists have narrowed the possibilities for a final theory of physics down to two a-priori equally likely possibilities. The main difference between them is that Theory 1 predicts that the universe is a billion times bigger than Theory 2 does. In particular, assuming the universe is relatively homogeneous (which both theories agree about), Theory 2 predicts that there are going to be about a billion times as many sentient observers in the universe. So the physicists are planning on building a massive particle accelerator to distinguish the two theories – a project that will cost many billions of dollars. Now, the philosophers come along and, say, that Theory 2 is the right one to within a billion-to-one confidence, since, conditioned on Theory 2 being correct, we're a billion times more likely to exist in the first place. The question is whether the philosophers should get the Nobel Prize in Physics for their "discovery."

Of course, what the philosophers are assuming here is the Self-Indication Assumption. So here's where things stand with the SSA and SIA. The SSA leads to the Doomsday Argument, while the SIA leads to Presumptuous Philosophers. It seems like, whichever one you believe, you get a bizarre consequence.

Finally, if we want to combine the anthropic computing idea with the Doomsday Argument, then there's the Adam and Eve puzzle. Suppose that Adam and Eve are the first two observers, and that what they'd really like is to solve an instance of an **NP**-complete problem, say, 3SAT. To do so, they pick a random assignment, and form a very clear intention beforehand that if the assignment happens to be satisfying, then they won't have any kids, whereas if the assignment is not satisfying, then they will go forth and multiply. Now, let's assume the SSA. Then, conditioned on having chosen an unsatisfying assignment, how likely is it that they would be an Adam and Eve in the first place, as opposed to one of the vast number of future observers? If we assume that they'll ultimately have (say) 2^{2n} descendants, then the probability would seem to be at most 2^{-2n+1}. Therefore, conditioned upon the fact that they *are* the first two observers, the SSA predicts that, with overwhelming probability, they will pick a satisfying assignment. If you're a hardcore Bayesian, you can take your pick between SSA and SIA and swallow the consequences either way!

19 Free will

So, in this chapter, we're going to ask – and hopefully answer – this question of whether there's free will or not. If you want to know where *I* stand, I'll tell you: I believe in free will. Why? Well, the neurons in my brain just fire in such a way that my mouth opens and I say I have free will. What choice do I have?

Before we start, there are two common misconceptions that we have to get out of the way. The first one is committed by the free will camp, and the second by the anti-free-will camp.

The misconception committed by the free will camp is the one I alluded to before: if there's no free will, then none of us are responsible for our actions, and hence (for example) the legal system would collapse. Well, I know of only one trial where the determinism of the laws of physics was *actually* invoked as a legal defense. It's the Leopold and Loeb trial in 1926.[1] Have you heard of this? It was one of the most famous trials in American history, next to the OJ trial. So, Leopold and Loeb were these brilliant students at the University of Chicago (one of them had just finished his undergrad at 18), and they wanted to prove that they were Nietzschean supermen who were so smart that they could commit the perfect murder and get away with it. So they kidnapped this 14-year-old boy and bludgeoned him to death. And they got caught – Leopold dropped his glasses at the crime scene.

They were defended by Clarence Darrow – the same defense lawyer from the Scopes Monkey Trial, considered by some to be the greatest defense lawyer in American history. In his famous closing address, he actually made an argument appealing to the determinism

---

[1] See, for example, http://law2.umkc.edu/faculty/projects/ftrials/leoploeb/leopold.htm

of the universe. "Who are we to say what could have influenced these boys to do this? What kind of genetic or environmental influences could've caused them to commit the crime?" (Maybe Darrow thought he had nothing to lose.) Anyway, they got life in prison instead of the death penalty, but apparently, it was because of their age, and not because of the determinism of the laws of physics.

Alright, what's the problem with using the nonexistence of free will as a legal defense?

Student: The judge and the jury don't have free will either.

Scott: Thank you! I'm glad someone got this immediately, because I've read whole essays about this, and the obvious point never gets brought up.

The judge can just respond, "The laws of physics might have predetermined your crime, but they *also* predetermined my sentence: DEATH!" (In the US, anyway. In Canada, maybe 30 days' jail term...)

Actually, I've since found a couplet by Ambrose Bierce that makes the point very eloquently:

"There's no free will," says the philosopher;
"To hang is most unjust."
"There is no free will," assent the officers;
"We hang because we must."

Alright, that was the misconception of the free will camp. Now on to the misconception of the *anti*-free-will camp. I've often heard the argument that says that, not only is there no free will, but the very concept of free will is incoherent. Why? Because *either* our actions are determined by something, or else they're not determined by anything, in which case they're random. In neither case can we ascribe them to "free will."

For me, the glaring fallacy in the argument lies in the implication Not Determined $\Rightarrow$ Random. If that were correct, then we couldn't have complexity classes like **NP** – we could only have **BPP**.

The word "random" means something specific: it means you have a probability distribution over the possible choices. In computer science, we're able to talk perfectly coherently about things that are nondeterministic, but not random.

Look, in computer science we have many different sources of nondeterminism. Arguably, the most basic source is that we have some algorithm, and we don't know in advance what input it's going to get. If it were always determined in advance what input it was going to get, then we'd just hardwire the answer. Even talking about algorithms in the first place, we've sort of inherently assumed the idea that there's some agent that can freely choose what input to give the algorithm.

Student: Not necessarily. You can look at an algorithm as just a big compression scheme. Maybe we do know all the inputs we'll ever need, but we just can't write them in a big enough table, so we write them down in this compressed form.

Scott: OK, but then you're asking a technically *different* question. Maybe there's no efficient algorithm for some problem such that there is an efficient compression scheme. All I'm saying is that the way we use language – at least in talking about computation – it's very natural to say there's some transition where we have this set of possible things that could happen, but we don't know which is going to happen or even have a probability distribution over the possibilities. We would like to be able to account for all of them, or maybe *at least one* of them, or the majority of them, or whatever other quantifier we like. To say that something is either determined or random is leaving out whole swaths of the Complexity Zoo.[2] We have lots of ways of producing a single answer from a set of possibilities, so I don't think it's logically incoherent to say that there could exist transitions in the universe with several allowed possibilities over which there isn't even a probability distribution.

---

[2] http://www.complexityzoo.com

Student: Then they're determined.

Scott: What?

Student: According to classical physics, everything is determined. Then, there's quantum mechanics, which is random. You can always build a probability distribution over the measurement outcomes. I don't think you can get away from the fact that those are the only two kinds of things you can have. You can't say that there's some particle that can go to one of three states, but that you can't build a probability distribution over them. Unless you want to be a frequentist about it, that's something that just can't happen.

Scott: I disagree with you. I think it does make sense. As one example, we talked about hidden-variable theories. In that case, you don't even have a probability distribution over the future until you specify *which* hidden-variable theory you're talking about. If we're just talking about measurement outcomes, then yes, if you know the state that you're measuring and you know what measurement you're applying, quantum mechanics gives you a probability distribution over the outcomes. But if you don't know the state or the measurement, then you don't even get a distribution.

Student: I know that there are things out there that aren't random, but I don't concede this argument.

Scott: Good! I'm glad someone doesn't agree with me.

Student: I disagree with your argument, but not your result that you believe in free will.

Scott: My "result"?

Student: Can we even *define* free will?

Scott: Yeah, that's an excellent question. It's very hard to separate the question of whether free will exists from the question of what the definition of it is. What I was trying to do is, by saying what I think free will is *not*, give some idea of what the concept seems to refer to. It seems to me to refer to some transition in the state of the universe where there are several possible outcomes, and we can't even talk coherently about a probability distribution over them.

Student: Given the history?

Scott: Given the history.

Student: Not to beat this to death, but couldn't you at least infer a probability distribution by running your simulation many times and seeing what your free will entity chooses each time?

Scott: I guess where it becomes interesting is, what if (as in real life) we don't have the luxury of repeated trials?

NEWCOMB'S PARADOX

So let's put a little meat on this philosophical bone with a famous thought experiment. Suppose that a super-intelligent Predictor shows you two boxes: the first box has $1000, while the second box has *either* $1 000 000 or nothing. You don't know which is the case, but the Predictor has already made the choice and either put the money in or left the second box empty. You, the Chooser, have two choices: you can either take the second box only, or both boxes. Your goal, of course, is money and not understanding the universe.

Here's the thing: the Predictor made a prediction about your choice before the game started. If the Predictor predicted you'll take only the second box, then he put $1 000 000 in it. If he predicted you'll take both boxes, then he left the second box empty. The Predictor has played this game thousands of times before, with thousands of people, and has never once been wrong. Every single time someone picked the second box, they found a million dollars in it. Every single time someone took both boxes, they found that the second box was empty.

First question: why is it obvious that you should take both boxes? Right: because *whatever's* in the second box, you'll get $1000 more by taking both boxes. The decision of what to put in the second box has already been made; your taking both boxes can't possibly affect it.

Second question: why is it obvious that you should take only the second box? Right: because the Predictor's never been wrong! Again and again you've seen one-boxers walk away with $1 000 000, and two-boxers walk away with only $1000. Why should this time be any different?

This paradox was popularized by a philosopher named Robert Nozick in 1969.[3] There's a famous line from his paper about it: "To almost everyone, it is perfectly clear and obvious what should be done. The difficulty is that these people seem to divide almost evenly on the problem, with large numbers thinking that the opposing half is just being silly."

There's actually a third position – a boring "Wittgenstein" position – which says that the problem is simply incoherent, like asking about the unstoppable force that hits the immovable object. If the Predictor actually existed, then you wouldn't have the freedom to make a choice in the first place; in other words, the very fact that you're debating which choice to make implies that the Predictor can't exist.

Student: Why can't you get out of the paradox by flipping a coin?
Scott: That's an excellent question. Why can't we evade the paradox using probabilities? Suppose the Predictor predicts you'll take only the second box with probability p. Then he'll put \$1 000 000 in that box with the same probability p. So your expected payoff is

$$1\,000\,000p^2 + 1\,001\,000p(1 - p) + 1000(1 - p)^2$$
$$= 1\,000\,000p + 1000(1 - p),$$

leading to exactly the same paradox as before, since your earnings will be maximized by setting $p = 1$. So my view is that randomness really doesn't change the fundamental nature of the paradox at all.

To review, there are three options: are you a one-boxer, a two-boxer, or a Wittgenstein?

Student: Is it really meaningless if you replace the question "what do you choose to do" with "how many boxes will you take?" It's not so

---

[3] R. Nozick, Newcomb's problem and two principles of choice. In *Essays in Honor of Carl G. Hempel*, ed. N. Rescher, Synthese Library, Dordrecht, the Netherlands. (1969), pp. 114–115.

296 QUANTUM COMPUTING SINCE DEMOCRITUS

much that you're *choosing*; you're reflecting on what you would in fact do, whether or not there's choice involved.

Scott: That is, you're just predicting your own future behavior? That's an interesting distinction.

Student: How good of a job does the Predictor have to do?

Scott: Maybe it doesn't have to be a *perfect* job. Even if he only gets it right 90% of the time, there's still a paradox here.

Student: So by the hypothesis of the problem, there's no free will and you have to take the Wittgenstein option.

Scott: Like with any good thought experiment, it's never any fun just to reject the premises. We should try to be good sports.

I can give you my own attempt at a resolution,[4] which has helped me to be an intellectually fulfilled one-boxer. First of all, we should ask what we really mean by the word "you." I'm going to define "*you*" to be anything that suffices to predict your future behavior. There's an obvious circularity to that definition, but what it means is that whatever "you" are, it ought to be closed with respect to predictability. That is, "you" ought to coincide with the set of things that can perfectly predict your future behavior.

Now let's get back to the earlier question of how powerful a computer the Predictor has. Here's you, and here's the Predictor's computer. Now, you could base your decision to pick one or two boxes on anything you want. You could just dredge up some childhood memory and count the letters in the name of your first-grade teacher or something and based on that, choose whether to take one or two boxes. In order to make its prediction, therefore, the Predictor has to know absolutely *everything* about you. It's not possible to state a priori what aspects of you are going to be relevant in making the decision. To me, that seems to indicate that the Predictor has to solve what one might call a "you-complete" problem. In other words, it seems the Predictor needs to run a simulation of

---

[4] After giving these lectures in 2006, I learned that Radford Neal independently proposed similar ideas. See R. M. Neal, Puzzles of anthropic reasoning resolved using full non-indexical conditioning, http://www.cs.toronto.edu/~radford/ftp/anth.pdf

you that's so accurate it would essentially bring into existence another copy of you.

Let's play with that assumption. Suppose that's the case, and that now you're pondering whether to take one box or two boxes. You say, "alright, two boxes sounds really good to me because that's another $1000." But here's the problem: when you're pondering this, you have no way of knowing whether you're the "real" you, or just a simulation running in the Predictor's computer. If you're the simulation, and you choose both boxes, then that actually *is* going to affect the box contents: it will cause the Predictor not to put the million dollars in the box. And that's why you should take just the one box.

Student: I think you could predict very well *most* of the time with just a limited dataset.

Scott: Yeah, that's probably true. In a class I taught at Berkeley, I did an experiment where I wrote a simple little program that would let people type either "f" or "d" and would predict which key they were going to push next. It's actually very easy to write a program that will make the right prediction about 70% of the time. Most people don't really know how to type randomly. They'll have too many alternations and so on. There will be all sorts of patterns, so you just have to build some sort of probabilistic model. Even a very crude one will do well. I couldn't even beat my own program, knowing exactly how it worked. I challenged people to try this and the program was getting between 70% and 80% prediction rates. Then, we found one student that the program predicted *exactly* 50% of the time. We asked him what his secret was and he responded that he "just used his free will."

Student: It seems like a possible problem with "you-completeness" is that, at an intuitive level, you is not equal to me. But then, anything that can simulate me can also presumably simulate you, and so that means that the simulator is both you and me.

Scott: Let me put it this way: the simulation has to bring into being a copy of you. I'm not saying that the simulation is *identical* to you. The simulation could bring into being many other things as well, so that the problem it's solving is "you-hard" rather than "you-complete."

Student: What happens if you have a "you-oracle" and then decide to do whatever the simulation *doesn't* do?

Scott: Right. What can we conclude from that? If you had a copy of the Predictor's computer, then the Predictor is screwed, right? But you don't have a copy of the Predictor's computer.

Student: So this is a theory of metaphysics which includes a monopoly on prediction?

Scott: Well, it includes a Predictor, which is a strange sort of being, but what do you want from me? That's what the problem stipulates.

One thing that I liked about my solution is that it completely sidesteps the mystery of whether there's free will or not, in much the same way that an **NP**-completeness proof sidesteps the mystery of **P** versus **NP**. What I mean is that, while it *is* mysterious how your free will could influence the output of the Predictor's simulation, it doesn't seem *more* mysterious than how your free will could influence the output of your own brain! It's six of one, half a dozen of the other.

One reason I like this Newcomb's Paradox is that it gets at a connection between "free will" and the inability to predict future behavior. Inability to predict the future behavior of an entity doesn't seem *sufficient* for free will, but it does seem somehow necessary. If we had some box, and if without looking inside this box, we could predict what the box was going to output, then we would probably agree among ourselves that the box doesn't have free will. Incidentally, what would it take to convince me that I don't have free will? If after I made a choice, you showed me a card that predicted what choice I was going to make, well that's the sort of evidence that seems both necessary and sufficient. And modern neuroscience does get close to this in some restricted situations. For example, in the famous experiments of Libet from the 1980s,[5] they'd attach electrodes to someone's brain and tell them that they could press a certain button whenever they

---

[5] B. W. Libet, Do we have free will? *Journal of Consciousness Studies*, 6 (1999), 47–57.

felt like it. A full second or more before the subject was conscious of making the decision to press the button, certainly before they physically moved their finger, you could see a so-called readiness potential forming in the pattern of neural firings. Now, that doesn't yet imply that one can actually *predict* when the subject will press the button: a crucial, rarely discussed gap in these experiments is that they failed to address how often the readiness potential formed *without* the subject pressing the button. On the other hand, more recent experiments – for example, those of Soon *et al.*[6] in 2008 – have used fMRI scans to predict which of *two* buttons the subject would press, somewhat better than chance (e.g., 60% of the time) and a few seconds before the subject was consciously aware of the decision. It's easy to overstate the significance of this sort of result: after all, we can also predict human decisions somewhat better than chance *without* fMRI, just by exploiting the fact that most humans tend to do the same things over and over! Conjurors, seducers, advertisers, and so on have known that since the dawn of history. On the other hand, it would be foolish to think that neuroscience's prediction abilities won't steadily improve. If so, then at some point, they might well force even diehard free-will believers to admit that at least *some* choices are much less "free" than they feel – or at least, that whatever is determining these choices acts earlier in time than it seems to subjective awareness.

If free will depends on an inability to predict future behavior, then it would follow from that free will somehow depends on our being *unique*: on it being impossible to copy us. This brings up another of my favorite thought experiments: the teleportation machine.

Suppose that, in the far future, there's a very simple way of getting to Mars – the Mars Express – in only 10 minutes. It encodes the positions of all the atoms in your body as information, then transmits it to Mars as radio waves, reconstitutes you on Mars, and (naturally) destroys the original. Who wants to be the first to sign up and buy

---

[6] C. S. Soon, M. Brass, H.-J. Heinze, and J.-D. Haynes, Unconscious determinants of free decisions in the human brain. *Nature Neuroscience*, 11 (2008), 543–45.

tickets? You can assume that destroying the original is painless. If you believe that your mind consists solely of information, then you should be lining up to get a ticket, right?

Student: I think there's a big difference between the case where you take someone apart then put them together on the other end, and the case where you look inside someone to figure out how to build a copy, build a copy at the end and then kill the original. There's a big difference between moving and copying. I'd love to get moved, but I wouldn't go for the copying.

Scott: The way moving works in most operating systems and programming languages is that you just make a copy then delete the original. In a computer, moving *means* copy-and-deleting. So, say you have a string of bits $x_1, \ldots, x_n$ and you want to move it from one location to another. Are you saying it matters whether we first copy all of the bits then delete the first string, or copy-and-delete just the first bit, then copy-and-delete the second bit and so on? Are you saying that makes a difference?

Student: It does if it's *me*.

Another Student: I think I'd just want to be copied, then based on my experiences decide whether the original should be destroyed or not, and if not, just accept that there's another version of me out there.

Scott: OK. So which of the two yous is going to make the decision? You'll make it together? I guess you could vote, but you might need a third you to break the tie.

Student: Are you a quantum state or a classical state?

Scott: You're ahead of me, which always makes me happy. One thing that's always really interested me about the famous quantum teleportation protocol (which lets you "dematerialize" a quantum state and "rematerialize" it at another location) is that in order for it to work, you need to measure – and hence destroy – the original state. But going back to the classical scenario, it seems even more problematic if you *don't* destroy the original than if you do. Then you have the problem of which one is "really" you.

Student: This reminds me of the many-worlds interpretation.

Scott: At least there, two branches of a wavefunction are never going to interact with each other. At most, they might interfere and cancel each other out, but here the two copies could actually have a conversation with each other! That adds a whole new layer of difficulties.

Student: So if you replaced your classical computer with a quantum computer, you couldn't just copy-and-delete to move something...

Scott: Right! This seems to me like an important observation. We know that if you have an unknown quantum state, you *can't* just copy it, but you *can* move it. So then the following question arises: is the information in the human brain encoded in *some* orthonormal basis? Is it copyable information or noncopyable information? The answer does not seem obvious a priori. Notice that we aren't asking if the brain is a quantum computer (let alone a quantum *gravity* computer a la Penrose), or whether it can factor 300-digit integers. Maybe Gauss could, but it's pretty clear that the rest of us can't. But even if it's only doing classical computation, the brain could still be doing it in a way that involves single qubits in various bases, in such a way that it would be physically impossible to copy important parts of the brain's state. There wouldn't even have to be much entanglement for that to be the case. We know that there are all kinds of tiny effects that can play a role in determining whether a given neuron will fire or not. So, how much information do you need from a brain to predict a person's future behavior (at least probabilistically)? Is all the information that you need stored in "macroscopic" variables like synaptic strengths, which are presumably copyable in principle? Or is some of the information stored microscopically, and possibly not in a fixed orthonormal basis? These are not metaphysical questions. They are, in principle, empirically answerable ones.

Now that we've got quantum in the picture, let's stir the pot a little bit more and bring in *relativity*. There's this argument (again, you can read whole PhD theses about all these things) called the *block-universe argument*. The idea is that somehow special relativity precludes the existence of free will. Here you are, and you're trying to decide whether to order pizza or Chinese take-out. Here's your friend, who's going to come over later and wants to know what you're going to order. As it happens, your friend is traveling close to the speed of light in your rest frame. Even though you perceive yourself agonizing over the decision, from her perspective, your decision has already been made.

Student: You and your friend are spacelike-separated, so what does that even *mean*?

Scott: Exactly. I don't really think, personally, that this argument says anything about the existence or nonexistence of free will. The problem is that it only works with spacelike-separated observers. Your friend can say, in principle, that in what she perceives to be her spacelike hypersurface, you've already made your decision – but she still doesn't know what you actually ordered! The only way for the information to propagate to your friend is from the point where you actually made the decision. To me, this just says that we don't have a total time-ordering on the set of events – we just have a partial ordering. But I've never understood why that should preclude free will.

I have to rattle you up somehow, so let's throw quantum, relativity, and free will all into the stew. There was a paper recently by Conway and Kochen called The Free Will Theorem,[7] which got a fair bit of press. So what is this theorem? Basically, Bell's Theorem (discussed in Chapter 12), or rather an interesting consequence of Bell's Theorem. It's kind of a mathematically obvious consequence, but still very interesting. You can imagine that there's no fundamental randomness in the universe, and that all of the randomness we observe in quantum mechanics and the like was just predetermined at the beginning of time. God just fixed some big random string, and whenever people make measurements, they're just reading off this one random string. But now suppose we make the following three assumptions.

---

[7] http://arxiv.org/abs/quant-ph/0604079

1. We have the free will to choose in what basis to measure a quantum state. That is, at least the detector settings are not predetermined by the history of the universe.

2. Relativity gives some way for two actors (Alice and Bob) to perform a measurement such that in one reference frame Alice measures first, and in another frame Bob measures first.

3. The universe cannot coordinate the measurement outcomes by sending information faster than light.

Given these three assumptions, the theorem concludes that there exists an experiment – namely, the standard Bell experiment – whose *outcomes* are also not predetermined by the history of the universe. Why is this true? Basically, because supposing that the two outcomes *were* predetermined by the history of the universe, you could get a local hidden-variable model, in contradiction to Bell's Theorem. You can think of this theorem as a slight generalization of Bell's Theorem: one that rules out not only *local* hidden-variable theories but also hidden-variable theories that obey the postulates of special relativity. Even if there were some nonlocal communication between Alice and Bob in their different galaxies, as long as there are two reference frames such that Alice measures first in one and Bob measures first in the other, you can get the same inequality. The measurement outcomes can't have been determined in advance, even probabilistically; the universe *must* "make them up on the fly" after seeing how Alice and Bob set their detectors. I wrote a review[8] of Steven Wolfram's book[9] a while ago where I mentioned this, as a basic consequence of Bell's Theorem that ruled out the sort of deterministic model of physics that Wolfram was trying to construct. I didn't call my little result the Free Will Theorem, but now I've learned my lesson: if I want people to pay attention, I should be talking about free will! Hence, this chapter.

---

[8] http://www.scottaaronson.com/papers/nks.pdf
[9] S. Wolfram, *A New Kind of Science*, Wolfram Media, 2002.

Actually, since I first wrote this chapter, the basic observation behind the Conway–Kochen "Free Will Theorem" has been used to great effect in quantum information science, to get protocols for generating so-called Einstein-certified random numbers. These are numbers that are *physically guaranteed* to be random, unless Nature resorted to faster-than-light communication to bias them, or did something equally drastic seeming (e.g., sent information backward in time). So, this is extremely different from the pseudorandomness we discussed in Chapter 8: here the numbers are *really* random assuming basic principles of physics, rather than apparently random assuming computational complexity hypotheses. You might ask: once we assume the current framework of physics (and in particular, quantum mechanics), isn't it *obvious* that we can generate true random numbers? Ah, but even then, suppose your quantum-mechanical random number generator wasn't working correctly, or was secretly tampered with by an adversary. What we want are numbers that might or might not pass some statistical test, but, if they *do* pass the test, we can conclude are random without knowing anything about the detailed physics of the devices that generated the numbers. Rather, all we want to assume is that the devices satisfied some *really* basic physical principles like locality.

At an intuitive level, it's not hard to understand how Bell's Theorem and the Conway–Kochen "Free Will Theorem" might give you this. I mean, the whole point of those results is that Alice and Bob do a certain experiment on entangled particles, and quantum mechanics predicts a result for the experiment that can't possibly be explained using local hidden variables. Instead, Alice and Bob's measurement outcomes must be genuinely probabilistic – with Nature "rolling the dice on the fly" at the instant of measurement – simply because that's the only way to explain the outcomes without Alice's choice of measurement conveying a signal to Bob or vice versa.

But there's a major problem here: namely, Alice and Bob need random numbers to perform a Bell-type experiment in the first place! For their choices of measurement have to be random as well. So it's far from obvious whether, by performing a Bell experiment, Alice and Bob can get more random bits *out* than they originally put *in*! And, in any case, the most we can hope for is *randomness expansion*: that is, a protocol by which Alice and Bob can convert n truly random bits into $m \geq n$ truly random bits, assuming no faster-than-light communication and so forth. Well, such randomness expansion is precisely what we now know to be achievable. The first result along these lines came from Pironio *et al.* (2010),[10] who (building on earlier ideas by Roger Colbeck) showed how to use Bell experiments to expand n random bits into n^2 nearly random bits. More recently, Vazirani and Vidick (2012)[11] showed how to get *exponential* randomness expansion, investing n random bits and getting out c^n for some $c > 1$. At the time of this writing, it remains open whether one can expand randomness by even more than an exponential amount in this way.

Years ago, I was at one of John Preskill's group meetings at Caltech. Usually, it was about very physics-y stuff and I had trouble understanding. But once, we were talking about a quantum foundations paper by Chris Fuchs, and things got very philosophical very quickly. Finally, someone got up and wrote on the board: "Free Will or Machine?" And asked for a vote. "Machine" won, seven to five.

I'll leave you with the following puzzle for next chapter: Dr. Evil is on his moon base, and he has a very powerful laser pointed at the Earth. Of course, he's planning to obliterate the Earth, being evil and all. At the last minute, Austin Powers hatches a plan, and sends Dr. Evil the following message: "Back in my lab here on Earth, I've created a replica of your moon base down to the last detail. The replica even

---

[10] S. Pironio, A. Acın, S. Massar, A. Boyer de la Giroday, D. N. Matsukevich, P. Maunz, S. Olmschenk, D. Hayes, L. Luo, T. A. Manning, and C. Monroe, Random numbers certified by Bell's theorem. *Nature*, **464** (2010), 1021–4. http://arxiv.org/abs/0911.3427

[11] U. Vazirani and T. Vidick, Certifiable quantum dice – or, true random number generation secure against quantum adversaries. In *Proceedings of Annual ACM Symposium on Theory of Computing* (2012), pp. 61–76. http://arxiv.org/abs/1111.6054

contains an exact copy of you. Everything is the same. Given that, you actually don't know if you're in your real moon base or in my copy here on Earth. So if you obliterate the Earth, there's a 50% chance you'll be killing yourself!" The puzzle is, what should Dr. Evil do? Should he fire the laser or not? (See here[12] for the paper about this.)

---

[12] Adam Elga, Defeating Dr. Evil with self-locating belief. http://philsci-archive.pitt.edu/1036/

20 Time travel

In the last chapter, we talked about free will, superintelligent predictors, and Dr. Evil planning to destroy the Earth from his moon base. Now I'd like to talk about a more down-to-earth topic: time travel. The first point I have to make is one that Carl Sagan made: we're all time travelers – at the rate of one second per second! Har har! Moving on, we have to distinguish between time travel into the distant future and into the past. Those are very different.

Travel into the distant future is by far the easier of the two. There are several ways to do it.

- Cryogenically freeze yourself and thaw yourself out later.
- Travel at relativistic speed.
- Go close to a black hole horizon.

This suggests one of my favorite proposals for how to solve **NP**-complete problems in polynomial time: why not just start your computer working on an **NP**-complete problem, then board a spaceship traveling at close to the speed of light and return to Earth to pick up the solution? If this idea worked, it would let us solve much more than just **NP**. It would also let us solve **PSPACE**-complete and **EXP**-complete problems – maybe even all computable problems, depending on how much speedup you want to assume is possible. So what are the problems with this approach?

Student: The Earth ages, too.

Scott: Yeah, so all your friends will be dead when you get back. What's a solution to that?

Student: Bring the whole Earth with you, and leave your computer floating in space.

Scott: Well, at least bring all your friends with you!

Let's suppose you're willing to deal with the inconvenience of the Earth having aged exponentially many years. Are there any other problems with this proposal? The biggest problem is, how much *energy* does it take to accelerate to relativistic speed? Ignoring the time spent accelerating and decelerating, if you travel at a v fraction of the speed of light for a proper time t, then the elapsed time in your computer's reference frame is

$$t' = \frac{t}{\sqrt{1 - v^2}}.$$

It follows that, if you want t' to be exponentially larger than t, then v has to be exponentially close to unity. There might already be fundamental difficulties with that, coming from quantum gravity, but let's ignore that for now. The more obvious problem is, you're going to need an exponential amount of energy to accelerate to this speed v. Think about your fuel tank, or whatever else is powering your spaceship. It's going to have to be exponentially large! Just for locality reasons, how is the fuel from the far parts of the tank going to affect you? Here, I'm using the fact that spacetime has a constant number of dimensions. (Well, and I'm also using the Schwarzchild bound, which limits the amount of energy that can be stored in a finite region of space: your fuel tank certainly can't be any denser than a black hole!)

Let's talk about the more interesting kind of time travel: the backward kind. If you've read any science fiction, you've probably heard the notion of *closed timelike curves (CTCs)*: regions of spacetime where *locally* it always looks like time is moving steadily forward with the laws of physics being perfectly obeyed, but *globally* you find that time has the topology of a loop, so that by going far enough into the future you re-encounter the present. So, basically just a fancier, more Einsteinian way to say "time travel into the past."

But can CTCs actually exist in Nature? This question has a very long history of being studied by physicists on weekends. It was discovered early on, by Gödel and others, that classical general relativity admits CTC solutions. All of the known solutions, however, have some element that can be objected to as being "unphysical." For example, some solutions involve wormholes, but that requires "exotic matter" having negative mass to keep the wormhole open.[1] They all, so far, involve either nonstandard cosmologies or else types of matter or energy that have yet to be experimentally observed. But that's just *classical* general relativity. Once you put quantum mechanics in the picture, it becomes an even harder question. General relativity is not just a theory of some fields in spacetime, but of spacetime itself, and so once you quantize it, you'd expect there to be fluctuations in the causal structure of spacetime. The question is, why *shouldn't* that produce CTCs?

Incidentally, there's an interesting metaquestion here: why have physicists found it so hard to create a quantum theory of gravity? The technical answer usually given is that, unlike (say) Maxwell's equations, general relativity is not renormalizable. But I think there's also a simpler answer, one that's much more understandable to a doofus layperson like me. The real heart of the matter is that general relativity is a theory of spacetime itself, and so a quantum theory of gravity is going to have to be talking about superpositions over spacetime and fluctuations of spacetime. One of the things you'd expect such a theory to answer is whether CTCs can exist. So quantum gravity seems "CTC-hard," in the sense that it's at least as hard as determining if CTCs are possible! And even I can see that this can't possibly be a trivial question to settle. Even if CTCs are impossible, presumably they're not going to be *proven* impossible without some far-reaching new insight. Of course, this is just one instantiation of a general problem: that no one really has a clear idea of what it means to treat spacetime itself quantum mechanically.

In the field I come from, it's never our place to ask if some physical object exists or not, it's to assume it exists and see what computations we can do with it. Thus, from now on, we'll assume

---

[1] For an accessible introduction to this topic, see K. Thorne, *Black Holes and Time Warps: Einstein's Outrageous Legacy*, W. W. Norton & Company, 1995 (reprint edition).

CTCs exist. What would the consequences be for computational complexity? Perhaps surprisingly, I'll be able to give a clear and specific answer to that.

So how would you exploit a CTC to speed up computation? First, let's consider the naïve idea: compute the answer, then send it back in time to before your computer started.

From my point of view, this "algorithm" doesn't work even considered on its own terms. (It's nice that, even with something as wacky as time travel, we can definitively rule certain ideas out!) I know of at least two reasons why it doesn't work.

Student: The universe can still end in the time you're computing the answer.

Scott: Yes! Even in this model where you can go back in time, it seems to me that you still have to quantify how much time you spend in the computation. The fact that you already have the answer at the beginning doesn't change the fact that you still have to do the computation! Refusing to count the complexity of that computation is like maxing out your credit card, then not worrying about the bill. You're going to have to pay up later!

Student: Couldn't you just run the computation for an hour, go back in time, continue the computation for another hour, then keep repeating until you're done?

Scott: Ah! That's getting toward my second reason. You just gave a slightly less naïve idea, which also fails, but in a more interesting way.

Student: The naïve idea involves iterating over the solution space, which could be uncountably large.

Scott: Yeah, but let's assume we're talking about an **NP**-complete problem, so that the solution space is finite. If we could merely solve **NP**-complete problems, we'd be pretty happy.

Let's think some more about the proposal where you compute for an hour then go back in time, compute for another hour then go back again and so on. The trouble with this proposal is that it doesn't take seriously that you're *going back in time*. You're treating time as a spiral, as some sort of scratchpad that you can keep erasing and writing over, but you're not going back to some other time, you're going back to *the* time that you started from. Once you accept that this is what we're talking about, you immediately start having to worry about the Grandfather Paradox (i.e., where you go back in time and kill your grandfather). For example, what if your computation takes as input a bit b from the future, and produces as output a bit $\neg b$, which then goes back in time to become the input? Now when you use $\neg b$ as input, you compute $\neg\neg b = b$ as output, and so on. This is just the Grandfather Paradox in a computational form. We have to come up with some account of what happens in this situation. If we're talking about CTCs at all, then we're talking about something where this sort of behavior can happen, and we need some theory of what results.

My own favorite theory was proposed by David Deutsch[2] in 1991. His proposal was that, if you just go to quantum mechanics, the problem is solved. Indeed, quantum mechanics is overkill: it works just as well to go to a classical *probabilistic* theory. In the latter case, you have some probability distribution $(p_1, \ldots, p_n)$ over the possible states of your computer. Then the computation that takes place within the CTC can be modeled as a Markov chain, which transforms this distribution to a different one. What should we impose if we want to avoid Grandfather Paradoxes? Right: that the output distribution should be the same as the input one. We should impose the requirement that Deutsch calls *causal consistency*: the computation within the CTC must map the input probability distribution to itself. In deterministic physics, we know that this sort of consistency can't always be achieved – that's just another way of stating the Grandfather Paradox. But as soon as we go to probabilistic theories, well, it's a basic fact that every Markov chain has at least one stationary distribution. In this case of the Grandfather Paradox, the unique solution

---

[2] David Deutsch, Quantum mechanics near closed timelike lines. *Physical Review D* 44 (1991), 3197–3217.

is that you're born with probability $\frac{1}{2}$, and *if* you're born, you go back in time and kill your grandfather. Thus, the probability that you go back in time and kill your grandfather is $\frac{1}{2}$, and hence you're born with probability $\frac{1}{2}$. Everything is consistent; there's no paradox.

One thing that I like about Deutsch's resolution is that it immediately suggests a model of computation. First, we get to choose a polynomial-size circuit $C: \{0, 1\}^n \rightarrow \{0, 1\}^n$. Then Nature chooses a probability distribution D over strings of length n such that $C(D) = D$, and gives us a sample y drawn from D. (If there's more than one fixed point D, then we'll suppose to be conservative that Nature makes her choice adversarially.) Finally, we can perform an ordinary polynomial-time computation on the sample y. We'll call the complexity class resulting from this model $\mathbf{P_{CTC}}$.

Student: Shouldn't we be talking about $\mathbf{BPP_{CTC}}$, since $\mathbf{P}$ doesn't have access to any randomness, whereas with closed timelike curves you have to have a distribution?

Scott: That's a tricky question–even with a fixed-point distribution, we can still require the CTC computer to produce a deterministic *output* (so that in essence, randomness is only used to avoid the Grandfather Paradox and not for any other purpose). On the other hand, if you relax that requirement and let the answer have some probability of error, it turns out that you get the same complexity class. That is, one can show that $\mathbf{P_{CTC}} = \mathbf{BPP_{CTC}} = \mathbf{PSPACE}$.

What can we say about this class? My first claim is that $\mathbf{NP} \subseteq \mathbf{P_{CTC}}$; that is, CTC computers can solve $\mathbf{NP}$-complete problems in polynomial time. Do you see why? More concretely, suppose we have a Boolean formula ϕ in n variables, and we want to know if there's a satisfying assignment. What should our circuit C do?

Student: If the input is a satisfying assignment, spit it back out?

Scott: Good. And what if the input isn't a satisfying assignment?

Student: Iterate to the next assignment?

Scott: Right! And go back to the beginning if you've reached the last assignment.

We'll just have this loop over all possible assignments, and we stop as soon as we get to a satisfying one. Assuming there *exists* a satisfying assignment, the only stationary distributions will be concentrated on satisfying assignments. So when we sample from a stationary distribution, we'll certainly see such an assignment. (If there are no satisfying assignments, then the stationary distribution is uniform.)

We're assuming that Nature gives us this stationary distribution for free. Once we set up the CTC, its evolution *has* to be causally consistent to avoid grandfather paradoxes. But that means Nature has to solve a hard computational problem to *make* it consistent! That's the key idea that we're exploiting.

Related to this algorithm for solving **NP**-complete problems is what Deutsch calls the "knowledge creation paradox." The paradox is best illustrated through the movie *Star Trek IV*. The Enterprise crew has gone back in time to the present (meaning to 1986) in order to find a humpback whale and transport it to the twenty-third century. But to build a tank for the whale, they need a type of plexiglass that hasn't been invented yet. So in desperation, they go to the company that *will* invent the plexiglass, and reveal the molecular formula to that company. They then wonder: how *did* the company end up inventing the plexiglass? Hmmmm...

Note that the knowledge creation paradox is a time travel paradox that's *fundamentally different* from the Grandfather Paradox, because here there's no actual logical inconsistency. This paradox is purely one of computational complexity: somehow this hard computation gets performed, but where was the work put in? In the movie, somehow this plexiglass gets invented without anyone ever having taken the time to invent it!

As a side note, my biggest pet peeve about time travel movies is how they always say, "Be careful not to step on anything, or you might change the future!" "Make sure this guy goes out with that girl like he was supposed to!" Dude – you might as well step on anything you want. Just by disturbing the air molecules, you've already changed everything.

OK, so we can solve **NP**-complete problems efficiently using time travel. But can we do more than that? What is the actual computational power of CTCs? I claim that, certainly, $\mathbf{P_{CTC}}$ is contained in **PSPACE**. Do you see why?

Well, we've got this exponentially large set of possible inputs $x \in \{0, 1\}^n$ to the circuit C, and our basic goal is to find an input x that eventually cycles around (that is, such that $C(x) = x$, or $C(C(x)) = x$, or...). For then we'll have found a stationary distribution. But finding such an x is clearly a **PSPACE** computation. For example, we can iterate over all possible starting states x, and for each one apply C up to 2^n times and see if we ever get back to x. Certainly, this is in **PSPACE**.

My next claim is that $\mathbf{P_{CTC}}$ is *equal* to **PSPACE**. That is, CTC computers can solve not just **NP**-complete problems, but all problems in **PSPACE**. Why?

Well, let $M_0, M_1, \ldots$ be the successive configurations of a **PSPACE** machine M. Also, let M_{acc} be the "halt and accept" configuration of M, and let M_{rej} be the "halt and reject" configuration. Our goal is to find which of these configurations the machine goes into. Note that each of these configurations takes a polynomial number of bits to write down. Then, we can define a polynomial-size circuit C that takes as input some configuration of M plus some auxiliary bit b. The circuit will act as follows:

$$C(\langle M_i, b \rangle) = \langle M_{i+1}, b \rangle$$
$$C(\langle M_{\mathrm{acc}}, b \rangle) = \langle M_0, 1 \rangle$$
$$C(\langle M_{\mathrm{rej}}, b \rangle) = \langle M_0, 0 \rangle.$$

So, for each configuration that isn't the accepting or rejecting configuration, C increments to the next configuration, leaving the auxiliary bit as it was. If it reaches an accepting configuration, then it loops back to the beginning and sets the auxiliary bit to 1. Similarly, if it reaches a rejecting configuration, then it loops back and sets the auxiliary bit to 0.

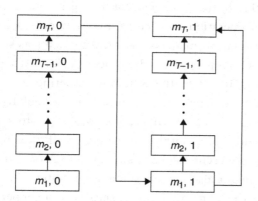

Now if we think about what's going on, we have two parallel computations: one with the answer bit set to 0, the other with the answer bit set to 1. If the true answer is 0, then the rejecting computation will go around in a loop, while the accepting computation will lead into that loop. Likewise, if the true answer is 1, it's the *accepting* computation that will go around in a loop. The only stationary distribution, then, is a uniform distribution over the computation steps with b set to the correct answer. We can then read off a sample and look at b, to find out whether the **PSPACE** machine accepts or rejects.

Thus, we can tightly characterize P_{CTC} as equal to **PSPACE**. One way to think about it is that having a CTC makes time and space equivalent as computational resources. In retrospect, maybe we should have expected that all along, but we still have to show it!

Now, there's an obvious question that we have to ask: what if we have a *quantum* computer acting inside the CTC? Obviously, we need to know the answer. How does this work? Now we have a polynomial-sized quantum circuit instead of a classical circuit, and we say that we have two sets of qubits: "CTC qubits" and "chronology-respecting qubits." We can do some quantum computation on both of them, but we're only really going to care about the CTC qubits.

At this point I need to introduce a concept that we haven't seen so far in this book, that of the *superoperator*. A superoperator is the most general type of operation allowed in quantum mechanics; it includes both unitary transformations and measurements as special cases. In fact, every superoperator can be *thought of* as just a giant unitary transformation, involving the system we're acting on as well as a second, "ancilla" system (which, in some cases, will look like it's "measuring" the first system). For this reason, superoperators don't actually change the rules of quantum mechanics at all: they're just a convenient way to represent the effects on system A of a unitary transformation that might also involve some other system B (which we don't care about at the moment). Roughly speaking, superoperators are to unitary transformations as mixed states are to pure states.

Mathematically, a superoperator is a function S that maps a mixed state (i.e., density matrix) ρ to another mixed state $S(\rho)$. We'll assume for simplicity that ρ and $S(\rho)$ live in the same number of dimensions, though it's possible to relax even that rule. Then the rule is that a superoperator must have the form

$$S(\rho) = \sum_k E_k \rho E_k^*$$

where

$$\sum_k E_k^* E_k = I$$

is the identity matrix.

Exercises for the non-lazy reader: Prove that superoperators always map valid mixed states (that is, hermitian positive semi-definite matrices with trace 1) to other valid mixed states. Give an example of a superoperator that (unlike a unitary transformation) can map a pure state to a mixed state. For more of a challenge, prove that every unitary transformation possibly involving an ancilla system gives rise to some superoperator, and conversely that every superoperator can be realized by a unitary transformation possibly involving an ancilla system.

So, to return to the subject of CTCs, if we start with a global unitary transformation on both the CTC and the causality-respecting qubits, then "trace out" (or ignore) the causality-respecting ones, we're left with some induced superoperator S that acts on the CTC qubits. Then, Nature will adversarially find a mixed state ρ that is a fixed point of S: i.e., such that $S(\rho) = \rho$. It's not always possible to find a pure state $\rho = |\psi\rangle\langle\psi|$ with that property, but by basic linear algebra (Deutsch worked out the details) there *is* always such a mixed state.

Exercise for the non-lazy reader: Prove this.

Here, ρ is a state just over the CTC qubits. The only real reason for the other qubits is that, without them, the superoperator would always be unitary, in which case the maximally mixed state I would always be fixed point. And that would trivialize the model.

As a general principle, quantum computers can simulate classical ones, and (as is easily shown) it's no different when we throw in CTCs. So we can certainly say that $\mathbf{BQP_{CTC}}$ contains **PSPACE**. But what's an *upper* bound on $\mathbf{BQP_{CTC}}$?

EXPSPACE would certainly work. Can you give a better upper bound?

So we're given an n-qubit superoperator (specified implicitly by a circuit), and we want to find a fixed point of it. This is basically a linear algebra problem. We know that you can do linear algebra in time polynomial in the dimension of the Hilbert space, which in this case is 2^n. This implies that we can simulate $\mathbf{BQP_{CTC}}$ in **EXP**. So we now have that $\mathbf{BQP_{CTC}}$ is somewhere between **PSPACE** and **EXP**. In my survey paper on "NP-complete problems and physical reality,"[3] pinning this down further was the main technical open problem!

---

[3] http://www.scottaaronson.com/papers/npcomplete.pdf

Around 2008, John Watrous and I were able to solve the problem.[4] Our result was that $\textbf{BQP}_{\text{CTC}} = \textbf{P}_{\text{CTC}} = \textbf{PSPACE}$. In other words, *if CTCs existed, then quantum computers would be no more powerful than classical ones.*

Student: Do we know anything about other classes with closed timelike curves? Like $\textbf{PSPACE}_{\text{CTC}}$?

Scott: That one is going to be **PSPACE** again. On the other hand, you can't just take any complexity class and append a $_{\text{CTC}}$ to it. You have to say what that means, and for some classes (like **NP**) it won't even make any sense.

In the last part of the chapter, I can give you a little hint of why $\textbf{BQP}_{\text{CTC}} \subseteq \textbf{PSPACE}$. Given a superoperator S that's described by a polynomial-size quantum circuit, which maps n qubits to n qubits, our goal is to compute a mixed state ρ such that $S(\rho) = \rho$. We won't be able to write down ρ explicitly (it would be far too large to fit in a **PSPACE** machine's memory), but all we're really aiming to do is to simulate the result of some polynomial-time computation that could have been performed on ρ.

Let $\text{vec}(\rho)$ be the "vectorization" of ρ (a vector with 2^{2n} components, one for each matrix entry of ρ). Then there exists a $2^{2n} \times 2^{2n}$ matrix M such that, for all ρ, $S(\rho) = \rho$ if and only if $M\,\text{vec}(\rho) = \text{vec}(\rho)$. In other words, we can just expand everything out from matrices to vectors, and then our goal is to find a $+1$ eigenvector of M.

Define $P := \lim_{z \to 1}(1 - z)(I - zM)^{-1}$. Then by Taylor expansion

$$
\begin{aligned}
MP &= M \lim_{z \to 1}(1 - z)(I + zM + z^2 M^2 + \cdots) \\
&= \lim_{z \to 1}(1 - z)(M + zM^2 + z^2 M^3 + \cdots) \\
&= \lim_{z \to 1}(1 - z)/z(zM + z^2 M^2 + z^3 M^3 + \cdots) \\
&= \lim_{z \to 1}(1 - z)/z[(I - zM)^{-1} - I] \\
&= \lim_{z \to 1}(1 - z)/z(I - zM)^{-1} \\
&= \lim_{z \to 1}(1 - z)(I - zM)^{-1} \\
&= P.
\end{aligned}
$$

[4] S. Aaronson and J. Watrous, Closed timelike curves make quantum and classical computing equivalent. In *Proceedings of the Royal Society A*, **465** (2009), 631–647. http://arxiv.org/abs/0808.2669

In other words, P projects onto fixed points of M. For all v, $M(Pv) = (Pv)$.

So now all we need to do is start with some arbitrary vector v – say, vec(I) where I is the maximally mixed state – and then compute:

$$Pv = \lim_{z \to 1}(1 - z)(I - zM)^{-1}v.$$

But how do we do apply this matrix P in **PSPACE**? Well, we can apply M in **PSPACE** since it's just a polynomial-time quantum computation. But what about taking a matrix inverse? Here, we borrow something from computational linear algebra. Csanky's algorithm, proposed in the 1970s, lets us compute the inverse of an $n \times n$ matrix not merely in polynomial time, but by a circuit of depth $\log^2 n$. Similar algorithms are actually used in practice today, for example, when doing scientific computing with lots of parallel processors. Now, "shifting everything up" by an exponential, we find that it's possible to invert a $2^{2n} \times 2^{2n}$ matrix using a circuit of size $2^{O(n)}$ and depth $O(n^2)$. But computing the output of an exponential-size, polynomial-depth circuit (which is described to us implicitly) is a **PSPACE** computation – in fact it's **PSPACE**-complete. As a final step, one can take the limit as $z \to 1$ using algebraic rules, and some further tricks due to Beame, Cook, and Hoover.[5]

Obviously, I'm skipping a lot of details.

There's an additional point an additional point that needs to be argued: that this P always projects onto the vectorization of a density matrix. If you look at the power series above, each individual term maps a vectorization of a density matrix onto another such vectorization, so the sum has to project onto vectorizations of density matrices as well. (Well, you might worry about the normalization, but that works out also.)

Since I first wrote this chapter in 2006, there have been some interesting further developments in the tale of CTC computation – so, I now feel like I should "travel back in time" to report about them!

---

[5] P. Beame, S. A. Cook, and H. J. Hoover, Log depth circuits for division and related problems. *SIAM Journal on Computing*, **15**:4 (1986), 994–1003.

First, a debate erupted in the quantum computing community about whether Deutsch's causal consistency model is really the "right" way to think about CTCs. It started with a paper by Bennett *et al.*[6] who pointed out that Deutsch's framework fails to respect the "statistical interpretation of mixed states." In other words, if you feed a state $\rho = (\rho_1 + \rho_2)/2$ as input to a CTC computer, the result might *not* be the same as if you feed ρ_1 with probability $\frac{1}{2}$ and ρ_2 with probability $\frac{1}{2}$. The problem is particularly severe if you imagine that the input to the CTC is just one half of some larger entangled state – in that case, there's *no* well-defined prescription for what the CTC computer should do. On the one hand, you could argue that this is completely unsurprising: after all, the whole *point* of a CTC computer would be to solve hard problems by breaking the linearity of quantum mechanics or even classical probability theory! And when you break linearity, you're asking for *precisely* this sort of ill-definedness. On the other hand, it is indeed pretty unpleasant to have one's face rubbed in the ill-definedness.

So, what do Bennett *et al.* propose as an alternative? Their prescription is that, if you want to talk about CTCs at all, then you need to assume that what happens inside the CTC isn't causally affected by anything in the entire rest of the universe. And thus, the output states of CTCs could be useful as "quantum advice states" (see Chapter 14), but not as anything more than that. So Bennett and others' analog of the complexity class **BQP**$_{\mathrm{CTC}}$ is actually a subclass of **BQP/qpoly**. My own reaction is that, sure, you can do this, but it basically amounts to defining CTCs out of existence! In other words, while Deutschian CTCs are indeed "diseased" in serious ways, this seems to me like a perfect example of a medicine that ends the disease only by killing the patient. If we remove CTCs from the dynamics – if we stipulate that Nature can hand you certain static "advice states," which you

6 C. H. Bennett, D. Leung, G. Smith, and J. A. Smolin, Can closed timelike curves or nonlinear quantum mechanics improve quantum state discrimination or help solve hard problems? *Physical Review Letters* **103** (2009), 170502. http://arxiv.org/abs/0908 .3023

can *interpret* (if you like) as the fixed points of superoperators, but you don't get to specify your *own* superoperator S and have Nature find a fixed point of S for you – then one can ask, in what sense are we still talking about CTCs at all?

A second major salvo in the CTC wars came with a paper by Lloyd *et al.*[7] in 2009. Unlike Bennett *et al.*, these authors didn't want to "define CTCs out of existence," but they gave a formal model for how they work that was extremely different from Deutsch's. Putting a pure state $|\psi\rangle$ into a closed timelike curve basically just means that you apply some transformation to $|\psi\rangle$, then you perform a projective measurement, then you *postselect* on getting back the same state $|\psi\rangle$ that you started with. *If* the postselection succeeds, then you're allowed to say that $|\psi\rangle$ has "travelled through time and met up with its past self." So, this gives rise to a complexity class that's contained in **PostBQP**, or postselected quantum polynomial time. Indeed, it's not hard to show that you get *exactly* **PostBQP**, which, by my **PostBQP = PP** theorem (see Chapter 18), means you get exactly **PP**, believed to be larger than **NP** but properly contained in **PSPACE**. Lloyd *et al.* actually argue that their model is more "reasonable" than Deutsch's, because Deutsch's lets you solve **PSPACE**-complete problems with polynomial resources, whereas theirs "merely" lets you solve **PP**-complete problems! On the other hand, there's also a clear sense in which their model is *less* reasonable: namely, there can easily be postselected measurements that succeed with probability zero. (For example, if you start with a qubit in state $|0\rangle$, then apply a NOT to it, then measure in the $\{|0\rangle, |1\rangle\}$ basis, you'll *never* find it back in its initial state.) For this reason, the model of Lloyd *et al.* can't be said to "resolve the grandfather paradox" in the same way Deutsch's does. Indeed, the only way to deal with grandfather paradoxes is to assume that small errors always cause postselected measurements to succeed

[7] S. Lloyd, L. Maccone, R. Garcia-Patron, V. Giovannetti, and Y. Shikano, The quantum mechanics of time travel through post-selected teleportation. *Physical Review* D, **84** (2011), 025007. http://arxiv.org/abs/1007.2615

with nonzero probability, the analog of the old idea that "if you go back in time and try to kill your grandfather, you'll always find that the gun jams, or something else mysteriously prevents you." (More about that shortly.)

My own view is that Lloyd *et al.* are talking less about CTCs themselves than about certain postselected quantum-mechanical experiments that "simulate" or "model" CTCs. (Indeed, one feature of the model of Lloyd *et al.* is that, at least with small numbers of qubits and moderately large postselection success probabilities, you can actually *do* the requisite experiments. The experiments were in fact done,[8] leading both to the entirely predictable results and to the entirely predictable misunderstandings by the popular press, which dutifully reported that physicists had now experimentally demonstrated a quantum time machine.)

Probably the biggest change in my own thinking about CTC computation came as a result of understanding a point that Deutsch had discussed in his original CTC paper. Inexcusably, though, I overlooked this point until much later, when I gave a talk about the **BQP**$_\text{CTC}$ = **PSPACE** theorem, and the philosopher of science Tim Maudlin (who was in the audience) forced me to come to terms with the point. The point is the following: even if (1) the laws of physics let us implement any polynomial-size circuit C we want, and (2) finding the fixed point of an arbitrary polynomial-size circuit is a **PSPACE**-complete problem, that *still* doesn't directly imply that we could use CTCs to solve **PSPACE**-complete problems.

The problem is that *the simulation of the abstract circuit C by the "real" laws of physics, even if it works fine in a non-CTC world, might not preserve the property that finding fixed points is **PSPACE**-complete.* In other words, the laws of physics that we're using to implement C might always allow an "out" – for example, an asteroid destroying the computer or the computer mysteriously never turning on – that maintain causal consistency inside the CTC without

---

[8] http://arxiv.org/abs/1005.2219

ever needing to run C. (This, of course, is the computational analog of "the gun jamming" when you go back in time to try to kill your grandfather.) If so, then when you run your CTC computer, you might always just get one of these spurious, easy-to-find, computationally uninteresting fixed points.

Now, you might object that even in ordinary life, *without* time travel, there's always the possibility of an asteroid hitting our computer, or some other unforeseen calamity causing our "real" computation to diverge from our abstract mathematical model of it! Yet we normally don't take that obvious fact to have relevance for complexity theory, or to mean that the laws of physics *don't* support universal computation at all. So why is the situation different with CTCs in the picture? Because now we're doing something new and exotic: asking Nature to find a *fixed point* of a given physical evolution, but not specifying *which* fixed point. That being so, if there are "doofus" fixed points lying around – ones that *don't* correspond to any fixed point of the original circuit C being simulated, and *don't* require solving any hard computational problem – then why shouldn't Nature be lazy and choose one of those, rather than one of the "hard" fixed points? If so, then in the presence of CTCs, "mysterious" computer failures would be the norm rather than an exotic aberration.

To solve this problem, one would need to show that finding a fixed point of the universe's *actual* evolution equations – given by the Standard Model, quantum gravity, or whatever – is a **PSPACE**-complete problem (and also that it's possible in principle to set up the requisite initial states). Crucially, it's *not* enough here to point out that the laws of physics are Turing universal, because it's easy to construct toy examples of "physical laws" that are Turing universal, yet for which fixed points are easy to find. (To illustrate, imagine that every physical system contained a "control bit" b, and that the universe ran a universal computation if $b = 1$ or applied the identity map if $b = 0$. Such a universe would be as capable of universal computation as ours is, yet it could always return doofus fixed points by setting $b = 0$.) What my and Watrous's result showed was simply that there *exist* computationally efficient laws for which finding fixed points is a hard computational problem, but it remains open whether our actual universe's laws are among them.

Interestingly, Deutsch's view is that CTCs must *not* enable the solution of hard computational problems. For if they did, then they would violate what Deutsch calls the "Evolutionary Principle": the principle that "knowledge can only come into existence via evolutionary processes" (or, translated into computer-science terms, that **NP**-complete and similar problems shouldn't be "solvable as if by magic"). Thus, Deutsch would say that the final laws of physics, whatever they are, will necessarily admit these doofus fixed points, thereby preventing Nature from having to solve a **PSPACE**-complete problem to ensure consistency around a CTC. Personally, I find this a strange way to argue. If CTCs existed, it's obvious that they would force us to reevaluate pretty much everything we thought we understood about space, time, causality, and more. What on earth makes Deutsch so confident that the Evolutionary Principle would survive such an upheaval, when so many other basic-seeming intuitions would not? Conversely, why not uphold the Evolutionary Principle, and much else besides, by simply conjecturing that *CTCs can't exist* – a conjecture that seems perfectly compatible with everything we know?

As usual, I'll end with a puzzle for next chapter. Suppose you can only fit a single bit at a time through a CTC. You can make as many CTCs as you like, but you can only send one bit through each, not a polynomial number of bits. (After all, we don't want to be extravagant!) In this alternate model, can you solve **NP**-complete problems in polynomial time?

21 Cosmology and complexity

Puzzle from last chapter: What can you compute with "narrow" CTCs that only send one bit back in time?

Solution: let x be a chronology-respecting bit, and let y be a CTC bit. Then, set $x := x \oplus y$ and $y := x$. Suppose that $\Pr[x = 1] = p$ and $\Pr[y = 1] = q$. Then, causal consistency implies $p = q$. Hence, $\Pr[x \oplus y = 1] = p(1 - q) + q(1 - p) = 2p(1 - p)$.

So we can start with p exponentially small, and then repeatedly amplify it. We can thereby solve **NP**-complete problems in polynomial time (and indeed **PP** ones also, provided we have a quantum computer).

I'll start with the "New York Times model" of cosmology – that is, the thing that you read about in popular articles until fairly recently – which says that everything depends on the density of matter in the universe. There's this parameter Ω which represents the mass density of the universe, and if it's greater than unity, the universe is closed. That is, the matter density of the universe is high enough that, after the Big Bang, there has to be a Big Crunch. Furthermore, if $\Omega > 1$, spacetime has a spherical geometry (positive curvature). If $\Omega = 1$, the geometry of spacetime is flat and there's no Big Crunch. If $\Omega < 1$, then the universe is open, and has a hyperbolic geometry. The view was that these are the three cases.

Today, we know that this model is wrong in at least two ways. The first way it's wrong is of course that it ignores the cosmological constant. As far as astronomers can see, space is roughly flat. That is, no one has detected a nontrivial spacetime curvature at the scale of the universe. There could be some curvature, but if there is, then it's pretty small. The old picture would therefore lead you to think that the universe must be poised on the brink of a Big Crunch: change the matter density just a tiny bit, and you could get a spherical universe that collapses or a hyperbolic one that expands forever. But in fact, the universe is not anywhere *near* the regime where there would be a Big Crunch. Why are we safe? Well, you have to look at what the energy density of the universe is made up of. There's matter, including ordinary matter as well as dark matter, there's radiation, and then there's the famous *cosmological constant* detected a decade ago, which describes the energy density of empty space. Their (normalized) sum Ω seems to equal unity as far as anyone can measure, which is what makes space flat, but the cosmological constant Λ is not zero, as had been assumed for most of the twentieth century. In fact, about 70% of the energy density of the observable universe (in this period of time) is due to the cosmological constant.

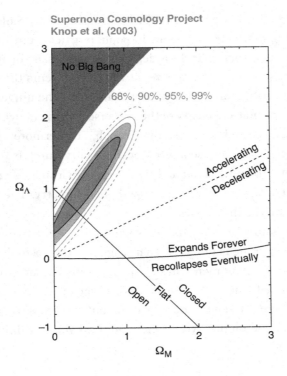

Along the diagonal black line is where space is flat. This is where the energy densities due to the cosmological constant and matter sum to unity. In the previous view, there was no cosmological constant, and space was flat, and so we're at the intersection of the two solid black lines. You can see the other solid black line slowly starts curving up. If you're above this line, then the universe expands forever, whereas if you're below this line, then the universe recollapses. So if you're at the intersection, then you really are right at the brink between expanding and collapsing. But, given that 70% of the energy density of the universe is due to Λ, you can see that we're somewhere around the intersection of the diagonal line with the inner oval – i.e., nowhere near where we recollapse.

But that's only one thing that's wrong with the simple "spherical/flat/hyperbolic" trichotomy. Another thing wrong with it is that the geometry of the universe and its topology are two separate questions. Just assuming the universe is flat doesn't imply that it's infinite. If the universe had a constant *positive* curvature, that would imply it was finite. Picture the Earth; on learning that it has a constant positive curvature, you would conclude it's round. I mean, yes, it could curve off to infinity where you can't see it, but assuming it's homogeneous in curvature, mathematically it has to curve around in either a sphere or some other more complicated finite shape. If space is flat, however, that doesn't tell you whether it's is finite or infinite. It could be like one of the video games where when you go off one end of the screen, you reappear on the other end. That's perfectly compatible with geometric flatness, but would correspond to a closed topology. The answer, then, to whether the universe is finite or infinite, is unfortunately that we don't know. (For more, see this paper[1] by Cornish and Weeks.)

Student: But with positive curvature, you could have something that tapers off infinitely like a paraboloid.

---

[1] N. J. Cornish and J. R. Weeks, Measuring the shape of the universe. *Notices of the American Mathematical Society* (1998). http://arxiv.org/abs/astro-ph/9807311

Scott: Yes, but that wouldn't be *uniform* positive curvature. Uniform means that the curvature is the same everywhere.

Student: It seems like what's missing in all these pictures so far is *time*. Are we saying that time started at some fixed point, or that time goes all the way back to negative infinity?

Scott: All of these pictures assume that there was a Big Bang, right? All of these are Big Bang cosmologies.

Student: So if time started at some finite point, then time is finite. But relativity tells us that there's really no difference between space and time, right?

Scott: No, it doesn't tell us that. It tells us that time and space are interrelated in a nontrivial way, but time has a different *metric signature* than space. As an aside, this is one of my pet peeves. I actually had a physicist ask me once how **P** could be different from **PSPACE** since "relativity tells us that time and space are the same." Well, the point is that time has a negative signature. This is related to the fact that you can go backward and forward in space, but you can only go forward in time. We talked in the last chapter about CTCs. The point about CTCs is that they would let you go backward in time and as a consequence, time and space really *would* become equivalent as computational resources. But as long as you can only go one direction in time, it's not the same as space.

Student: So can we go far in space enough to loop around?

Scott: If your arm was long enough, could you stretch it out in front of you and punch yourself in the back of your head? As I was saying, the answer is that we don't know.

Student: As far as the spread of mass is concerned, I think that people believe that is finite, because of the Big Bang.

Scott: That's a misconception about the Big Bang. The Big Bang is not something that happens at one point in space; the Big Bang is the event that creates spacetime itself. The standard analogy is that the galaxies are little spots on a balloon, and as the balloon expands, it's not that the spots are rushing away from each other, it's that the balloon is getting bigger. If spacetime is *open*, then it could well be that instead of just having a bunch of matter crowded around, you've actually got an infinite amount of matter at the moment of the Big Bang. As time goes by, the infinite universe gets stretched out, but at any point in time, it would still go on infinitely. If you look at our local horizon, we see things rushing away from each other, but that's just because we can't go past that horizon and see what's beyond it. So the Big Bang isn't some explosion that happened at some time and place; it's just the beginning of the whole manifold.

Student: But then shouldn't the mass/energy not spread out faster than the speed of light?

Scott: That's another great question; I'm glad to have something I can actually explain! Within a fixed reference frame, you can have two points appearing to recede from each other faster than light, but the reason they appear to recede is just that the intervening space is expanding. Indeed, the empirical fact is that faraway galaxies do rush away from each other faster than light. What's limited by the speed of light is the speed with which an ant can move along the surface of the expanding balloon – *not* the expansion speed of the balloon itself.

Student: So would it be possible to *observe* an object moving away faster than the speed of light?

Scott: Well, if some light was emitted a long time ago (say, shortly after the Big Bang), then by the time that light reaches us, we may be able to infer that the galaxy the light came from must *now* be receding away from us faster than the speed of light.

Student: Can two galaxies move *toward* each other faster than the speed of light?

Scott: In a collapse, yes.

Student: How do we avoid all the old paradoxes that come with allowing objects to move faster than the speed of light?

Scott: In other words, why doesn't faster-than-light expansion or contraction cause causality problems? See, this is where I start having to defer to people who *actually* understand GR. But let me take a shot: there are certainly possible geometries of spacetime – for example, those involving wormholes, or Gödel's rotating universe – that *do* have causality problems. But what about the actual geometry we live in? Here, things are just receding away from each other, which is not something you can actually use to send signals faster than light. What you *can* get, in our geometry, are objects that are so far away from each other that naïvely they should "never have been in causal contact," but nevertheless seem like they must have been. So, the hypothesis is that there was a period of rapid inflation in the extremely early universe, so that objects could reach equilibrium with each other and only *then* be causally separated by inflation.

So what is this cosmological constant? Basically, a kind of antigravity. It's something that causes two given points in spacetime to recede away from each other at an exponential rate. What's the obvious problem with that? As the Woody Allen character's mother told him, "Brooklyn is not expanding." If this expansion is such an important force in the universe, why doesn't it matter within our own planet or galaxy? Because on the scale that we live, there are other forces like gravity that are constantly counteracting the expansion. Imagine two magnets on the surface of a slowly expanding balloon: even though the balloon is expanding, the magnets still stick together. It's only on the scale of the entire universe that the cosmological constant is able to win over gravity.

You can talk about this in terms of the scale factor of the universe. Let's measure the time t since the beginning of the universe in the rest frame of the cosmic background radiation (the usual trick). How "big" is the universe as a function of t? Or to put it more carefully, given two test points, how has the distance between them changed as a function of time? The hypothesis behind inflation is that at the very beginning – at the Big Bang – there's this enormous exponential growth for a few Planck times. Following that, you've got some expansion, but also have gravity trying to pull the universe together. It works out there that the scale factor increases as $t^{2/3}$. Ten billion years after the Big Bang, when life is first starting to form on Earth, the cosmological constant starts winning out over gravity.

After this, it's just exponential all the way, like in the very beginning but not as fast.

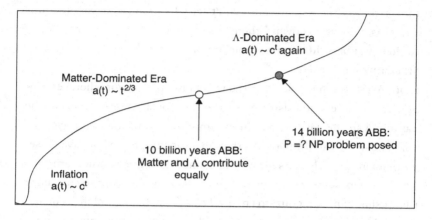

It's an interesting question as to why we should be alive at a time when the cosmological constant is 70% and matter is 30% of the energy density. Why shouldn't one of them be almost all and the other negligible? Why should we be living in the small window where they're both of the same order of magnitude? One argument you can make is the anthropic one: if we were in a later epoch, then there'd maybe be two or three of us here, and the rest of us would be outside of the cosmological horizon. The universe would be a much thinner place.

So that's how physicists would describe the cosmological constant, but how I would describe it is just the inverse of the number of bits that can ever be used in a computation! More precisely:

$$\text{max bits} = \frac{3\pi}{\Lambda}.$$

In Planck units, the cosmological constant is about 10^{-121}, and so we find that 10^{122} is about the maximum number of bits that could ever be used in a computation in the physical world. (We're going to get later to what exactly we mean by the maximum number of bits that can be involved in a computation.) How do we get to that interpretation of the cosmological constant?

Student: What's the definition of the cosmological constant?

Scott: It's the vacuum energy. Again, this is physics. People don't *define* things, they *observe* them. They don't actually know what this vacuum energy is, they just know it's there. It's an energy of empty space, and could have many different possible origins.

Student: An average?

Scott: Well, yes, but it seems to be very close to constant wherever people can measure it and also seems to be very constant over time. No one has found any deviation from the assumption that it's the same everywhere. One way to think of it is that, in a vacuum, there's always these particle/antiparticle pairs forming and annihilating each other. Empty space is an extremely complicated thing! So maybe it's not so surprising that it should have a nonzero energy. Indeed, the hard problem in quantum field theory is *not* to explain why there's a cosmological constant, but rather to explain why it isn't 10^{120} times larger than it is! A naïve quantum field theory argument gives you a prediction that the entire universe should just blow apart in an instant.

Student: So is this Ω_Λ?

Scott: No, Ω_Λ is the fraction of the total energy density that's composed of the cosmological constant. So that also depends on the matter density, and unlike Λ itself it can change with time.

To see what any of this has to do with computation, we have to take a detour into the holographic bound. This is one of the few things that seems to be *known* about quantum gravity, with the string theorists and loop quantum gravity people actually agreeing. Plus it's a bound, which is a language I speak. My treatment will follow a nice survey paper[2] by Bousso. I'm going to make this assigned reading, but only for the physicists. We saw way back that there's this Planck area $\ell_p^2 = G\hbar/c^3$. You can get it by combining a bunch of physical constants together until the units cancel such that you get length squared. Planck himself did that back around 1900. This is clearly very

---

[2] R. Bousso, The holographic principle. *Reviews of Modern Physics*, **74** (2002), 825–874, http://arxiv.org/abs/hep-th/0203101

deep, because you're throwing together Newton's constant, Planck's constant and the speed of light and you're getting an area scale which is on the order of 10^{-69} m^2.

The holographic bound says that, in any region of spacetime, the amount of entropy that you can put in the region – or up to a small constant, the number of bits you can store in it – is at most the surface area of the region measured in Planck units divided by 4. This is the surprising part: the number of bits you can store doesn't grow with the volume, it grows with the surface area. I can show you a derivation of this (or rather, what the physicists take to be a derivation).

Student: Does the derivation tell you why you divide by 4 and not, say, 3?

Scott: The string theorists believe they have an explanation of that. It's one big success that they like to lord over other quantum gravity approaches! For the loop quantum gravity people, the constant comes out wrong and they have to adjust it by hand by what they call the Immirzi parameter. (Note added: since 2006, there have been claims by the LQG camp to have solved this problem.)

The rough intuition is that, if you try to build a cube of bits (say, a hard disk) and keep making it bigger and bigger, then it's eventually going to collapse to a black hole. At that point, you can still put more bits in it, but when you do that, the information just sort of gloms onto the event horizon in a way that people don't fully understand. But however it happens, from that point on, the information content is just going to increase like the surface area.

To "derive" this, the first ingredient we need is the so-called Bekenstein bound. Bekenstein was the guy who back in the 1970s realized that black holes should have an entropy. Why? If there's no entropy and you drop something into a black hole, it disappears, which would seem to violate the Second Law of Thermodynamics. Furthermore, black holes exhibit all sorts of *unidirectional* properties: you can drop something in a black hole but you can't get it out, or you can merge two black holes and get a bigger one but then you can't split one black hole into multiple smaller black holes. This unidirectionality is extremely reminiscent of entropy. This is obvious in retrospect; even someone like me can see it in retrospect.

So what is this Bekenstein bound? It says that in Planck units, the entropy S of any given region satisfies

$$S \leq 2\pi kER/\hbar c$$

where k is Boltzmann's constant, E is the energy of the region, and R is the radius of the region (again, in Planck units). Why is this true? Basically, this formula combines π, Boltzmann's constant, Planck's constant and the speed of light. It has to be true. (I'm learning to think like a physicist. Kidding!) Seriously, it comes from a thought experiment where you drop some blob of stuff into the black hole and figure out how much the temperature of the black hole must increase (using physics we won't go into), and then use the relation between temperature and entropy to figure out how much the entropy of the black hole must have increased. You then apply the Second Law and say that the blob you dropped in must have had *at most* the entropy gained by the black hole. For otherwise, the total entropy of the universe would have decreased, contradicting the Second Law.

Student: Doesn't the area go like the square of the radius?
Scott: It does.
Student: Then why should R appear in the Bekenstein bound and not R^2?
Scott: We're getting to that!

That's fact one. Fact two is the *Schwarzschild bound*, which says that the energy of a system can be at most proportional to its radius. In Planck units, $E \leq R/2$. This is again because, if you were to pack matter/energy more densely than that, it would eventually collapse to a black hole. If you want to build a hard disk where each bit takes a fixed amount of energy to represent, then you can make a one-dimensional Turing tape which could go on indefinitely, but if you tried to make it even two dimensional, then when it became big enough it would collapse to a black hole. The radius of a black hole is proportional to its mass (its energy) by this relationship. You could say that a black hole gives you the most bang for your buck in terms of having the most energy in a given radius. So black holes are maximal in at least two senses: they have the most energy per radius and *also* the most entropy per radius.

Now, if you accept these two facts, then you can put them together:

$$S \leq 2\pi ER \leq \pi R^2 = A/4.$$

That is, the entropy of any region is at most the surface area in Planck units divided by 4. As for explaining why we divide A by 4, in effect we've reduced the problem to explaining why $E \leq R/2$. The π goes away since the surface area of a sphere is $4\pi R^2$.

There actually is a problem with the holographic bound as I've stated it – it clearly fails in some cases. One of them would be a closed spacetime. Let's say that space is closed – if you go far enough in one direction you appear back in another direction – and let's say that this region here can be at most proportional to the surface area. But how do I know that this is the inside? There's a joke where a farmer hires a mathematician to build a fence in as efficient a fashion as possible – that is, to build a fence with the most area inside given some perimeter. So the mathematician builds a tiny circle of fence, steps inside and declares the rest of the Earth to be outside. Maybe the whole rest of the universe is the inside! Clearly, the amount of entropy in the entire rest of the universe could be more than the surface area of this tiny little black hole, or whatever else it is. In general, the problem with the holographic bound is that it is not "relativistically covariant." You could have the same surface area, and in one reference frame the holographic bound is true, whereas in another it might fail.

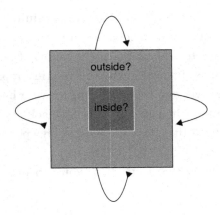

Anyway, it appears that Bousso and others have essentially solved these problems. The way they do it is by looking at "null hypersurfaces," which are made up of paths traced by photons (geodesics). These are relativistically invariant. So the idea is that you have some region, and you look at the light rays emanating from the surface of the region. Then, you define the inside of the region to be the direction in which the light rays are converging upon each other. One advantage of doing it this way is that you can switch to another reference frame, but these geodesics are unchanged. On this account, the way you should interpret the holographic bound is as upper bounding the amount of entropy you could see in the region if you could travel from the surface inward at the speed of light. In other words, the entropy being upper bounded is the entropy you would see along these null hypersurfaces. Doing it this way *seems* to solve the problems.

So what does any of this have to do with computation? You might say that if the universe is infinite, then clearly in principle you could perform an arbitrarily long computation. You just need enough Turing machine tape. What's the problem with that argument?

Student: The tape would collapse to a black hole?

Scott: As I said, you could just have a one-dimensional tape, and that could be extended arbitrarily.

Student: What if the tape starts receding away from you?

Scott: Right! Your bits are right there, then after you turn your back for just a few tens of billions of years, they've receded beyond your cosmological horizon due to the expansion of the universe.

The point is, it's not enough just to have all of these bits available in the universe somewhere. You have to be able to control all of them – you have to be able to set them all – and then you need to be able to access them later while performing a computation. Bousso formalizes this notion with what he calls a "causal diamond," but I'd just call it a computation with an input and an output. The idea is you have some starting point P and some endpoint Q, and then you look at the intersection of the forward light cone of P and the past light cone of Q. That's a causal diamond.

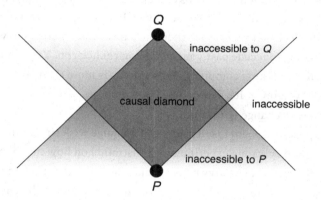

The idea is that for any experiment we could actually perform – any computation we could actually do – we're going to have to have some starting point of the experiment, and some end point where you collect the data (read the output). What's relevant isn't the total amount of entropy in the universe, but just the total amount of entropy that can be contained in one of these causal diamonds. So now, Bousso has this other paper[3] where he argues that if you're in a de Sitter space – that is, a space with a cosmological constant, like

---

[3] R. Bousso, Positive vacuum energy and the N-bound. *Journal of High Energy Physics*, 0011:038 (2000). http://arxiv.org/abs/hep-th/0010252

the space we seem to live in – then, the amount of entropy that can be contained in one of these causal diamonds is at most $3\pi/\Lambda$. That's why, in our universe, there's the bound of around 10^{122} bits. The point is that the universe is expanding at an exponential rate, and so a point that's at the edge of our horizon now will be, after another 15 billion years or so (another age of the universe), a constant factor as far away as it is now.

Student: So where do you place P and Q to get that number?

Scott: You could put them anywhere. You're maximizing over all P and Q. That's really the key point here.

Student: Then where does the maximum occur?

Scott: Well, pick P wherever you like, then pick Q maybe a couple tens of billions of years in its causal future. If you don't wrap your computation up after 20 billion years or so, then the data at the other end of your memory is going to recede past your cosmological horizon. You can't actually build a working computer whose radius is more than 20 billion light years or whatever. It's depressing, but true.

Student: Does Λ change with time?

Scott: The prevailing belief is that it doesn't change with time. It might, but there are pretty strong experimental constraints on how much. Now the proportion Ω_Λ of the energy density taken up by Λ, that is changing. As the universe gets more and more dilute, the proportion of the energy taken up by Λ gets bigger and bigger, even though Λ itself stays the same.

Student: But the radius of the universe is changing.

Scott: Yes. In our current epoch, we get to see a larger and larger amount of the past as light reaches us from farther and farther away. But once Λ starts winning out over matter, the radius of the observable universe will reach a steady state of 10 billion light years or whatever it is.

Student: Why is it 10 billion light years?

Scott: Because that's the distance such that something that far away from you will appear to be receding away from you at the speed of light, if there's no countervailing influence of gravity.

Student: So it's just a coincidence that that distance happens to be about the current size of the observable universe?

Scott: Either a coincidence or something deeper that we don't fully understand yet!

This is fine, but I promised you that I'd talk about computational complexity. Well, if the holographic bound combined with the cosmological constant put a finite upper bound on the number of bits in any possible computation, then you might argue that we can only solve problems that are solvable in constant time! And you might feel that in some sense, this trivializes all of complexity theory. Fortunately, there's an elegant way out of that: we say that now we're interested in asymptotics not just in n (the size of the input), but in $1/\Lambda$. Forget for now that Λ has a known (tiny) value, and think of it as a varying parameter – then complexity theory comes back! Taking that point of view, let me make the following claim: suppose the universe is $(1 + 1)$ dimensional (that is, one space and one time dimension) and has cosmological constant Λ. Then the class of problems that we can solve is contained in DSPACE($1/\Lambda$): the class of problems solvable by a deterministic Turing machine using $\sim 1/\Lambda$ tape squares. In fact it's *equal* to DSPACE($1/\Lambda$), depending on what assumptions you want to make about the physics. Certainly, it at least contains DSPACE($1/\sqrt{\Lambda}$).

First of all, why can't we do *more* than DSPACE($1/\Lambda$)?

Well, to be more formal, let me define a model of computation that I'll call the Cosmological Constant Turing machine. In this model, you've got an infinite Turing machine tape, but now at every time step, between every two squares, there's an independent probability Λ of a new square forming with a '$*$' symbol in it. As a first pass, this seems like a reasonable model for how Λ would affect computation. Now, if your tape head is at some square, the squares at a distance $1/\Lambda$ will appear to be receding away from the tape head at a rate of one square per time step on average. So, you can't hope to ever journey to those squares. Every time you step toward them, a new square will probably be born in the intervening space. (You can think of the speed of light in this model as one tape square per time step.) So, the class of problems you can solve will be contained in DSPACE$(1/\Lambda)$, since you can always just record the contents of the squares that are within $1/\Lambda$ of the current position of the tape head, and ignore the other squares.

But can we actually achieve DSPACE$(1/\Lambda)$? You might imagine a very simple algorithm for doing so. Namely, just think of your $1/\Lambda$ bits as a herd of cattle that keep wandering away from each other. You have to keep lassoing them together like a cosmological cowboy. In other words, your tape head will just keep going back and forth, compressing the bits together as they try to spread out while simultaneously performing the computation on them. Now, the question is, can you actually lasso the bits together in time $O(1/\Lambda)$? I haven't written out a proof of this, but I don't think it's possible in less than $\sim 1/\Lambda^2$ time with a standard Turing machine head (one without, e.g., the ability to delete tape squares). On the other hand, certainly you can lasso $\sim 1/\sqrt{\Lambda}$ bits in $O(1/\Lambda)$ time. You can therefore compute DSPACE$(1/\sqrt{\Lambda})$. I conjecture that this is tight.

A second interesting point is that in two or more dimensions you don't get the same picture. In two dimensions, the radius still doubles on a timescale of about $1/\Lambda$, but even to visit all the bits that need to be lassoed now takes on the order of $1/\Lambda^2$ time. And so we can ask if there is something you can do on a 2-D square grid in time $1/\Lambda$ which you couldn't do in time $1/\Lambda$ on a 1-D tape. You've got this $1/\Lambda^2$ space here, and intuitively you'd think that you can't make use of more than $1/\Lambda$ of the tape squares in $1/\Lambda$ time, but it's not clear if that's actually true. Of course, for added fun, you can also ask all of these questions for quantum Turing machines.

The other thing you can ask about is *query complexity* in this model. For example, what if you lost your keys and they could be anywhere in the universe? If your keys are somewhere within your cosmological horizon, and your space has one dimension, then in principle you can find them. You can traverse the entire space within your horizon in time $O(1/\Lambda)$. But in two dimensions, the number of locations you can check before most of the observable universe has receded is only like the square root of the number of possible locations. You can pick some faraway place to go, take a journey there, and by the time you come back the region has doubled in size.

In the quantum case, there's actually a way out: use Grover's algorithm! Recall that Grover's algorithm lets us search a database of N items in only $\sqrt{N}$ steps. So it would seem that this would let us search a 2-D database of size on the order of the observable universe. But there's a problem. Think about how Grover's algorithm actually works. You've got these query steps interleaved with the amplitude amplification steps. In order to amplify amplitudes, you've got to collect all the amplitudes in one place, so that you can perform the Grover reflection operation. If we think about some quantum robot searching a 2-D database having dimension $\sqrt{N} \times \sqrt{N}$, then you only need to do $\sqrt{N}$ iterations of Grover's algorithm, since there's only N items in the database, but each iteration takes $\sqrt{N}$ time, since the robot has to gather the results of all the queries. That's a problem, because we don't seem to get any benefit over the classical case. Thus, the proposed solution for searching a database the size of the universe doesn't seem to work. It *does* seem to give us some advantage in three dimensions. If you think of a 3-D hard disk, here the side length is $N^{1/3}$, so we would need $\sqrt{N}$ Grover iterations taking $N^{1/3}$ time each, giving a total time of $N^{5/6}$. At least that's somewhat better than N. As we add more dimensions, the performance would get closer to $\sqrt{N}$. For example, if space had 10 large dimensions, then we'd get a performance of $N^{12/22}$.

In a paper[4] I wrote with Andris Ambainis years ago, what we did is we showed that you can use a recursive variant of Grover's algorithm to search a 2-D grid using time of order $\sqrt{N} \log^{3/2} N$.

[4] S. Aaronson and A. Ambainis, Quantum search of spatial regions. *Theory of Computing*, 1 (2005), 47–79. http://www.scottaaronson.com/papers/ggtoc.pdf

For three or more dimensions, the time order is simply $\sqrt{N}$. I can give some very basic intuition as to how our algorithm works. What you do is use a divide-and-conquer strategy: that is, you divide your grid into a bunch of smaller grids. Then you can keep dividing the subgrid into smaller subgrids, and appoint regional Grover's algorithm commanders for each subgrid.

Even, as a first step, let's say that you search each row separately. Each row only takes $\sqrt{N}$ time to search, and then you could come back and collect everything together. You can then do a Grover search of the $\sqrt{N}$ rows, taking $N^{1/4}$ time, giving a total time of $N^{3/4}$.

That's the first way of solving the problem. Later, other people discovered a simpler and better way to solve the problem, using quantum random walks. But the bottom line is that, given a 2-D database the size of the universe, you actually can search it for a marked item before it recedes past the cosmological horizon. You can only do one search, or at best a constant number of searches, but at least you can find one thing you're really desperate for.

22 Ask me anything

To remind you, this book is based on a course I taught in 2006. On the last day of class, I followed the great tradition pioneered by Richard Feynman, in which the last class should be one where you can ask the teacher anything. Feynman's rule was that you could ask about anything except politics, religion, or the final exam. In my case, there *was* no final exam, and I didn't even make politics or religion off-limits. This chapter collects some of the questions people asked me, together with my responses.

Student: Do you often think about using computer science to limit or give us a hint about physical theories? Do you think that we'll be able to discover physical theories which give more powerful models than quantum computation?

Scott: Is **BQP** the end of the road, or is there more to be found? That's a fantastic question, and I wish more people would think about it. I'm being a bit of a politician here and not answering directly, because obviously the answer is "I don't know." I guess the whole idea with science is that if we don't know the answer, we don't try to sprout one out of our butt or something. We try to base our answers on something. So, everything we know is consistent with the idea that quantum computing *is* the end of the road. Greg Kuperberg had an analogy I really liked. He said that there are people who keep saying that we've gone from classical to quantum mechanics so what other surprises are in store? But maybe that's like first assuming the Earth is flat, and then on discovering that it's round, saying who knows, maybe it has the topology of a Klein bottle. There's a surprise in a given direction, but once you've assimilated it, there may not be any further surprise in that same direction.

The Earth is still as round as it was for Eratosthenes. We talked before about the strange property of quantum mechanics that it seems like a very brittle theory. Even general relativity, you could imagine putting in torsion or other ways of playing around with it. But quantum mechanics is very hard to fool around with without making it inconsistent. Of course, that doesn't prove that there's nothing beyond it. To people in the 1700s, it probably looked like you couldn't twiddle around much with Euclidean geometry without making it inconsistent. But on the other hand, the mere fact that something is conceivable doesn't imply that we ought to spend time on it. So, are there actual ideas about what could be beyond quantum mechanics?

Well, there are these quantum gravity proposals where it looks like you don't even have unitarity – people can't even get the probabilities to sum to unity. The positive spin on that would be "Woohoo! We found something beyond quantum mechanics!" The negative spin would be that these theories (as they currently stand) are just nonsense, and when quantum gravity people finally figure out what they're doing, they'll have recovered unitarity. And then there are phenomena that seem to change our understanding of quantum mechanics a little bit. One of these is the black hole information loss problem:

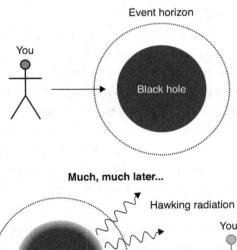

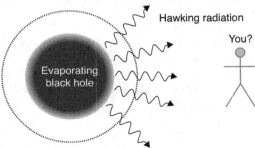

So here's you falling into a black hole. The basic problem is that all the information about you – if you fall into the black hole – is supposed to come out later as Hawking radiation. If the physics outside the event horizon is unitary, that information would *have* to come out. We don't know exactly how the information comes out, though. If you do a semiclassical calculation, it seems like only completely thermal noise coming out. However, most physicists (even Hawking) now believe that, if we *really* understood what was going on, then we'd see that the information comes out.

The trouble is, once you're in the black hole, you're not even near the event horizon. You're headed straight for the singularity. On the other hand, if the black hole is going to be leaking out information, then it seems like the information should somehow be *on* the event horizon or very close to it. This is especially so since we know that the *amount* of information in the black hole is proportional to the surface area. But from your perspective, you're just somewhere in the interior. So it seems like the information has to be in two places at once.

Anyway, one proposal that people like Gerard 't Hooft and Lenny Susskind have come up with is that, yes, the information *does* get "duplicated." On its face, that would seem to violate unitarity, and specifically the No-Cloning Theorem. But on the other hand, how would you ever *see* both copies of the information? If you're inside the black hole, then you're never going to see the outside copy. You can imagine that if you're really desperate to find out if the No-Cloning Theorem is violated – so desperate you'd sacrifice your life to find out – you could first measure the outside copy, then jump in to the black hole to look for the inside copy. But here's the funny thing: people actually calculated what would happen if you tried to do this, and they found that you'd have to wait a very long time for the information to come out as Hawking radiation, and by the time one copy comes out via Hawking radiation, the other copy is already at the singularity. It's like there's some kind of censorship that acts to keep you from seeing both copies at once. So from any one observer's perspective, it's as if unitarity is maintained. So it's funny that there are these little things that seem like they might cause a conflict with quantum mechanics or lead to a more powerful model of computation, but when you really examine them, it no longer seems like they do.

Since I first wrote this chapter in 2006, there have been very exciting further developments on the black-hole front. Among other things, there are new arguments that, contrary to decades of doctrine, it might *not* be true that an observer falling into a black hole would "see nothing special" as they passed the event horizon, and would only start seeing crazy quantum-gravity effects in the fraction of a second before being annihilated at the singularity. Instead, quantum gravity might be needed even to predict what such an observer would see *at* the event horizon!

A first indication of this came from work by the string theorist Samir Mathur, on the so-called fuzzball picture of black holes.[1] Mathur was motivated by string theory's "AdS/CFT correspondence," which *defines* certain quantum theories of gravity in D spatial dimensions by first constructing ordinary quantum field theories in $D - 1$ spatial dimensions, and then arguing that D-dimensional quantum gravity is just a "dual description" of the lower-dimensional quantum field theory. If AdS/CFT is correct, then at least in string theory, black holes *must* be describable by perfectly ordinary, unitary, reversible quantum mechanics – which then implies that the infalling bits of information *must* somehow come out in the Hawking radiation. The problem is that this abstract argument doesn't explain *how* the bits make their way out – or even how it's *possible* for them to get out, given Hawking's semiclassical calculation suggesting that they can't. So, Mathur set out to calculate what happens in some string-theory "model scenarios" that capture at least *some* aspects of physical black holes. What he found – or claims to have found – is that the "region

---

[1] See, for example, S. D. Mathur, The fuzzball proposal for black holes: an elementary review. *Fortschritte der Physik*, **53** (2005), 793–827. http://arxiv.org/abs/hep-th/0502050, http://arxiv.org/abs/1208.2005, http://www.physics.ohio-state.edu/~mathur/faq2.pdf

of quantum gravity weirdness" does *not* remain a little Planck-sized nugget at the singularity, but instead grows in size until it's a complicated "fuzzball," filling the entire region inside the event horizon. So in this picture, the reason why the bits can come out in the Hawking radiation is fundamentally the same as the reason why the bits describing a lump of coal can come out when you burn the coal: namely, because the bits are *there* on the surface!

Now, Mathur explicitly *wasn't* saying that a large observer falling into a black hole would see anything special at the event horizon – indeed, he conjectures that there's an "approximate dual description," valid for realistic-sized observers, in which those observers would continue through the event horizon all the way to the singularity, just as predicted by classical general relativity. This description would be valid *despite* the fact that, in some sense, the "real physics" would be taking place at the surface of the fuzzball, the place we used to call the event horizon.

Recently, however, there have been arguments[2] that an observer *would* encounter something special at the event horizon – that, in fact, the observer would smack into a "firewall" there and burn up, long before getting anywhere close to the singularity! Or at least, that if that isn't what would happen for "young" black holes, it *is* what would happen for "old" black holes, ones that have already radiated away at least half of their bits in Hawking radiation. I can't reproduce the argument for this prediction in any detail, but it's based on a modified version of Hawking's information loss paradox. At the time of this writing (January 2013), the whole field seems to be in a state of confusion over the "firewall" business, with even some of the experts changing their minds on a month-to-month basis.

Whatever the exact outcome, I confess to "relief" about the new developments, since they support the vague, inchoate feeling I

---

[2] See, for example, A. Almheiri, D. Marolf, J. Polchinski, and J. Sully, *Black Holes: Complementarity or Firewalls?* http://arxiv.org/abs/1207.3123 and http://blogs.discovermagazine.com/cosmicvariance/2012/09/27/guest-post-joe-polchinski-on-black-holes-complementarity-and-firewalls/

had ever since learning about the black hole information problem: namely, that *there must be something "physically special" happening at the event horizon, regardless of what classical general relativity might say to the contrary.* I mean, consider the perspective of an observer Alice *outside* a black hole, who watches what happens as her dimwitted friend Bob jumps into the hole. It's well known that, because light takes longer and longer to escape the closer you get to the event horizon, Alice will never actually *see* Bob recede past the event horizon. Instead, Bob will seem to Alice to get closer and closer to the horizon without ever going past. According to the modern view, the quantum information corresponding to Bob will actually get "smeared" and "pancaked" all across the event horizon at incredible speed. Then, if she's willing to wait 10^{70} years or so, Alice will see that event horizon that Bob was pancaked onto slowly evaporate in a mist of Hawking radiation – a mist whose information content is just a constant multiple of the area of the event horizon in Planck units. Again in the modern view, if Alice collects and pieces together the Hawking radiation painstakingly enough, then she can in principle recover the very "Bob qubits" that fell in. Now, given all this, I ask you: is it really plausible to describe the event horizon as a perfectly ordinary place with no funky quantum-gravity effects – to say that any new physics must be confined to a tiny singularity? I say no – and physicists seem more and more to agree!

But even if we accept that, the question remains of whether there's also a "complementary" perspective – namely, Bob's perspective – in which he continues past the event horizon without incident, living for perhaps a few more hours (in the case of a super-massive black hole, like the one at the center of our galaxy) before dying violently at the singularity. Maybe there is such a perspective, maybe there isn't, maybe there is but it's only approximate. Curiously, though, it's not obvious that the question of what Bob "experiences" after passing the event horizon even belongs in the domain of science! For whatever Bob experiences or doesn't experience, there's no possible way for him to *communicate* it to the rest of us outside the black hole. It's true that the *information* about Bob will eventually come out in subtle correlations between the photons of Hawking radiation. But the process that produced those photons could've been just as well described by Alice's "complementary" perspective – the perspective where Bob got pancaked on the event horizon and never made it past! In that case, Alice wouldn't need to make any reference to Bob's "experiences" after crossing the horizon. So, in what sense did Bob's final hours of subjective awareness – the hours between crossing the event horizon and hitting the singularity – actually "exist"? Only Bob knows!

Of course, you could argue that this is not so different from the situation that we're *all* in, all the time, with regard to minds other than our own. Philosophically, Alice can't be absolutely certain that there's "anything that it's like" to be Bob, even if Bob is sitting across from her in a Cleveland apartment building rather than hurtling toward the singularity of a black hole. I'd say that, as often the case, what physics does is simply to "take us full circle," forcing us to see an ancient philosophical puzzle in a new light – here, through the possibility of two complementary descriptions, one where Bob gets squashed to a Planck-length-thin pancake and one where he lives for a few hours more.

Setting aside Bob's subjective experience, what all the modern views on black holes seem to agree on is that there's no need to modify quantum mechanics even a little. Yes, black holes are a weird and wonderful laboratory for the principles of quantum mechanics, but evidence seems to be mounting that they ultimately don't *challenge* those principles, any more than any other physical object. But if so – if even these most extreme, most gravitational objects in the universe don't overturn quantum mechanics – then it becomes much harder to imagine what *could* overturn QM. Something in cosmology? In the very early universe? In the link between mind and brain? Well, maybe, but we might have to come to terms to the possibility that QM is fundamentally *true*.

And this, finally, brings me to the point of this long digression, and the occasion for wrapping it up. Figuratively speaking, physicists have by now journeyed to the ends of the universe, yet failed to turn up *any* phenomena that would make the complexity class encompassing our computational abilities any larger or smaller than **BQP**, Bounded-Error Quantum Polynomial-Time. That's not to say this can never happen – just that **BQP** has proved an extremely formidable opponent.

Anyway, looking out to physics is the "obvious" way to address the question of what could be out there beyond **BQP**. But a second way to address it is to look internally, within complexity theory. In other words, we can ask, from a purely mathematical standpoint, what reasonable-looking complexity classes *are* there above **BQP**, which some future theory of physics could plausibly give rise to.

When we ask this question, the first thing we notice is that, of the computational models that give us more than **BQP**, most of them give us *vastly* more: they let us solve **NP**-complete problems in polynomial time, and often even **PP**-complete and **PSPACE**-complete problems. This is true, for example, if we add in nonlinearities, posts-elected measurements, or CTCs. And, of course, these models are all logical possibilities – but to me, they seem not merely too fantastical, but too *boring*! In the past, Nature has always been more wily than this; she's always found ways to give us some of what we want but not all of it. So, suppose we want to believe that there's something more powerful than quantum computing, but that still can't solve **NP**-complete problems in polynomial time. Then, how much "room" is there for such a model? We do have some problems that seem to be easier than **NP**-complete, but that are still too hard to efficiently solve with a quantum computer. Two examples are Graph Isomorphism and approximate shortest vector. Very "close" to **NP**-complete, but probably not quite there, seem to be the problems of inverting one-way functions and distinguishing random from pseudorandom functions.

Years ago I came up with one example of a computational model (discussed in Chapter 12), where you get to see the entire history of a hidden variable during the course of a quantum computation.

I gave evidence that in this model you do get more than with ordinary quantum computing – for example, you get Graph Isomorphism and approximate shortest vector – but still you don't get the **NP**-complete problems. On the other hand, my model was admittedly rather artificial. So maybe there *is* one more dramatic step before you get to **NP**-complete – I'm not sure.

Student: How can you say "one step"? You can theoretically always contrive a problem between any other two problems.
Scott: Of course, but here's the point: no one was interested in quantum computing when Bernstein and Vazirani discovered you could solve the Recursive Fourier Sampling problem. People only became interested when it was found that you could solve problems that were *previously* considered to be important, like factoring. So if we judge our hypothetical new model by the same standard, and ask what problems it can solve that we already think are important, there are arguably not that many of them between factoring and **NP**-complete. So again, there could be some new model that gets you slightly beyond **BQP** – maybe it lets you solve Graph Isomorphism, or the Hidden Subgroup Problem for a few more non-abelian groups – but, at least in our current picture, there's only a limited amount of "room" between **BQP** and the **NP**-complete problems.
Student: Where would you ever get an oracle?
Scott: You just define it. Let A be an oracle...
Student: That's a bit of an issue.
Scott: It is, it is. It's strange to me that only computer scientists get this kind of flak for using the techniques that they have to answer questions. Like physicists say that they're going to do some calculation in the perturbative regime. "Oh! Of course, what else would you do? These are deep and difficult problems." Of course, you're going to do what works. Computer scientists say that we can't yet prove that $P \neq NP$, but we'll study it in the relativized world. "That's cheating!" It just seems obvious that you just start with the kind of results that you can prove and work from there. One objection that could be made against oracle results in the past would be that some of them were just trivial. Some of them

essentially just amounted to restatements of the question. But these days, we've got some very nontrivial oracle separations. I mean, I can tell you in very concrete terms what an oracle result's good for. About every month or so, I see another paper on the arXiv solving **NP**-complete problems in polynomial time on a quantum computer. This must be the easiest problem in the world. Often these papers are very long and complicated. But if you know about oracle results, you don't have to read the papers. That's a very useful application. You can say if this proof works, then it also works relative to oracles, but that can't be the case, because we know of an oracle where it's false. Of course, that probably won't convince the author, but it will at least convince you.

As another example, I gave this oracle relative to which **SZK** (Statistical Zero-Knowledge) is not in **BQP**. In other words, finding collisions is hard for a quantum computer. Sure enough, as the years go by, I see these papers that talk about how to find collisions with a constant number of queries on a quantum computer, and without reading the paper I can say no, this has to fail, because it's not doing anything nonrelativizing. So, oracles are there to tell you what approaches not to try. They direct you toward the nonrelativizing techniques that we know we're eventually going to need.

Student: What complexity class are you?

Scott: I'm not even all of **P**. I'm not even **LOGSPACE**! Especially if I haven't had much sleep.

Student: What's the complexity class for creativity?

Scott: That's an excellent question. I was thinking about it just this morning. Someone asked me if humans have an oracle in their head for **NP**. Well, maybe Gauss or Wiles did. But for most of us, finding proofs is a very hit-or-miss business. You can change your perspective and it seems pathetic that after three billion years of natural selection and after this time building up civilizations, all the wars and everything else, we can solve a *few* instances of SAT – but if you switch to the Riemann Hypothesis or Goldbach's Conjecture instances, suddenly we can't solve those.

ASK ME ANYTHING 353

When it comes to proving theorems, you're dealing with a very special case of an **NP**-complete problem. You aren't just taking some arbitrary formula of size polynomial in n, you're taking some fixed question of fixed size and asking, does this have a proof of size n? So you're uniformly generating these instances for whatever length proof you're looking for. But even for this sort of problem, the evidence is *not* good that we have some sort of general algorithm for solving them. A few people decided to forsake their social lives and spend their whole lives in this monastic existence, thinking about math problems. Finally, they've managed to succeed on a few problems and sometimes even win Fields Medals for that. But there's still this huge universe of problems that everyone knows about and no one can solve. So I would say that, before reaching for Penrose-style speculations about human mathematical creativity transcending computation, we should first make sure the data actually supports the hypothesis that humans are good at finding proofs. I'm not convinced that it does.

Now, it's clear, that in certain cases, we are very good at finding patterns or taking a problem that looks to be hard and decomposing it into easier subproblems. In many cases, we're much better at that than any computer. We can ask, "why is that?" That's a very big question, but I think part of the answer is we've got a billion-year head start. We've got the advantage of a billion years of natural selection giving us a very good toolbox of heuristics for solving certain kinds of search problem. Not all of them and not all the time, but in some cases, we can do really well. Like I said, I believe that **NP**-complete problems are not efficiently solvable in the physical universe, so I believe that there can never be a machine that can just prove any theorem efficiently, but there could certainly be machines that would take advantage of the same kind of creative insight that human mathematicians have. They don't have to beat God, they just have to beat Andrew Wiles. That could be an easier problem, but it takes us outside of the scope of complexity theory and into AI.

Student: So even if there's no way to solve **NP**-complete problems in polynomial time, human mathematicians could still be rendered obsolete?

Scott: Sure. And after the computers take over from us, maybe they'll worry that *they'll* be out of a job once some **NP** oracle comes along.

Student: Bell inequalities seem to be an important tool in studying the limitations of quantum mechanics. We know what happens if we have completely nonlocal boxes, but what happens (say, to computational complexity) if we allow correlations just above what, say, quantum entanglement gives?

Scott: That's a good question, and there are people who have been thinking about it.

To provide a bit of context, there's this important result called *Tsirelson's inequality*,[3] which you can think of as "the quantum version of the Bell inequality." The Bell inequality says that Alice and Bob can win a certain game, called the CHSH game, at most 75% of the time in a classical universe, but can win ~85% of the time if they share entangled qubits. Now, Tsirelson's inequality says that, even *with* entangled qubits, there's still a limit to what Alice and Bob can do: they can't win the CHSH game more than ~85% of the time, despite the fact that even winning 100% of the time still wouldn't let them send signals faster than light. So one might say the limits imposed by quantum mechanics are a bit stronger than they "had to be" – in particular, stronger than the limits imposed by the no-signalling principle.

Now, around a decade ago, a trend started of studying hypothetical "superquantum" theories, which would violate Tsirelson's inequality, but which *still* wouldn't allow any faster-than-light communication. The simplest way to do that is just to postulate the existence of so-called "nonlocal boxes": magical devices that let Alice and Bob win the CHSH game, say, 95% of the time instead of just 85%.

---

[3] See http://en.wikipedia.org/wiki/Tsirelson's_bound

You can then study how other issues are affected by these boxes. For example, Brassard et al.[4] (building on an earlier result of Wim van Dam[5]) showed that, if you have a good enough nonlocal box (if the error is small enough), then it makes communication complexity trivial (i.e., all communication problems can be solved with just a single bit).

The fundamental problem is that you can imagine Tsirelson's bound is violated – that is, you can imagine that there are these nonlocal correlations stronger than anything allowed by quantum mechanics – but saying that still doesn't give us a *model of computation*. I mean, what are the allowed operations? What's the space of possible states, that gives rise to the possibility of nonlocal boxes? If we had answers to *those* questions, then we could begin to think about computational complexity in these hypothetical worlds.

Student: Do you see there being a bit more clearing up of the complexity classes? We just keep getting more and more.

Scott: To me, that's like asking a chemist if she sees a clearing up of the Periodic Table. Is nitrogen going to collapse with helium? In our case, it's a little bit better than for the chemist, since we can expect a collapse of *some* classes. For example, we hope and expect that **P**, **RP**, **ZPP**, and **BPP** are going to collapse. We hope and expect that **NP**, **AM**, and **MA** are going to collapse. **IP** and **PSPACE** already collapsed. So yeah, there are collapses, but we also know that there are other pairs of classes that can't collapse. We know, for example, that **P** is different from **EXP**, which immediately tells you that either **P** has to be different from **PSPACE** or **PSPACE** has to be different from **EXP**, or both. So not everything can collapse. That shouldn't really be surprising.

[4] G. Brassard, H. Buhrman, N. Linden, A. A. Methot, A. Tapp, and F. Unger, A limit on nonlocality in any world in which communication complexity is not trivial. *Physical Review Letters* 96 (2006), 250401. http://arxiv.org/abs/quant-ph/0508042

[5] W. van Dam, Implausible consequences of superstrong nonlocality. (2005). http://arxiv.org/abs/quant-ph/0501159

Now, maybe complexity theory took a wrong turn when it gave everything this string of random-looking capital letters as its name – I appreciate how they can look to people like codenames or inside jokes. But really, we're just talking about different notions of computation. Time, space, randomness, quantumness, having a prover around. There are as many complexity classes as there are different notions of computation. So, the richness of the complexity zoo just seems like an inevitable reflection of the richness of the computational world.

Student: Do you think that **BPP** will collapse with **P**?
Scott: Oh, yeah. Absolutely. We have not just one but several reasonable-looking circuit lower bound conjectures where we know that, if they're true, then **P** = **BPP**. I mean, there were people who realized even in the 1980s that **P** should equal **BPP**. Even then, Yao pointed out that, if you had good enough cryptographic pseudorandom number generators, then you could use them to derandomize any probabilistic algorithm, hence **P** = **BPP**. Now, what people managed to do in the 1990s is to get the same conclusion with weaker and weaker assumptions.

Besides that, there's also an "empirical" case, in that two of the most spectacular results in complexity theory in the last decade were the AKS primality test showing that primality testing is in **P**, and Reingold's result that searching an undirected graph is in deterministic logspace. So, this program of taking specific randomized algorithms and derandomizing them has had considerable success. It sort of increases one's confidence that, if we were smart enough or knew enough, then this would probably work for other **BPP** problems as well. You can also look at a specific case, like derandomizing polynomial identity testing, and maybe this is a good example to illustrate the point.

The question is, if you've got some polynomial like $x^2 - y^2 - (x + y)(x - y)$, *is it identically zero?* In this case, the answer is yes. But you could have some very complicated polynomial identity involving variables raised to very high powers, and then it's not obvious how you would check it efficiently even with a computer. If you tried to expand everything out, you'd get an exponential number of terms.

Now, we do know of a fast randomized algorithm for this problem: namely, just plug in some random values (over some random finite field) and see whether the identity holds or not. The question is whether this algorithm can be *derandomized*. That is, is there an efficient deterministic algorithm to check whether a polynomial is identically zero? If you bang your head against this problem, you quickly get into some very deep questions in algebraic geometry. For example, can you come up with some small list of numbers, such that, given any polynomial $p(x)$ described by a small arithmetic formula, all you have to do is plug in the numbers in that list, and if $p(x) = 0$ for every x in the list, then it's zero everywhere? That seems like it *should* be true, because all you should have to do is pick some "generic" set of numbers to test which is much larger than the size of the formula for p. For example, if you find that $p(1) = 0$, $p(2) = 0, \ldots, p(k) = 0$, then either p must be zero, or else it must be evenly divisible by the polynomial $(x - 1) \ldots (x - k)$. But is there *any* nonzero multiple of $(x - 1) \ldots (x - k)$ that can be represented by an arithmetic formula of size much smaller than k? That's really the crucial question. If you can prove that no such polynomial exists, then you'll give a way to derandomize polynomial identity testing (a major step towards proving $\mathbf{P = BPP}$).

Student: What do you think the chances are that three Indian mathematicians will come up with an elementary proof?

Scott: I think it's gonna take at least four Indian mathematicians! We know today that if you prove good enough circuit lower bounds, then you can prove $\mathbf{P = BPP}$. But Impagliazzo and Kabanets also proved a result in the other direction: if you want to derandomize, you're going to *have* to prove circuit lower bounds. To me that gives some explanation as to why people haven't succeeded yet in proving that $\mathbf{P = BPP}$. It's all because we don't know how to prove circuit lower bounds. The two problems are almost – though not quite – the same.

Student: Does $P = BPP$ imply that $NP = MA$?

Scott: Almost. If you derandomize **PromiseBPP**, then you derandomize **MA**. No one has any idea of how to derandomize **BPP** that wouldn't also derandomize **PromiseBPP**.

Student: How would you answer an intelligent design advocate? Without getting shot?

Scott: You know, I'm genuinely not sure. It's one of those cases where there might be anthropic selection going on. If someone could be persuaded by evidence on this question, then wouldn't he or she already have been? I think we have to concede that there are people for whom the most important thing about a belief isn't whether it's true, but rather some other properties of the belief, such as its role in a community. So they're playing a different game where beliefs are judged by a different standard. It's like you're a basketball player on a football field.

Student: Is complexity theory relevant to the evolution versus intelligent design controversy?

Scott: To the extent that you need complexity theory, it's all sort of trivial complexity theory. For example, just because we believe that **NP** is exponentially hard doesn't mean that we believe that every particular instance (say, evolving a working brain or a retina) has to be hard.

Student: When Steven Weinberg came to talk at the Perimeter Institute, the question was asked, "where does God fit into all of this?" His answer was to just dismiss religion as an artifact of our evolution that now has no value, and that we'd eventually grow out of it. Do you agree with him?

Scott: So I think that there are several questions here.

Student: You're being a politician.

Scott: Look, this is a hot topic, with books like Richard Dawkins's *The God Delusion*[6] . . .

---

[6] Mariner Books, 2008 (reprint edition)

Student: Was it a good book?

Scott: Yes. Dawkins is always amusing, and he's at his absolute best when he's ripping into bad arguments like a Rottweiler. Anyway, one way to think about it is that the world would clearly be a better place if there were no wars, or for that matter if there were no lawyers and no one sued anyone else. And there are those who want to turn that idea into an actual political program. I'm not talking about people who oppose specific wars like the one in Iraq for specific reasons, but absolute pacifists. And the obvious problem with their position is a game-theoretic one. Yes, the world would be a better place with no armies, but the other guys have an army.

It's clear that religion fills some sort of role for people; otherwise, it wouldn't have been so ubiquitous for thousands of years or resisted very significant efforts to stamp it out. For example, maybe people who believe God is on their side are braver in battle. Or maybe religion is one of the factors (besides more obvious factors) that induces men and women to get married and have lots of babies, and is therefore adaptive just from a Darwinian point of view. Years ago I was struck by an irony: in contemporary America, you've got these stereotypical coastal elites who believe in Darwinism and often live by themselves well into their thirties or forties, and then you've got these stereotypical heartland folks who reject Darwinism but marry young and have 7 kids, 49 grandkids and 343 great-grandkids. So then it's not really a contest between "Darwinists" and "anti-Darwinists"; it's just a contest between Darwinian theorists and Darwinian practitioners!

If this idea is right – that is, if religion has played this role throughout history of helping inspire people to win wars, have more babies, etc. – then the question arises, how are you ever going to counter religion except with a competing religion?

Student: I'm sure that's what people are thinking about when they decide whether or not to believe in a religion.

Scott: I'm not saying it's conscious, or that people are thinking it through in these terms. Maybe a few are, but the point is they don't *have* to in order for it to describe their behavior.

Student: We can have lots of kids without accepting a religion if we want to.

Scott: Sure, we *can*, but *do* we on average? I don't know the numbers offhand, but it does tend to be true in modern society that religious people have more children on average.

Now, there's another key factor, which is that sometimes *irrationality can be supremely rational*, because it's the only way of proving to someone else that you're committed to something. Like if someone shows up at your doorstep and asks for $100, you're much more likely to give it to him if his eyes are bloodshot and he looks really irrational – you don't know what he's going to do! The only way that this is actually effective is if the show of irrationality is *convincing*. The person can't just feign, or you'll see through it. He has to be really, really irrational and show that he's ready to get revenge on you no matter what. If you believe that the person's going to defend his honor to the death, you're probably not going to mess with him.

So the theory is that religion is a way of committing yourself. Someone might *say* that he believes in a certain moral code, but others might figure talk is cheap and not trust him. On the other hand, if he has a long beard and prays every day and really seems to believe that he'll have an eternity in hellfire if he breaks the code, then he's making this very expensive commitment to his belief. It becomes much more plausible that he means it. So in this theory, religion functions as a way of publicly advertising a commitment to a certain set of rules. Of course, the rules might be good or they might be terrible. Nevertheless, this sort of public commitment to obeying a set of rules, backed up with supernatural rewards and punishments, seems like an important element of how societies organized themselves for thousands of years. It's why rulers trusted their subjects not to rebel, men trusted their wives to stay faithful, wives trusted their husbands not to abandon them, etc., etc.

So, I feel like these are the sorts of game-theoretic forces that Dawkins and Hitchens and the other antireligion crusaders are up against, and that they maybe don't sufficiently acknowledge in their writing. What makes it easier for them, of course, is that their opponents can't just come out and say, "yes, *of course* it's all a load of hooey, but here are the important social functions it serves!" Instead, religious apologists often resort to arguments that are easily demolished (at least since the days of Hume and Darwin) – since their *real* case, though considerably stronger, is one that's hard for them to make openly!

In summary, maybe it's true that humans (if we survive long enough) will eventually outgrow religion, now that we have something better to fill religion's explanatory role. But before that will happen, I think that *at the least* we'll need to better understand the social functions that religion played for most of history and still plays in most of the world, and maybe come up with alternative social mechanisms to solve the same sorts of problem.

Student: I was just thinking of if there's another case where irrationality might be preferred over rationality.

Scott: Where to begin?

Student: Especially if you have incomplete information. Like if you have a politician who's committed and won't change his ideals later on, you can feel more assured that he'll do what he said he would.

Scott: Because he has conviction. He believes in what he says. To most voters, that matters more than the actual content of the beliefs.

Student: I'm not sure that's best for the public interest.

Scott: Right, that's the problem! But how do you defeat people who have mastered the mechanisms of rational irrationality? By saying, "no, look here, you've got your facts wrong"? Which game are you playing?

Or take another example: a singles bar. The ones who succeed are the ones best able to convince themselves (at least temporarily) of certain falsehoods: "I'm the hottest guy/girl here." This is a very clear case where irrationality seems to be rational in some sense.

Student: The standard example is if you're playing Chicken with someone, it's advantageous to you if you break your steering wheel so it can't turn.

Scott: Exactly.

Student: Why is computer science not a branch of physics departments?

Scott: The answer to that isn't philosophical, it's historical. Computer scientists back in the day were either mathematicians or electrical engineers. People who would have been computer scientists when there wasn't such a department went into either math or electrical engineering. Physics had its plate full with other things, and to get into physics you had to learn this enormous amount of other stuff which maybe wasn't directly relevant if you just wanted to hack around and write programs, or if you wanted to think theoretically about computation. Paul Graham has said that computer science is not so much a unified discipline as a collection of people thrown together by accident of history, like Yugoslavia.[7] You've got the "mathematicians," the "hackers," and the "experimentalists," and we just throw them all together in the same department and hope they sometimes talk to each other. But I do think (and this is a clichéd thing to say) that the boundaries between CS, math, physics, and so on are going to look less and less relevant, more and more like a formality. It's clear that there's a terrain, but it's not clear where to draw the boundaries.

---

[7] P. Graham, Hackers and painters. http://www.paulgraham.com/hp.html

Index

#P (complexity class), 253–257, 260, 286, 288
1-norm, 112–113, 117, 119, 146
2-norm, 112–113, 117, 119, 146
2SAT, 70
3-coloring, 191–193
3SAT, 59–64, 70, 194, 203–204, 219, 250, 289

AC⁰ (complexity class), 259–261
ACC⁰ (complexity class), 260–261
Adleman, Leonard, 87–88, 91, 102, 139, 283
AdS/CFT, 346
advice, 83–90, 202, 211–215, 217, 320
Agrawal, Manindra, 77, 88
Aharonov, Dorit, 223
Aharonov, Yakir, 211
Ahn, Louis von, 36
Ajtai, Miklos, 100, 106, 258–259
algebrization, 258, 260
Alice, 69–70, 103, 127–128, 130–131,
        176–178, 209–210, 303–305,
        348–349, 354
ALL (complexity class), 213–214
Allen, Woody, 214
Alon, Noga, 239
AM (complexity class), 189, 245, 249, 265,
        355
Amazon, 102
Ambainis, Andris, 210, 239, 341
amplitudes, 28, 71, 109, 114–116, 119–123,
        125, 131, 139, 146–148, 160, 179,
        201, 217, 220, 223, 283, 285, 341
analog computer, 217, 222, 224
Anderson, Pamela, 164
anthropic principle, 169, 230, 266, 276–279,
        282, 286, 289, 331, 358
anyon, 226
Appel, Kenneth, 37, 187
Aristotle, 1–2
arithmetization, 250, 258

Arkhipov, Alex, 287–288
Arora, Sanjeev, 51
Arthur, 188–189, 203, 253–255, 257, 265
Axiom of Choice, 14–16, 26–28

Babai, Laszlo, 188, 206
Babbage, Charles, 33
Baez, John, 276
Banach-Tarski paradox, 15
Barak, Boaz, 51
Bayesianism, 229, 232
Bayesians, 5, 205, 232–233, 267, 271, 276,
        289
Bayes's Theorem, 228, 232, 266–268, 274,
        276
BB84 states, 127–128
Beame, Paul, 319
beamsplitter, 287
Beigel-Reingold-Spielman Theorem, 285
Bekenstein, Jakob, 32, 333–334
Bekenstein bound, 333–334
Bell, John, 162, 171, 176, 198
Bell inequality, 6, 109, 171–172, 176,
        302–305, 354
Bennett, Charles, 127, 130, 137, 145–146,
        149, 320–321
Bernstein, Ethan, 139, 142–143, 145, 149,
        224, 351
Bierce, Ambrose, 291
Big Bang, 85, 184, 212, 325–326, 328–330
Big Crunch, 325–326
birthday paradox, 196–197
black box, 29, 141–142, 146, 153, 205, 248
black-box group, 205
black hole, 32, 202, 221–222, 307–308,
        333–336, 344–349
block universe, 301
Blum, Lemore, 98
Blum, Manuel, 98
Blum-Blum-Shub generator, 98
Blum Speedup Theorem, 49, 82, 85

Bob. *See* Alice
Bohm, David, 183–185
Bohmian mechanics, 5, 183, 201
Bohr, Niels, 5, 202, 216
Boltzmann, Ludwig, 167, 334
Boltzmann's constant, 334
Boolean formula, 29, 64, 70, 79, 85–86, 138, 212, 247, 250, 255, 312
Boolean function, 84, 133, 213, 231, 239, 241, 250, 256, 284
BosonSampling, 287
Bostrom, Nick, 267, 272–274, 288
Bousso, Raphael, 32, 332, 336–337
BPP (complexity class), 79–82, 84, 86–91, 132, 135, 138, 140, 142, 144, 147, 188–189, 219, 246–247, 258, 279–282, 286, 291, 355–358
BPP_path (complexity class), 279–282, 286
BPP/poly (complexity class), 214
BPP/qpoly (complexity class), 211–214, 217, 320
BQP_CTC (complexity class), 317–318, 320, 322
Brassard, Giles, 127, 130, 145, 196–197, 355
Bremner, Michael, 297
Buhrman, Harry, 214
Busy Beaver, 28, 42–43

Caesar cipher, 94
Cantor, Georg, 12
CAPTCHAs, 36
cardinality, 11–12, 14, 16, 18
Carmichael numbers, 77
Carter, Brandon, 273
causal consistency, 311, 320, 322, 325
causal diamond, 337
cellular automaton, 99
Chaitin, Gregory, 213
Chalmers, David, 42
Chernoff bound, 73–74, 80
chess, 38, 254–255
Chinese Roon, 38–39
Chiribella, Giulio, 131
Choice, Axiom of, 14–16, 26–28
Christiano, Paul, 129
CHSH game, 354
Chuang, Isaac, 149
Church, Alonzo, 31
Church-Turing Thesis, 31–33
ciphertext, 94–97, 105, 107

circuit, 54, 60–61, 135–136, 145, 156, 189, 241, 243, 247–248, 255–257, 259–262, 312, 314, 316–319, 322–323, 356–357
CircuitSAT, 60–62, 107
Clique, 63
closed timelike curves, 308–324, 328, 350
Cohen, Paul, 26
Colbeck, Roger, 305
Completeness Theorem, 18, 24–25
computable, 19, 22, 28, 30–31, 42, 66, 84, 90, 96, 101, 149, 219, 259, 284, 307
computational complexity, 39, 44, 51, 71, 94, 132, 156, 186, 202, 240, 248, 287, 304, 310, 313, 339, 354–355
computational learning theory, 229, 232–233, 236, 238
computational zero-knowledge proof, 193
concept class, 231, 233–236, 239–240, 242
coNP (complexity class), 64–67, 86, 88, 193, 247, 249, 253, 263
consciousness, 40–41, 53, 151, 155–157, 276
consistency, 23, 25–27, 152, 248, 311
constant-depth circuits, 145, 242, 259–260
contextuality, 71, 171, 175
continuum, 2, 16, 26–27
Continuum Hypothesis, 14, 16, 26–28
Controlled-NOT gate, 125, 130, 135–136, 283
Conway, John, 302, 304
Cook, Stephen, 58–59, 203, 319
Cook-Levin Theorem, 62, 203
Cook reduction, 58–59
Copenhagen interpretation, 5, 201–202
Coppersmithm Don, 49
coRP (complexity class), 81, 88
correlation, 71, 172
cosmological constant, 325–327, 330–332, 337, 339
cosmological horizon, 331, 337–338, 340, 342
countable, 8, 12, 16, 18, 23
Cramér's Conjecture, 76
creationists, 163, 358–359
Crépeau, Claude, 130
Crick, Francis, 52
cryptography, 74, 89, 93–95, 97, 100, 102, 107–108, 127, 241, 244, 248
Csanky's algorithm, 319
CZK (complexity class), 193

Dam, Win van, 355
Darrow, Clarence, 290–291
Darwin, Charles, 361
Darwinians, 214–215, 257, 359
Davies, Paul, 359
Dawkins, Richard, 358–359, 361
decoherence, 160, 163, 165–170, 218, 224–225
Deep Blue, 37
Democritus, 1–4, 147
density matrix, 316
dequantization, 91
derandomization, 84, 88–91, 193, 247, 282, 356–358
Deutsch, David, 5, 147–148, 201, 245, 311–313, 317, 320–322, 324
diagonalization, 84, 256, 260–261
digital commitment, 192
Doomsday Argument, 272–277, 288–289
double-slit experiment, 184
DQP (complexity class), 195, 198–199
Dr. Evil paradox, 306
Drucker, Andrew, 214
Dwork, Cynthia, 100, 106
Dyakonov, Michel, 224

eigenvalue, 121, 208, 237
eigenvector, 208, 318
Einstein, Albert, 122, 160, 171, 175, 304
Einstein-Podolsky-Rosen channels, 130
Elga, Adam, 306
ELIZA, 35
entanglement, 71, 159, 163, 177, 220, 264, 301, 354
entropy, 32, 90, 166–169, 185, 236, 333–337
EPR (Einstein-Podolsky-Rose) pair, 122, 124, 130, 171, 262
Eratosthenes, 343
Euclid, 105
Euclidean norm. See 2-norm
event horizon, 221–222, 333, 344–349
evolution (biological), 34, 38, 111, 358–359.
    See also natural selection
Evolutionary Principle, 324
EXP (complexity principle), 55, 70, 138–139, 262, 307, 317, 355
expectation, 72
exponential time, 54–55, 64, 70, 89, 138, 240, 260, 263
EXPSPACE (complexity principle), 317

Extended Church-Turing Thesis, 31, 217, 219
Extended Riemann Hypothesis, 76

factoring, 57, 64–65, 77, 91, 98–100, 105–106, 140, 146–147, 157, 192, 218–219, 244–245, 286, 351
fat-shattering dimension, 240
fault-tolerance, 165, 218, 223, 225–226
Fermat's Last Theorem, 18
Fermat's Little Theorem, 18
Feynman, Richard, 57, 139–140, 283, 343
Feynman path integral, 283
Fields Medal, 27, 353
finite field, 205, 250, 253, 255, 258, 357
firewall, 347
first-order logic, 8–10, 18
fixed point, 312, 317, 319, 321–324, 328
fMRI, 159, 299
FOCS (Foundations of Computer Science), 62
forcing, 26
Fortnow, Lance, 248–249, 254
Four-Color Theorem, 37, 40–41, 63, 187
Fourier Checking, 145
Franzén, Torkel, 18
Fredkin, Ed, 244
free will, 290–291, 293–294, 296–299, 301–303, 307
Free Will Theorem, 302, 304, 307
Frege, Gottlob, 8, 52, 187
Friedberg, Richard, 30
Fuchs, Chris, 131, 202, 305
Fundamental Theorem of Algebra, 252
fuzzball, 346–347

garbage, 206
Gauss, Cal Friedrich, 57, 157, 301, 352
gavagai, 228
general relativity, 110, 222, 277, 308–309, 344, 347–348
geodesics, 336
Geometric Complexity Theory (GCT), 261
Gill, John, 78–79
GMW (Goldreich-Micali-Wigderson) protocol, 190
Gödel, Kurt, 18–19, 22, 26, 150–152, 156, 187, 308, 329. See also Completeness Theorem; Incompleteness Theorem
God's Coin Toss, 265, 267, 273–274

Goldbach's Conjecture, 21, 213, 252
Goldilocks Principle, 277
Goldreich, Oded, 97, 190, 217, 241
Goldwasser, Shafi, 241
Google, 37, 230
Gott, Richard, 273
Gottesman, Daniel, 135, 227
Gottesman-Knill Theorem, 135
Graham, Paul, 362
Grandfather Paradox, 311–313, 321–323
Graph Isomorphism, 186, 195, 198–199, 219, 350–351
Graph Nonsomorphism, 193–194
Grochow, Joshua, 261
group non-membership problem, 205, 208, 212
Grover's algorithm, 109, 146, 197, 199, 287, 341–342
grue, 228

Hadamard gate, 133, 135, 138, 198, 207, 264, 284
Haken, Wolfgang, 37, 187
halting problem, 21–22, 29–31, 42, 45, 47, 84, 136, 154, 156, 213
Hamiltonian, 204, 220, 287
hard on average, 99–100
Hardy, Lucien, 131
Hartmanis, Juris, 47
Håstad, Johan, 101, 241
hat problem, 91–93
Hawking, Stephen, 345
Hawking radiation, 222, 344–349
Heisenberg, Werner, 216
Hempel, Carl, 242
Hidden Subgroup Problem, 351
hidden variables, 166, 169–171, 177–178, 185–186, 195, 197, 302, 359
hidden-variable theories, 160, 170–179, 181, 183, 185, 195, 198–199, 293, 303
Hilbert, David, 14
Hilbert space, 28, 169, 184–185, 210, 317
Hitchens, Christopher, 361
Holevo's Theorem, 209–221
holographic bound, 221–222, 332–333, 335–336, 339
Hoof, Gerard 't, 345
Hoover, H.J., 319
Høyer, Peter, 196–197

Hume, David, 229, 361
Hume's Problem of Induction, 228
hypercomputation, 31

Immirzi parameter, 333
Impagliazzo, Russell, 87, 89, 91, 101, 241, 247, 357
Incompleteness Theorem, 19, 22–25, 41, 150–151
independent (random variables), 73
indeterminism, 160
information-theoretically secure, 94–95
inteference, 71
intelligent design, 358
interactive proof, 186, 190, 246, 249, 255, 257–258, 260, 262–263
interference, 71, 114–115, 143, 148–149, 160, 164, 168, 207, 220
International Obfuscated C Code Contest, 51
ion trap, 204
IP (complexity class), 193, 249, 253–254, 258, 262–263, 265
irrationality, 360–362

Jennings, Ken, 37
Jeopardy!, 37
Jozsa, Richard, 130, 287

Kabanets, Valentine, 91, 357
Kahn, David, 94
Kant, Immanuel, 218
Karloff, Howard, 249
Karp, Richard, 58–59, 83, 86
Karp-Lipton Theorem, 86, 88
Karp reduction, 58–59
Kasparov, Garry, 37
Kayal, Neeraj, 77, 88
Kearns, Michael, 230
Kempe, Julia, 204
Kitaev, Alexei, 134, 204, 262–263
Kleene, Stephen, 30
Kochen, Simon, 171–172, 174–175, 302, 304
Kochen-Specker Theorem, 171–172, 174–175
Kolmogorov, Andrei, 71
Kripke, Saul, 52–53
Kuperberg, Greg, 208–209, 214, 343
Kurzweil, Ray, 158

Ladner's Theorem, 63, 65
large cardinals, 23
lattice, 99, 106–107
Laughlin, Robert, 223
Leibniz, Gottfried, 33, 187
Leslie, John, 270
Leucippus, 1
Leung, David, 320
Levin, Leonid, 57–59, 101, 203, 217, 220, 227, 241
Libet, Benjamin, 298
linearity, 73, 123, 125, 220–221, 223–224, 320
linearity of expectation, 73
linear-optical quantum computers, 287
linear programming, 54
Lipton, Richard, 83, 86
Lloyd, Seth, 124, 321–322
Löb's Theorem, 26
Loebner, Hugh, 36
LOGSPACE (complexity class), 352, 356
loop quantum gravity, 332–333
Lovelace, Ada, 33
Löwenheim–Skolem Theorem, 18
Luby, Michael, 101, 241
Lund, Carsten, 249, 255
Lutomirski, Andy, 209

MA (complexity class), 188–189, 203, 249, 255–256, 265, 281, 355, 358
Maldacena, Juan, 221
Many-Worlds Interpretation, 201, 300
Map Colorability, 62
Markov, A.A., 73
Markov chain, 311
Markov's inequality, 73–74
Mathur, Samir, 346–347
matrix multiplication, 49
Maudlin, Tim, 322
Max-Flow/Min-Cut Theorem, 180–181
maximally mixed state, 164, 167, 215, 317, 319
measurement, 4
mediocrity principle, 273
Merlin. See Arthur
Mertens, Stephen, 51
metamathematics, 10
metaphysics, 269, 298
Micali, Silvio, 190, 241
Miller, Gary, 77

mixed states, 115–117, 121–122, 125–126, 131, 164, 167, 215, 222, 236, 238, 316–320
model (logic), 9, 18, 23–25
modus ponens, 9
Monte Carlo simulation, 9, 74
Moore, Christopher, 51
Moshkovitz, Dana, 194
Muchnik, A.A., 30
Mulmuley, Ketan, 261
multiverse, 147–149, 167–169, 179

Nagasawa, Masao, 182
Naor, Moni, 260
natural proofs, 250, 259
natural selection, 352–353. See also evolution
Nayak, Ashwin, 210, 239
Neal, Radford, 296
Neumann, Jon von, 33, 52, 74, 93, 167, 223, 226
Newcomb's Paradox, 294, 298
NEXP (complexity class), 70, 258, 260
Nielsen, Michael, 149
Nisan, Noam, 87, 249
Nobel Prize, 140, 225, 283, 289
No-Cloning Theorem, 126–128, 131, 158, 345
nondeterminism, 292
nonlocal boxes, 354–355
nonlocality, 171
nonrelativizing, 209, 246–247, 255, 257, 352
nonuniformity, 82–83, 87–90
Nozick, Robert, 295
NP (complexity class), 56–67, 70, 79, 82, 85–86, 88, 91, 98–99, 137, 145–146, 156, 188–189, 194, 203–204, 241, 245–246, 248, 255, 257–258, 261, 281–282, 291, 298, 307, 318, 321, 351–352, 354–355, 358
NP ∩ coNP (complexity class), 64–66
NP-complete, 59–60, 62–65, 67, 70, 79, 85–88, 100, 108, 124, 145, 154, 190–191, 198–199, 203, 212, 219, 227, 241, 250, 279, 289, 307, 310, 312–314, 317, 324–325, 350–354
NP-completeness, 29, 42, 45, 58, 62, 100, 194, 298
NP-hard, 58
NP-intermediate, 219

NSA (National Security Agency), 95, 102
null hypersurfaces, 336
NUMB3RS, 59
Number Field Sieve, 105, 108

Obama, Barack, 246
observable universe, 156, 201, 326, 338–339, 341
Occam's Razor, 230, 235–236
one-time pad, 94
one-way function, 36, 101–103, 190–191, 193, 241, 350
oracle, 29–30, 58–60, 67, 82, 137, 142, 144–145, 154, 156, 199, 208–209, 245–246, 248–249, 254–255, 257–258, 281, 298, 351–352, 354
ordinal numbers, 13–14
Otter (computer program), 37

P (complexity class), 45, 54–57, 62–64, 66–67, 70, 83–85, 88–91, 98–100, 131, 140, 145, 156, 241, 244–248, 255, 257–258, 261, 286, 288, 305, 312, 328, 351–352, 355–358
Packing, 63
PAC learning, 230, 235
Papadimitriou, Christos, 51
passive optical elements, 287
PCP (Probabilistically Checkable Proof), 193–194
PCP Theorem, 194
P_{CTC} (complexity class), 312, 314–315, 318
Peano Arithmetic, 23–24, 27
Peano axioms, 9–10
Peikert, Chris, 106
Penrose, Roger, 33, 41, 150–156, 158, 187, 301, 353
Peres, Asher, 103, 130
permanent (matrix), 288
perturbation, 248, 351
PH. See polynomial hierarchy
Pinker, Steven, 244
plaintext, 94–97, 105, 107
Planck area, 221, 332
Planck scale, 2, 32, 185
Planck time, 330
Plato, 330
Platonism, 152, 186–187, 200
Poincaré Conjecture, 18

polynomial hierarchy, 66–67, 82, 86, 88, 144–145, 193, 245, 254, 282–283, 286–288
polynomial identity testing, 356–357
polynomials, 129, 250–252, 254, 286
polynomial time, 54–61, 63–68, 70, 77–85, 87–88, 90, 96–98, 100–102, 104–106, 124, 136–138, 140, 154, 188, 192–193, 195, 199, 203, 206, 208, 219, 241, 247–248, 253, 255, 258, 262, 279–283, 287, 307, 312, 318–319, 321, 325, 350, 352, 354
Post, Emil, 30
PostBPP (complexity class), 280–283, 286–288
PostBQP (complexity class), 214, 282–283, 286–288, 321
postselection, 214, 280–284, 286, 321–322
POVM (positive operator-valued measurement), 236–237
PP (complexity class), 78–79, 138–140, 189, 212–213, 254–257, 262, 282–286, 321, 325, 350
P/poly (complexity class), 83–88, 255, 258, 260
Pratt, Vaughn, 88
Preskill, John, 165, 223, 305
Presumptuous Philosophers, 288–289
primality testing, 54, 57, 77–78, 88, 356
Prime Number Theorem, 76
primes, 21, 64, 75–77, 250
Principle of Deferred Measurement, 283
private-key cryptosystems, 102
probabilities, 4, 7, 28, 71–72, 109–113, 115, 121, 123, 146, 162, 168, 174, 181–183, 201, 220, 238, 266–267, 275–276, 281, 295, 322, 344
Problem of Induction, 228
PromiseMA (complexity class), 257
promise problem, 196, 203–204, 257
pseudorandom function, 241, 259–260, 350
pseudorandom generator, 89–90, 96–101, 241, 247, 356
PSPACE (complexity class), 55–56, 139, 193, 214, 245, 253–254, 258, 262–263, 265, 307, 312, 314–315, 317–319, 321–322, 324, 328, 350, 355
public-key cryptography, 102, 128–129

public-key quantum money, 129
P versus NP question, 56–57, 91, 156, 246,
        258, 261, 298

QAM (complexity class), 165
QCMA (complexity class), 208–209
QIP (complexity class), 262–265
QIP[2] (complexity class), 262–265
QMA (complexity class), 203–204, 208–209,
        212, 214, 265
QMA-complete, 203
QMAM (complexity class), 203
qualia, 34
quantifier, 34
quantitative epistemology, 200
quantum advice, 211–212, 214–215, 217, 320
quantum advice state, 212, 215
quantum computer, 19, 65, 82, 105–107, 134,
        138–140, 144–145, 147–149, 153,
        157, 159, 165, 195, 203, 211–212,
        218, 222–223, 225–226, 246, 282,
        287–288, 301, 315, 325, 350, 352
quantum computing, 3, 6, 8, 28, 31–32, 64,
        71, 140, 144–145, 147–149, 157,
        163, 195, 199, 202–203, 217–220,
        225–227, 237, 244, 284, 286–287,
        320, 343, 350–351
Quantum Cook-Levin Theorem, 203
quantum fault-tolerance, 223
quantum gravity, 32, 150–151, 301, 308–309,
        323, 332–333, 344, 346–347
quantum interactive proof, 262
quantum key distribution, 127, 159
quantum mechanics, 3–6, 27–28, 44, 71,
        109–112, 115–119, 121, 123–127,
        129, 131–132, 146–148, 150,
        156–158, 160–164, 168–170, 172,
        178, 180, 182, 184, 201–202, 217,
        220, 223, 225, 236–238, 286, 293,
        302, 304, 309, 311, 316, 320–321,
        343–344, 346, 349, 354–355
quantum money, 128–129
quantum oracle, 208
quantum robot, 341
quantum state, 5, 115, 117, 123, 125–126,
        148, 163–164, 171–172, 176,
        201–203, 206–210, 214–215, 217,
        221, 236–241, 265, 282, 300–301,
        303, 320

quantum state tomography, 237–238
quantum teleportation. See teleportation
qubit, 114, 121–122, 125, 127–130, 132–139,
        143, 159, 163–165, 167, 198,
        202–204, 207–212, 214, 220–221,
        223, 225–227, 237–240, 264,
        283–284, 287, 301, 316–318,
        321–322, 348, 354
query complexity, 197, 340

Rabin, Michael, 77, 88, 105
randomized algorithm, 74, 77–81, 88–89,
        143, 188, 247, 356–357
randomness, 71, 74–78, 82, 87–90, 98, 166,
        247, 295, 302, 305, 312, 356
random walk, 206–207, 340
rational numbers, 12, 16, 119
Razborov, A.A., 259–260
Recursive Fourier Sampling, 142–145, 351
reduction, 42, 58–59, 64, 97, 100–102, 106
Regev, Oded, 100, 106–107, 204
Reingold, Oded, 256, 260, 285–286
relativize, 30, 246, 258
reliability of memory, 162, 168
religion, 5, 343, 358–361
Riemann Hypothesis, 26, 76, 124, 213, 352
Robbins Conjencture, 37
Rommnet, Mitt, 246
RP (complexity class), 80–81, 355
RSA, 102–103, 105, 107–108, 124
Rudich, Steven, 259–260
Rule 110, 99
Russel, Bertrand, 8, 15, 52

Sagan, Carl, 307
Sahai, Amit, 196
sample distribution, 233
sample space, 230–232, 234
Santhanam, Rahul, 257–258
Saxena, Nitin, 77, 88
Schöning, Uwe, 254
Schrödinger, Erwin, 3, 131, 181–182, 225
Schwarzschild bound, 334
Scopes Monkey Trial, 290
Searle, John, 33, 39
Second Incompleteness Theorem, 23–24, 151
Second Law of Thermodynamics, 124,
        166–167, 333–334
self-checking programs, 255

Self-Indication Assumption, 274, 289
self-printing program, 51
Self-Sampling Assumption, 274, 289
semidifinite programming, 262
sets, 8, 11–12, 14–15
set theory, 8, 10–11, 18, 23, 26–27, 29, 63, 151–152
Shamir, Adi, 254
Shannon, Claude, 74, 95, 166
shattering, 234
Shepherd, Dan, 287
Shi, Yaoyun, 134, 197
Shor's algorithm, 65, 106, 140–142, 144, 146–147, 201, 245, 287
Shortest Vector Problem, 99–100, 106
Shub, M., 98
Sipser, Michael, 82, 91, 248–249, 258
Smith, Graeme, 320
Smolensky, Roman, 259
Smolin, John A., 320
Solovay, Robert, 88, 134, 258
Solovay-Kitaev Theorem, 134
SPACE($f(n)$), 47, 54
Space Hierarchy Theorem, 49
special relativity, 111, 175, 301, 303
Specker, Ernst, 171, 174–175
Stable Marriage Problem, 67–68
Standard Model, 102, 323
Star Trek, 313
stationary distribution, 311, 313–315
statistical zero-knowledge proof, 186, 193, 195–196, 352
Stearns, Richard, 47
Stern-Gerlach experiment, 176
STOC (Symposium on Theory of Computing), 62
stochastic matrix, 113, 171–172, 179, 181–182
Stohers, Andrew, 49
Strassen, Volker, 49, 88
string theory, 332
strong AI, 33, 38, 41, 155
superoperator, 316–318, 321
superposition, 3–6, 133, 137, 141, 161–162, 165, 206–208, 212, 263–264
Susskind, Lenny, 345
Szemerédi, Endre, 264
SZK (complexity class), 193, 195, 352

Tapp, Alain, 196–197
Tarski, Alfred, 15, 37, 45–46
TC^0 (complexity class), 260
teleportation, 129, 131, 158–159, 299–300, 321
Thorne, Kip, 309
Threshold Theorem, 223–224, 226
time-constructibility, 48
time dilation, 221
TIME($f(n)$), 47, 49, 54
Time Hierarchy Theorem, 47, 49
time travel, 307–308, 310, 313–314, 321, 323
Toda's Theorem, 254, 286
transistor, 223
transition probabilities, 162, 174, 181–183
trapdoor one-way function, 103, 108
Tsirelson's inequality, 354–355
Turing, Alan, 18, 20–22, 29–36, 38, 47, 74, 94, 138, 151–156
Turing Award, 47
Turing degree, 30–31
Turing equivalent, 29
Turing machine, 20, 28–31, 42, 47, 55, 59, 61, 78, 83–84, 87, 136, 138, 153, 156, 200, 219, 336, 339–340
Turing reducible, 29
Turing Test, 35–38, 155

union bound, 72–73, 87, 141, 233
unitarity, 222, 344–345
unitary matrix, 113–114, 134, 171–172, 179, 182
universal programmable computers, 21

vacuum energy, 169, 332, 337
Vadhan, Salil, 196
Valiant, Leslie, 230, 232–233, 253
Vassilevska Williams, Virginia, 49
Vazirani, Umesh, 139, 142–143, 145, 149, 210, 224, 230, 238, 305, 351
VC dimension, 234–236, 239–240
Vidick, Thomas, 128, 305
Vinodchandran, N.V., 256–257
von Neumann, John. See Neumann, Jon von
von Neumann trick, 93

Watrous, John, 128, 205–207, 212, 262–263, 318, 323
Watson (Jeopardy computer), 37

Watson, James, 52
wavefunction, 160, 164, 167, 183, 185, 201, 301
weak measurement, 211
Weinberg, Steven, 123, 358
Weizebaum, Joseph, 35
well ordered, 13–15
Wiesner, Stephen, 127–128
Wigderson, Avi, 87, 89, 91, 190, 247, 258
Wiles, Andrew, 109, 352–353
Williams, Ryan, 260
Winograd, Shmuel, 49
Witten, Edward, 221
Wittgenstein, Ludwig, 295–296

Wolfram, Stephen, 99, 303
Wootters, William, 123, 130
worst-case/average-case equivalence, 99–100

Yahoo, 37
Yao, Andy, 90, 102, 356
you-complete, 296–297

Zeilinger, Anton, 225
Zermelo-Fraenkel axioms, 11, 14, 23
Zermelo-Fraenkel set theory, 26
zero-knowledge proof, 189–193, 195
ZPP (complexity class), 81, 88, 355

Ω (constant), 213

Printed in the United States
By Bookmasters